Lecture Notes in Physics

Edited by H. Araki, Kyoto, J. Ehlers, München, K. Hepp, Zürich
R. Kippenhahn, München, H. A. Weidenmüller, Heidelberg
and J. Zittartz, Köln

173

Stochastic Processes in Quantum Theory and Statistical Physics

Proceedings of the International Workshop
Held in Marseille, France, June 29 – July 4, 1981

Edited by S. Albeverio, Ph. Combe, and M. Sirugue-Collin

Springer-Verlag
Berlin Heidelberg New York 1982

Editors

Sergio Albeverio
Mathematisches Institut, Ruhr-Universität
D-4630 Bochum

Philippe Combe
Madeleine Sirugue-Collin
Centre de Physique Théorique, CNRS
Marseille-Luminy, B.P. 907
F-13288 Marseille Cedex 9

ISBN 3-540-11956-6 Springer-Verlag Berlin Heidelberg New York
ISBN 0-387-11956-6 Springer-Verlag New York Heidelberg Berlin

Depuis quelques années, la théorie des probabilités a acquis un statut important en physique mathématique, alors que jusque là, ses concepts comme ses méthodes n'avaient joué qu'un rôle marginal, ceci même en mécanique statistique classique, qui est un domaine naturel pour l'application de telles idées. La formulation de Gibbs de la mécanique statistique a certainement joué un rôle dans ce phénomène, dans la mesure où elle fournissait un outil presque automatique pour calculer les moyennes que les physiciens pouvaient confronter avec l'expérience. D'autre part, la théorie des probabilités n'était pas alors développée comme elle l'est actuellement et son enseignement n'avait pas dans les études académiques une place importante. Cette situation n'a pas changé lors du développement de la mécanique quantique. En effet, s'il y a une interprétation probabiliste de la mécanique quantique, il n'y a pas un espace de probabilité sous-jacent.

La réintroduction massive des idées probabilistes s'est faite tout d'abord dans les années soixante en mécanique statistique classique; plus tardivement en théorie des champs, quand on a identifié la théorie des champs euclidienne à une théorie de mécanique statistique classique. Aujourd'hui, ces notions ont diffusé même chez des physiciens moins proches de la physique mathématique, ceci dans la mesure où l'on a reconnu que les probabilités sont un outil très puissant dans l'étude des systèmes à grand nombre de degrés de liberté. Pour ne prendre qu'un exemple, la notion de configuration typique et les algorithmes probabilistes sont maintenant largement utilisés chez les physiciens qui calculent sur ordinateur les propriétés des modèles de théorie de jauge sur réseau. On peut s'attendre à ce que d'autres concepts probabilistes deviennent déterminants dans la description qualitative et quantitative des systèmes complexes.

A ce point, on peut envisager pour les probabilités et les processus stochastiques un rôle analogue à celui joué par l'analyse classique dans le développement de la mécanique au XVIII ème siècle. Cependant, il faut remarquer qu'il y a aujourd'hui une situation assez singulière : le langage probabiliste ne s'applique pas directement aux théories quantiques dans l'espace-temps réel. On utilise en effet deux schémas conceptuels différents, l'un pour parler de la réalité, l'autre pour faire des calculs dans la région euclidienne. Il est intéressant de se demander si cette dichotomie va se résoudre un jour, et comment. On peut rappeler qu'il y a eu dans l'histoire des sciences un illustre précédent : celui du système copernicien qui a été utilisé, bien avant sa reconnaissance comme description de la réalité, comme un instrument efficace de calcul. Notre question n'est pas complètement académique car une interprétation réaliste du langage euclidien est en principe possible : la mécanique stochastique en offre un exemple.

Dans les travaux qui suivent, on trouvera représentés différents aspects de la diffusion du langage probabiliste dans la physique théorique et mathématique contemporaine. On y trouvera aussi des conceptions alternatives des probabilités. Nous espérons que ce recueil pourra donner une orientation dans une situation qui est en plein mouvement.

G. JONA-LASINIO   M. SIRUGUE

ACKNOWLEDGEMENTS

The Workshop took place at C.I.R.M., Marseille, from June 29
to July 4, 1981.

We would like to thank all participants and contributors for creating a
very stimulating scientific atmosphere. The Workshop would not have been possible
without the interest and support of various persons and institutions. Our special
thanks are due to Professor André Aragnol, Director of the C.I.R.M., Professor
Mohammed Mebkhout, Doyen de la Faculté des Sciences de Luminy, Professor Guy
Pouzard, Président de l'Université de Provence, as well as the Centre National
de la Recherche Scientifique, the U.E.R. de Physique Paris VII, and the University
of Paris VII.

We are very grateful to Maryse Cohen-Solal for her very invaluable and
generous help throughout all stages of the Workshop as well as to Professor
Jean-Marie Souriau and the Centre de Physique Théorique, C.N.R.S., Luminy, for
the material facilities which were put at our disposal.

Marseille, July 1982

THE ORGANIZING COMMITTEE :

S. ALBEVERIO, Ph. COMBE, F. GUERRA, R. HOEGH-KROHN,

G. JONA-LASINIO, G. RIDEAU, R. RODRIGUEZ, M. SIRUGUE, M. SIRUGUE-COLLIN

- CONTENTS -

QUANTUM THEORY AND NON-KOLMOGOROVIAN

PROBABILITY

Luigi ACCARDI
Dipartimento di Matematica
UNIVERSITA' DI ROMA II
Torvergata - ROMA
ITALY

§ 1   QUANTUM THEORY AND NON–KOLMOGOROVIAN PROBABILITY

Quantum theory gave rise to a new statistical theory with some peculiar mathematical features such as the appearance of complex numbers, Hilbert spaces, ...

Some efforts have been devoted to understand wether the quantum statistical model - called in the following the $\mathcal{C}$-Hilbert space model - has some kind of intrinsic necessity or the whole quantum theory can be described within the framework of the classical probabilistic model - called in the following the Kolmogorovian model.

The author has recently proposed a new approach to this problem whose main idea can be schematically expressed as follows: the conditional probabilities relative to the values of a given set of observables are regarded as the basic objects of the theory; the possibility of describing a given set of statistical data (i.e. of conditional probabilities) within the framework of a given statistical model - kolmogorovian or not - can be checked by means of <u>statistical invariants</u> - characteristic of the probabilistic model and expressed uniquely in terms of the conditional probabilities.

Since the conditional probabilities can be experimentally evaluated, the choice among different probabilistic models is determined uniquely by the experimental data.

Two problems naturally arise at this point:

i)   describe the non kolmogorovian probabilistic models.

ii) compute the statistical invariants relative to a given set of conditional probabilities and to a given probabilistic model.

In the following we shall report on some recent progress concerning these problems.

The results obtained up to now are far from giving complete answers to these problems, even in the case of finite valued observables, but at least they allow to draw some definite conclusions concerning some old standing problems in the foundations of quantum theory in particular:

i)   the necessity to introduce non–kolmogorovian probabilistic models in order to describe the transition probabilities predicted by the quantum theory (and confirmed by many experiments).

ii) The necessity of using complex rather than real or quaternionic Hilbert spaces.

iii) The existence (at the moment only as a purely mathematical possibility) of sets of transition probabilities which cannot be described neither with a kolmogorovian nor with a $\mathcal{C}$-Hilbert space model.

iv) The existence of non–kolmogorovian models which are neither of $\mathcal{C}$-Hilbert space type nor known variants of them.

As shown in [3] these results also allow to clarify some controversial questions in the interpretation of quantum theory; we shall not discuss this point here, but only mention that, from the point of view of quantum probability the "superposition principle" of quantum theory is not considered as expressing a physical superposition of states, and all the experimental tests for this principle are interpreted as experimental evidences of the necessity of introducing nonkolmogorovian models in the probabilistic description of nature.

§ 2   Statistical invariants for some simple systems

In the following we will denote $A, B, C, \ldots$ some observable quantities and (a_α), (b_β), (c_γ), ... their values.

We assume that different indices α, β, γ, ... correspond to different values of the observable(non degeneracy) and that

$$\alpha, \ \beta, \ \gamma, \ \ldots = 1, \ldots, n \tag{2.1}$$

for some $n < +\infty$ independent on A, B, C, ... The basic statistical data concerning these observables are the quantum mechanical transition probabilities

$$P(A = a_\alpha \mid B = b_\beta); \ P(B = b_\beta) \mid C = c_\gamma); \ \ldots$$

We assume that

$$P(A = a_\alpha \mid B = b_\beta) > 0; \ \ldots \tag{2.2}$$

where, here and in the following, the dots after a relation mean that the some relation has to be understood for all observables A, B, C, ... and their values a_α, b_β, c_γ, ... The matrix $(P(A = a_\alpha \mid B = b_\beta))$ will be denoted $P(A|B)$.

__Definition (2.1)__ A __Kolmogorovian model__ for the set $\{ P(A|B) \}$ of transition probability matrices is defined by:

(i) a probability space (Ω, θ, μ)

(ii) for each of the observables $A, B, C, \ldots$ - a measurable partition (A_α), (B_β), (C_γ), ... of (Ω, θ) $(\alpha, \beta, \gamma, \ldots = 1, \ldots, n)$ such that for any couple of observables A, B and of values $a\alpha$, $b\beta$ one has:

$$P(A = a_\alpha \mid B = b_\beta) = \frac{\mu(A_\alpha \cap B_\beta)}{\mu(B_\beta)} \tag{2.3}$$

__Definition (2.2)__ A $\mathbb{C}$-Hilbert (resp. $\mathbb{R}$-Hilbert) space model for the set $\{P(A|B)\}$ of transition probability matrices is defined by:

(i) a complex (resp. real) Hilbert space H.

(ii) for each of the observables $A, B, C, \ldots$ - an orthonormal basis (ϕ_α^A), (ϕ_β^B), (ϕ_γ^C), ... of H $(\alpha, \beta, \gamma, \ldots = 1, \ldots, n)$ such that for any couple of observable A, B and of values a_α, b_β, one has:

$$P(A = a_\alpha \mid B = b_\beta) = |< \phi_\alpha^A, \phi_\beta^B >|^2 \tag{2.4}$$

Clearly the fulfillment of the symmetry conditions:

$$P(A = a_\alpha \mid B = b_\beta) = P(B = b_\beta \mid A = a_\alpha); \ \ldots \tag{2.5}$$

is necessary in order that the family $\{ P(A|B) \}$ admits a $\mathbb{C}$-Hilbert space model. In the following we shall deal only with transition probabilities which satisfy (2.2) that is, for any couple of observables A, B:

$$P(A|B) = {}^t P(B|A) \tag{2.6}$$

A first simple remark is the following:

__Proposition (2.3)__ For any $2 \leq n + \infty$, if the transition probability matrices between __two__ observables satisfy (2.6), then they admit a kolmogorovian model.

As we shall see later, Proposition (2.3) is false for __three__ (or more) observables even if condition (2.6) is satisfied. If condition (2.6) is not satisfied the statement is false even for two observables.

The analogue of Proposition (2.3) for $\mathbb{C}$-Hilbert space models does not hold for $n \geq 3$ in fact (cf. [5]) one has:

__Proposition (2.4)__ Let $n = 3$, and let A, B be two three valued observables such that

$$P(A|B) = {}^t P(B|A) = (p_{ij}); \ i, j = 1, 2, 3 \tag{2.7}$$

The transition matrices (2.7) admit a $\mathbb{C}$-Hilbert space model if and only if

4

$$\left| \sqrt{p_{11}p_{33}} - \sqrt{p_{21}p_{31}p_{32}} \right| \leq \sqrt{p_{22}}\,(1 + p_{31}) \leq \sqrt{p_{11}p_{33}} + \sqrt{p_{21}p_{31}p_{32}} \tag{2.8}$$

The problem of the existence of a $¢$-Hilbert space model for the transition probability matrix $P(A|B) = {}^{t}P(B|A)$ relative to two observables has been studied by M. Roos [7] , [8] who gave sufficient conditions for any finite n. Translated in our language (quite different from Roos' one) we can say that the investigations of this author are motivated by the attempt to check wether the experimentally measured transition probabilities in the decay of the $K°$ and $\bar{K}°$ mesons into $\Pi^{+}\Pi^{-}$ can be described by a non Kolmogorovian model which is not of $¢$-Hilbert space type.

Because of Proposition (2.3) the simplest case in which there might be a non trivial choice between Kolmogorovian and $¢$-Hilbert space models is the case of three two-valued observables. For the corresponding transition probability matrices we shall use the notations:

$$P = P(A|B) = \begin{bmatrix} p & 1-p \\ 1-p & p \end{bmatrix} = \begin{bmatrix} \cos^{2}\alpha/2 & \sin^{2}\alpha/2 \\ \sin^{2}\alpha/2 & \cos^{2}\alpha/2 \end{bmatrix}$$

$$Q = P(B|C) = \begin{bmatrix} q & 1-q \\ 1-q & q \end{bmatrix} = \begin{bmatrix} \cos^{2}\beta/2 & \sin^{2}\beta/2 \\ \sin^{2}\beta/2 & \cos^{2}\beta/2 \end{bmatrix} \tag{2.9}$$

$$R = P(C|A) = \begin{bmatrix} r & 1-r \\ 1-r & r \end{bmatrix} = \begin{bmatrix} \cos^{2}\gamma/2 & \sin^{2}\gamma/2 \\ \sin^{2}\gamma/2 & \cos^{2}\gamma/2 \end{bmatrix}$$

According to (2.2):

$$o < p,q,r < 1 \tag{2.10}$$

<u>Theorem (2.5)</u> The bi-stochastic matrices P, Q, R defined by (2.9) admit a Kolmogorovian model if and only if

$$| p + q - 1 | \leq r \leq 1 - | p - q | \tag{2.11}$$

<u>Theorem (2.6)</u> The following conditions are equivalent:

i)   The transition matrices P, Q, R admit a $¢$-Hilbert space model.

ii) $$- 1 \leq \frac{p + q + r - 1}{2\sqrt{pqr}} \leq 1 \tag{2.12}$$

iii) $[\sqrt{pq} - \sqrt{(1-p)(1-q)}\,]^{2} \leq r \leq [\sqrt{pq} + \sqrt{(1-p)(1-q)}\,]^{2}$ $\qquad$ 2.13

iv) $|\alpha - \beta| \leq \gamma \leq \alpha + \beta$; $\alpha + \beta + \gamma \leq 2\Pi$ $\qquad\qquad$ (2.14

<u>Theorem (2.7)</u> The following conditions are equivalent:

i)   The transition matrices P, Q, R admit a $\mathbb{R}$-Hilbert space model

ii) $$\frac{p + q + r - 1}{2\sqrt{pqr}} = + 1; \quad \text{or} \quad \frac{p + q + r - 1}{2\sqrt{pqr}} = - 1 \tag{2.15}$$

iii) α,β,γ satisfy one of the following three equalities:

$$\alpha + \beta + \gamma = 2\Pi; \quad 2\Pi - \alpha + \beta + \gamma = 2\Pi; \quad \alpha + \beta + 2\Pi - \gamma = 2\Pi \tag{2.16}$$

<u>Theorem (2.8)</u> The transition matrices P, Q, R admit a quaternion Hilbert space model if and only if they admit a $¢$-Hilbert space model.

Remark that condition (2.15) is the necessary and sufficient condition for the angles α, β, γ to be adiacent angles of a thriedron in $\mathbb{R}^3$. The first of the conditions (2.16) characterizes the flat thriedrons and the remaining two can be reduced to the first without changing the transition probabilities.

The inequality:

$$[\; \sqrt{pq} \; - \sqrt{(1-p)(1-q)}\;]^2 \le |p + q - 1| \le 1 - |p - q| \le [\sqrt{pq} + \sqrt{(1-p)(1-q)}\;]^2 \qquad (2.17$$

shows that the existence of a kolmogorovian model for P,Q,R always implies the existence of a $\mathbb{C}$-Hilbert space model. This implication however is very special to the 2-dimensional case in fact, as a conseuqnce of Proposition (2.4), in the 3-dimensional case this implication is false even for two observables (cf. [4] for an explicit example).

Matrices of the form (2.9) appear naturally in quantum theory as transition probability matrices of spin-1/2 systems. Thus, measuring the transition probabilities of a spin-1/2 system along three space directions whose angles α, β, γ are such that both inequalities (2.17) and (2.12) are strict, one could give an experimental proof of the inadeguacy of both the kolmogorovian and the $\mathbb{R}$-Hilbert space model to describe the results of quantum mechanical measurement. Theorem (2.8) shows that, at least in the simple case considered above, the transition probabilities alone do not allow to distinguish between complex and quaternion Hilbert space models. We conjecture that this statement has general validity.

If a given set of transition probabilities satisfies the conditions which guarantee the existence for them of a given probabilistic model, then changing the transition probabilities so that these conditions are still satisfied, the new set of transition probabilities will still admit the same model. In this sense we speak of <u>statistical invariants</u>.

§ 3.   Schwinger's algebra of measurements

As shown in § (2.), the (symmetric) transition probabilities relative to a <u>couple</u> of (finite valued) observables can always be described within the framework of the Kolmogorovian model. This means that for couples of observables the use of a Kolmogorovian model to describe the corresponding transition probabilities is not a purely statistical property – i. e. something which can be deduced directly fron the transition probabilities – but must be the expression of a further physical property of this couple of observables.

It is natural to conjecture that this physical property should be related to the incompatibility of the two observables in the quantum mechanical sense and, since the mutual incompatibility of two observables means that between them there is an indeterminacy principle of Heisenberg type, the problem arises wether it is possible to give a general statistical formulation of the indeterminacy principle with an intuitive physical significance and such that it imposes enough constraints on the mathematical model to imply the choice of Hilbert space models (or similar ones). To this goal a slight modification of Schwinger's "Algebra of measurements" turns out to be very useful.

Let A be an observable with values (a_α) ($\alpha = 1,\ldots, n < +\infty$ ·) and t be an instant of time. We associate the symbol $A_\alpha(t)$ to the idealized selective measurement, performed at time t, that from an ensemble of independent identically prepared systems accepts those for which the observable A takes the value a_α and rejects all the other ones (cf. [9], § (1.2)). Such a measurement will be called an _elementary filter_.

Denoting $A(s)$, $B(t)$ two measurement operations (not necessarily elementary filters) performed respectively at times s and t, we denote $A(s) \cdot B(t)$ $(s < t)$, the measurement operation consisting in performing in series the operations $A(s)$ and $B(t)$. If $A(s)$ and $B(s)$ are compatible measurement operations, the symbol $A(s)+B(s)$ will be associated to the realization in parallel of the two operations at the same instant s. With these notations the properties which characterize the elementary filters can be expressed with the symbolic notations:

$$\text{"}\lim_{\varepsilon \to 0}\text{"} \quad A_\alpha(t) A_{\alpha'}(t + \varepsilon) = \delta_{\alpha\alpha'} A_\alpha(t) \tag{3.1}$$

$$\sum_\alpha A_\alpha(t) = 1 \tag{3.2}$$

Where O and 1 are the filters defined in the obvious way and the "lim" in (3.1) has no precise methamtical meaning but only expresses the heuristic idea one has of first-kind measurements as those which if repeated "immediately after" yield the same result (for a discussion of these points cf. [6] , [9]).

In the following we will assume that all sequences of measurement operations considered are performed in a very short time, so that one can neglect all temporal effects with the exclusion of those specifying the time ordering of the operations. Consequently we omit the time label and write simply $A_\alpha B_\beta$ to mean $A_\alpha(t)B_\beta(t+\varepsilon)$ for _any_ $\varepsilon > 0$ $(\varepsilon \ll 1)$.

Finally if p is any number such that $0 < p < 1$, and A_α is the elementary filter associated to the event $A = a_\alpha$, the symbol pA_α will be associated to any measurement apparatus characterized by the following properties: i) if a system has been accepted by the apparatus pA_α then with certainty for this system $A = a_\alpha$. ii) from an ensemble of systems identically prepared by the condition $A = a_\alpha$ only the fraction p will be accepted by the apparatus, but on each system of the ensemble we cannot say wether it will be accepted or not. Applying to the elementary filters the operations described above, in a compatible way, we obtain symbols corresponding to a physically realizable operation which will be called _filters_. The operations of addition, multiplication and multiplication by a number in $[0,1]$ defined in the way described above among filters, enjoy the usual algebraic properties, but are not everywhere defined. With the adjunction of ideal elements (symbols not corresponding to any physical operation, like the addition of incompatible operations) one can embed the structure defined by these operations into an abstract associative algebra A over the reals called the _algebra of measurements_ associated to the given system.

Thus to any observable A of the system one can associate a partition of the identity in the algebra of measurements A, i.e. a family (A_α) of elements of A such that

$$A_\alpha A_{\alpha'} = \delta_{\alpha\alpha'} A_\alpha \tag{3.3}$$

$$\sum_\alpha A_\alpha = 1 \tag{3.4}$$

If A,B are two observables and (A_α), (B_β) the corresponding partitions of the identity in A, in the classical case one has:

$$A_\alpha B_\beta A_\alpha = A_\alpha B_\beta = B_\beta A_\alpha \tag{3.5}$$

corresponding to the fact that all classical measurements are compatible.

We single out the statistical essence of Heisenberg indeterminacy principle

in the statement that, if A and B are incompatible observables, the measurement
of any value of B introduces an indeterminacy in our knowledge of the values of A.
In the framework of Schwinger's algebra of measurements, we formulate this statement through the equality:

$$A_\alpha B_\beta A_\alpha = p_{\alpha\beta} A_\alpha \tag{3.6}$$

where $p_{\alpha\beta} = P(A = a_\alpha \mid B = b_\beta)$.

Equality (3.6) means that the presence of the filter B_β in the apparatus
$A_\alpha B_\beta A_\alpha$ doesn't yield any information, as it does in the classical case, on single
systems, but only on ensembles; more precisely: on the single system emerging from
the apparatus $A_\alpha B_\beta A_\alpha$ we can only say that $A = a_\alpha$, but on an ensemble of systems
prepared so that $A = a_\alpha$ we can say that exactly the fraction $p_{\alpha\beta}$ will be accepted
by the filter $A_\alpha B_\beta A_\alpha$.

Summing up; the statistical content of Heisenberg indeterminacy principle
can be formulated as a constraint imposed on the multiplication in the algebra of
measurements and involving the conditional probabilities among values of incompatible observables. This introduces a connection between the structure of the
algebra of measurements and the statistical properties of the system, the exploita-
tion of which will be the goal of the following sections.

Let us emphasize that condition (3.6) expresses only the statistical content
of the indeterminacy principle; the deeper physical content involves Planck's
constant and emerges when dealing with the explicit form of the transition probabilities.

Finally, in the above schematization the specific space-time location of the elementary filters has been neglected. However the structure theorems deduced from
this approach give rather clear indications on how to extend these ideas to more
elaborate and realistic models (this extension will not be discussed here).

§ 4.   Structure of the standard Heisenberg models

Motivated by the heuristic considerations in the preceeding section, let us
introduce the following definition

<u>Definition(4.1)</u> Let T be a set; $\{A(x): x \in T\}$ a family of observables such that $A(x)$
has values $(a_\alpha(x))$ with $\alpha = 1, \ldots, n$ and $n < +\infty$ independent on $x \in T$. Let for any
$x, y \in T$, $P(x,y) = (p_{\alpha\beta}(x,y))$ denote the transition probability matrix of $A(x)$ given
$A(y)$, i.e.

$$p_{\alpha\beta}(x,y) = P(A(x) = a_\alpha(x) \mid A(y) = a_\beta(y)) \tag{4.1}$$

An <u>Heisenberg model</u> for the family $\{P(x,y)\}$ of transition matrices is defined
by the assignment of:
i)   an R-algebra A (associative with identity 1)
ii) $\forall x \in T$, and for each value $a_\alpha(x)$ of $A(x)$ — an element $A_\alpha(x)$ of A such that, for
     any $x, y \in T$ and $\alpha, \beta = 1, \ldots, n$:

$$A_\alpha(x)A_{\alpha'}(x) = \delta_{\alpha\alpha'} A_\alpha(x) \tag{4.2}$$
$$\sum_\alpha A_\alpha(x) = 1 \tag{4.3}$$
$$A_\alpha(x) A_\beta(y) A_\alpha(x) = p_{\alpha\beta}(x,y) A_\alpha(x) \tag{4.4}$$

Just as in the kolmogorovian case, the existence of an Heisenberg model will impose some constraints on the transition probability matrices. Similarly it is clear that not any $\mathbb{R}$-algebra can be the algebra of measurements of an Heisenberg model. Thus we are faced with two problems:

<u>Problem I</u> Which $\mathbb{R}$-algebras can be the algebra of measurements of some Heisenberg model?

<u>Problem II</u> Under which conditions a family of transition probability matrices will admit on Heisenberg model?

We will solve Problems I and II for a particular class of Heisenberg models — called <u>standard</u> in the following — characterized by the property: for any $x, y \varepsilon T$ the products

$$A_\alpha(x) \, A_\beta(y) \quad ; \quad \alpha, \beta = 1, \ldots, n \tag{4.5}$$

are a linear basis of A over its center.

That is: denoting κ the center of A, the products (4.5) are a basis of A considered as κ-module.

The center of A has a clear physical meaning: the partitions of the identity over κ correspond to universally compatible observables (superselection rules) in the following we will see that it (or better, its minimal dimension over $\mathbb{R}$) has also a statistical significance.

We will first solve Problem (I), and for this it will be sufficient to consider the case when the set T has only two elements; then we will deduce the compatibility conditions which are necessary and sufficient for the existence of a standard Heisenberg model relative to an arbitrary set of transition probability matrices. We also investigate the problem of the uniqueness up to isomorphism of the standard Heisenberg model. Finally we characterize the usual quantum mechanical models within the larger class of the standard Heisenberg models.

If the set T has only two elements, then we are dealing with only two observables A and B with transition probability matrices

$$P(A|B) = (p_{\alpha\beta}(A|B) \quad ; \quad P(B|A) = (p_{\beta\alpha}(B|A)$$

An Heisenberg model for the transition matrices $P(A|B)$, $P(B|A)$ is defined by an $\mathbb{R}$-algebra A (associative with identity) and two partitions of the identity in A: (A_α), (B_β). If such a model is standard then, denoting x the center of κ, there will exist n^4 elements of κ — $\gamma^{\alpha'\beta'}_{\alpha\beta}$ — such that:

$$B_\beta A_\alpha = \sum_{\alpha'\beta'} \gamma^{\alpha'\beta'}_{\alpha\beta} A_{\alpha'} B_{\beta'} \tag{4.6}$$

The existence of $\gamma^{\alpha'\beta'}_{\alpha\beta} \varepsilon \kappa$ satisfying (4.6) is also a sufficient condition for the Heisenberg model relative to A, B to be standard provided that for any α, β:

$$\gamma \varepsilon \varepsilon \kappa; \quad \gamma A_\alpha B_\beta = 0 \;\Rightarrow\; \gamma = 0$$

and this property can always be assumed to hold.

The elements $\gamma^{\alpha'\beta'}_{\alpha\beta} \varepsilon \kappa$ satisfying (4.6) will be called the <u>structure constants of A</u> in the $A_\alpha B_\beta$-basis.

<u>Proposition (4.2)</u> Let A be an associative $\mathbb{R}$-algebra with center κ and let (A_α), (B_β) be partitions of the identity in A such that $\{A_\alpha B_\beta : \alpha, \beta = 1, \ldots, n\}$ is a κ basis of A. Then, if $(\gamma^{\alpha'\beta'}_{\alpha\beta})$ are the structure constants of A in the $A_\alpha B_\beta$-basis one has:

$$\sum_\alpha \gamma^{\alpha'\beta'}_{\alpha\beta} = \delta_{\beta\beta'} \quad ; \quad \forall \alpha' \tag{4.7}$$

$$\sum_{\beta} \gamma_{\alpha\beta}^{\alpha'\beta'} = \delta_{\alpha\alpha'} \quad ; \qquad \forall\ \beta' \tag{4.8}$$

$$\gamma_{\alpha'\beta}^{\alpha\beta'} \cdot \gamma_{\alpha''\beta'}^{\alpha\beta''} = \gamma_{\alpha'\beta}^{\alpha\beta''} \cdot \gamma_{\alpha''\beta'}^{\alpha'\beta''} \tag{4.9}$$

Conversely if κ is a commutative, associative $\mathbb{R}$-algebra with identity and $\gamma_{\alpha\beta}^{\alpha'\beta'} \in \kappa$ $(\alpha,\beta,\alpha',\beta' = 1,\ldots,n)$ are elements of κ satisfying (4.7), (4.8), (4.9), then there exists an $\mathbb{R}$-algebra A, associative with identity, and two partitions of the identity in A - (A_α), (B_β) - such that the products $A_\alpha B_\beta$ are a κ-basis of A with structure constants $\gamma_{\alpha\beta}^{\alpha'\beta'}$.

<u>Remark</u> Conditions (4.7), (4.8), (4.9) have been deduced using only associativity and the partitions of the identity. Hence they define a class of algebras strictly larger than the one associated to standard Heisenberg models. For example, the algebras associated to a kolmogorovian model for $P(A|B)$, $P(B|A)$ is included in this class and has structure constants

$$\gamma_{\alpha\beta}^{\alpha'\beta'} = \delta_{\alpha\alpha'}\, \delta_{\beta\beta'}$$

Introducing, as a further condition, the validity of the weak formulation of Heisenberg's principle discussed in the preceeding section:

$$A_\alpha B_\beta A_\alpha = p_{\alpha\beta} A_\alpha \quad ; \quad B_\beta A_\alpha B_\beta = p_{\beta\alpha} B_\beta \tag{4.10}$$

one can show that:

<u>Proposition (4.3)</u>   The (quantum mechanical) symmetry condition

$$P(\ A = a_\alpha\ |\ B = b_\beta) = P(B = b_\beta\ |\ A = a_\alpha) \tag{4.11}$$

is necessary for the existence of a standard Heisenberg model for the transition probability matrices $P(A|B)$, $P(B|A)$.

Assuming that condition (4.11) is satisfied and denoting $P = (p_{\alpha\beta})$ the bi-stochastic matrix defined by

$$P = P(A|B) = \text{traspose of } P(B|A) \tag{4.12}$$

the structure of the algebras associated with standard Heisenberg models is described by the following:

<u>Theorem (4.4)</u>   In the notations above the following conditions are equivalent:

i) A, (A_α), (B_β), κ define a standard Heisenberg model for the bi-stochastic matrix $P(A|B) = {}^tP(B|A)$ with structure constants $\gamma_{\alpha\beta}^{\alpha'\beta'}$.

ii) There exists a κ-valued matrix $U(A|B) = (u_{\alpha\beta}(A|B))$ such that

$$\gamma_{\alpha\beta}^{\alpha'\beta'} = \frac{u_{\alpha'\beta}(A|B)\, u_{\alpha\beta'}(A|B)}{u_{\alpha\beta}(A|B)\, u_{\alpha'\beta'}(A|B)}\ p_{\alpha\beta}(A|B) \tag{4.13}$$

$$\sum_{\alpha} \frac{p_{\alpha\beta}(A|B)}{u_{\alpha\beta}(A|B)} \cdot u_{\alpha\beta'}(A|B) = \delta_{\beta\beta'} \tag{4.14}$$

$$\sum_{\beta} u_{\alpha'\beta}(A|B)\ \frac{p_{\alpha\beta}(A|B)}{u_{\alpha\beta}(A|B)} = \delta_{\alpha\alpha'} \tag{4.15}$$

iii) There exists a κ-module H and two κ-bases (a_α), (b_β) of H such that the κ-linear rank one projections A_α, B_β onto $\mathbb{K}\cdot a_\alpha$, $\kappa\cdot b_\beta$, defined with respect to the (a_α), (b_β) bases, satisfy condition (4.10) above.

Motivated by Theorem (4.4) and by obvious quantum mechanical analogies, we introduce the following definition:

<u>Definition (4.5)</u>   Let $P = (p_{\alpha\beta})$ be a nxn bi-stochastic matrix, and let κ be an associative, commutative $\mathbb{R}$-algebra with unit. A κ-valued matrix $U = (u_{\alpha\beta})$ satisfying

$$\sum_{\alpha} (u_{\alpha\beta})^{*} u_{\alpha\beta'} = \delta_{\beta\beta'} \tag{4.16}$$

$$\sum_{\beta} u_{\alpha\beta} (u_{\alpha'\beta})^{*} = \delta_{\alpha\alpha'} \tag{4.17}$$

where

$$(u_{\alpha\beta})^{*} = P_{\alpha\beta}/u_{\alpha\beta} \tag{4.18}$$

will be called a κ-valued transition amplitude matrix for P (we always assume $u_{\alpha\beta} \neq 0$). Two κ-valued transition amplitude matrices for P are called equivalent if the standard Heisenberg models they define through theorem (4.4) have the same structure constants.

<u>Proposition (4.6)</u>   Two κ-valued transition amplitude matrices $(u_{\alpha\beta})$, $(v_{\alpha\beta})$ for a given bi-stochastic matrix P are equivalent if and only if there exist κ-valued functions $g_1(\alpha)$, $g_2(\beta)$ $(\alpha,\beta = 1,\ldots,n)$ such that:

$$u(\alpha,\beta) = g_1(\alpha) \, v(\alpha,\beta) g_2(\beta) \tag{4.19}$$

Starting from the above results the answer to Problem (II) can be formulated as follows:

<u>Theorem (4.7)</u>   Let T be any set; $\{P(x,y) : x,y \in T\}$ be a set of transition probability matrices satisfying the symmetry condition

$$p_{\alpha\beta}(x,y) = p_{\beta\alpha}(x,y) \; ; \quad \forall x,y \in T \; ; \; \forall \alpha\beta \tag{4.20}$$

κ – a commutative, associative $\mathbb{R}$-algebra with unit.

The following statements are equivalent:

i)   The family $\{P(x,y)\}$ admits a standard Heisenberg model with center κ, in the sense of Definition (4.1).

ii)   For each couple $x,y \in T$ there exists a κ-valued transition amplitude $U(x,y)$ for $P(x,y)$ such that the family $\{U(x,y)\}$ satisfies the conditions

$$U(x,y) \, U(y,z) = U(x,z) \; ; \; x,y,z \in T \tag{4.21}$$

where, by definition

$$U(x,x) = id \; ; \quad x \in T \tag{4.22}$$

iii)   There exists a κ-module H and for each $x \in T$, a x-basis $(a_\alpha(x))$ of H such that denoting, for each $x \in T$, $A_\alpha(x)$ the -linear rank one projection onto $\kappa \cdot a_\alpha(x)$ defined with respect to the $(a_\alpha(x))$ – basis, one has

$$A_\alpha(x) \, A_\beta(y) \, A_\alpha(x) = p_{\alpha\beta}(x,y) \, A_\alpha(x) \tag{4.23}$$

for any $x,y \in T$; and $\alpha,\beta = 1, \ldots, n$.

Writing equations (4.21), (4.22) in terms of the coefficients $(u_{\alpha\beta}(x,y))$, one easily recognizes in them a generalization of the quantum mechanical "principle of composite amplitudes" (cf. [2]) which therefore turns out to be mathematically equivalent to the weak form (4.23) of the Heisenberg principle – which has an immediate physical significance – plus the finite dimensionality condition (4.5) which apparently has not - but which can at least be considered as a "condition of minimal complexity" on the algebra of measurements.The principle of composite amplitudes appears thus as a compatibility condition on the transition probabilities $P(x,y)$ for the existence of a non-kolmogorovian model, just as the usual Chapman-Kolmogorov equation is a compatibility condition for the existence of a kolmogorovian model. Of course the existence of a set of κ-valued transition amplitudes satisfying (4.21), for a given set of transition probability matrices $\{P(x,y)\}$ is a property which can be expressed uniquely in terms of conditions on the probabilities $p_{\alpha\beta}(x,y)$ (cf. the examples in section (2.)). The explicit determination of such conditions might be difficult but, for a finite set T it is in principle always possible, while for an infinite set T can be made accessible with the introduction of some symmetry conditions (i. e. considering the action of a group on the index set T leaving the transition probabilities invariant).

Two standard Heisenberg models for the family of transition matrices $\{P(x,y)\}$ are called equivalent if, denoting $\{U(x,y)\}$, $\{V(x,y)\}$ the associated transition amplitudes $U(x,y)$, $V(x,y)$ are equivalent in the sense of Definition (4.5). Proposition (4.6) and the compatibility conditions (4.21), (4.22) imply then that this is the case if and only if there exists a function

$$c \; : \; (x,y)\epsilon \, TxT \longrightarrow c(x,y) \; \epsilon \; \kappa \qquad \text{satisfying}$$

$$c(x,y) \; c(y,z) \;\; = \;\; c(x,z) \qquad \forall x,y,z \; T \tag{4.24}$$

and, for each x,y T a κ-valued, diagonal matrix $D(x,y)$ such that

$$U(x,y) \;\; = \;\; c(x,y) \; D(x,y) \; V(x,y) \; D \; (y,x)^{-1} \tag{4.25}$$

The family $\{D(x,y)\}$ must satisfy the further set of compatibility conditions:

$$\frac{D(x,z)}{D(x,y)} \;\; = \;\; U(x,y) \; \frac{D(y,z)}{D(y,x)} \; U(y,z) \; \frac{D(z,x)}{D(z,y)} \; U(z,x) \tag{4.26}$$

which, for most families $\{U(x,y)\}$ imply that for each $x,y\epsilon \, T$, $D(x,y) = D(x)$ independently on y. Therefore in the generic case two families of transition amplitudes $\{U(x,y)\}$, $\{V(x,y)\}$ are equivalent if and only if for any $x,y\epsilon T$:

$$U(x,y) \;\; = \;\; c(x,y) \; D(x) \; V(x,y) \; D(y)^{-1} \tag{4.27}$$

with $c(x,y) \; \epsilon\kappa$ satisfying $(4 .24)$, and $\{D(x)\}$ - a family of diagonal matrices.

The usual quantum mechanical models can be thought as those standard Heisenberg models in which the involutions

$$u_{\alpha\beta}(x,y) \quad \rightarrow \quad p_{\alpha\beta}(x,y) \; / \; u_{\alpha\beta}(x,y)$$

do not depend on the statistics. More precisely:

<u>Proposition (4.8)</u> In the notations and assumptions of Theorem (4.7), the following statements are equivalent

i) There exists an involutive automorphism $j : \kappa \rightarrow \kappa$ $(j^2 = id)$ such that for each $x,y\epsilon T$ and for each $\alpha, \beta = 1,\ldots,n$:

$$j(u_{\alpha\beta}(x,y)) \;\; = \frac{P_{\alpha\beta}(x,y)}{u_{\alpha\beta}(x,y)}$$

ii) Denoting H^* the space of κ-linear functionals $H \rightarrow \kappa$, there exist and additive map $J : H \to H^*$ and an involutive automorphism $j : \kappa \to \kappa$, such that for each $x\ T$ and $\alpha = 1,\ldots,n$:

$$J(\lambda a_\alpha(x)) \;=\; j(\lambda)\, a_\alpha^*(x) \quad ; \quad \lambda\ \varepsilon\ \kappa$$

where $\{a_\alpha(x)\}$ is the κ-basis of H defined in Theorem (4.7) and $a^*_\alpha(x)$ denotes its dual basis ($\langle a^*_\alpha(x),\ a_{\alpha'}(x)\rangle = \delta_{\alpha\alpha'}$; where $\langle\ \cdot\,,\cdot\ \rangle$ is the κ-valued duality $\langle H^*, H\rangle$).

Bibliography

1) L. Accardi – Topics in quantum probability.
   Phys. Rep. __77__ (1981), 169–192

2) L. Accardi – Foundations of quantum probability.
   Rendiconti del Seminario Matematico. Torino 1982

3) L. Accardi – Non-Kolmogorovian models in probability theory.
   Talk given at the III Vilnius Conference on Probability and Mathematical Statistics. June 1981 (to appear)

4) L. Accardi, A. Fedullo. On the statistical meaning of complex members in quantum theory.
   Lett. Nuovo Cim. 1982

5) L. Accardi, A. Fedullo . Statistical invariants for probabilistic models of finite valued observables.
   to appear

6) B. Mielnik – Geometry of quantum states.
   Comm. Math. Phys. __9__(1968), 55–80

7) M. Roos – Studies of the principle of superposition in quantum mechanics.
   Comm. Phys.–Math. __33__ (1966), 3–17.

8) M. Roos – J. Math. Phys. 5,(1964), 1609; (1965) 6, 1354.

9) J. Schwinger – Quantum kinematic and dynamics.
   Academic Press (1970)

STOCHASTIC JUMP PROCESSES IN THE PHASE SPACE

REPRESENTATION OF QUANTUM MECHANICS

J. BERTRAND and G. RIDEAU
Laboratoire de Physique Théorique et Mathématique
Université Paris VII - 75251 Paris - FRANCE -

We want to describe the evolution of quantum observables in terms of stochastic jump processes, following a procedure similar to that set up by Maslov [1] and extended by Ph. Combe et al [2] . To characterize these observables, we shall use the representation by functions on phase space due to Wigner [3] , and shall restrict mainly to observables described by bounded continuous functions (belonging to $C(\mathbb{R}^2)$.

1. Definition of the system and evolution of observables

First, let us recall the Wigner-Weyl correspondence between operators $\hat{A}$ and "functions" $A(q,p)$ written as :

$$(1) \quad A(q,p) = 2\pi\hbar \int_{\mathbb{R}^2} dx\, dy\; e^{2i\pi(xp+yq)} Tr\left(e^{-2i\pi(x\hat{P}+y\hat{Q})}\hat{A}\right)$$

where

$$\hat{Q} = q \;,\; \hat{P} = -i\hbar\, \partial_q$$

Now, we consider a system with hamiltonian

$$\hat{H} = h(\hat{P}) + \mu\hat{V}$$

where h is a continuous function and the symbol $V(q,p)$ of $\hat{V}$ is the Fourier transform of a bounded complex measure $d\nu = e^{i\varphi}\, d\rho$, φ real, ρ non negative :

$$(2) \quad V(q,p) = \int_{\mathbb{R}^2} d\nu(y,x)\; e^{2i\pi(yq+xp)}$$

The evolution of the Heisenberg observables symbols $A(q,p,t)$ is given by the equation

$$(3) \quad \frac{dA(q,p,t)}{dt} = \frac{i}{\hbar} \left\{ h(p) + \mu V(q,p) , A(q,p,t) \right\}$$

where $\left\{ , \right\}$ denotes the Wigner-Moyal bracket.

Writing the perturbation series for this equation, we get for the Fourier transform $\tilde{A}(\tilde{q},p,t)$:

$$\tilde{A}(\tilde{q},p,t) = \int dq \, e^{-2i\pi q\tilde{q}} \, A(q,p,t)$$

the following expression :

$$\tilde{A}(\tilde{q},p,t) = \sum_{m=0}^{\infty} \left(\frac{i}{\hbar}\right)^m \int_0^t dt_m \cdots \int_0^{t_2} dt_1 \sum_{\varepsilon_j = \pm 1} \mu^m \int_{\mathbb{R}^{2m}} \prod_{i=1}^m \varepsilon_i \, d\nu(y_i, x_i)$$

$$(4)$$

$$\cdot \exp 2i\pi \left\{ \sum_{i=1}^m x_i \left[p - \sum_{j=i+1}^m \pi\hbar \varepsilon_j y_j + \pi\hbar \varepsilon_i (\tilde{q} - \sum_{j=i}^m y_j) \right] \right\}$$

$$\cdot \exp \frac{i}{\hbar} \left\{ \sum_{i=1}^{m+1} (t_i - t_{i-1}) \sum_{\eta = \pm 1} \eta \, h (p - \sum_{j=i}^{m+1} \pi\hbar \varepsilon_j y_j + \eta\pi\hbar(\tilde{q} - \sum_{j=i}^{m+1} y_j)) \right\}$$

$$\cdot \tilde{A}(\tilde{q} - \sum_{i=1}^m y_i , p - \sum_{i=1}^m \pi\hbar\varepsilon_i y_i , 0)$$

$$t_0 = 0 , t_{m+1} = t , y_{m+1} = 0$$

Now, we are going to define a stochastic process $Z(t)$ with respect to which (4) can be regarded as an expectation value.

<u>Definition of the stochastic process $Z(t)$</u>

$Z(t) \equiv (X(t), Y(t), \varepsilon(t))$ has trajectories in $\mathbb{R}^2 \times \mathbb{Z}$ which are constants except for a number N_t of jumps $Z \equiv (x_i, y_i, \varepsilon_i) , i = 1, \dots, m,$ $(x_i, y_i) \in \mathbb{R}^2 , \varepsilon_i = \pm 1$, occuring at stochastic instants $t_i \in [0, t[$ and which arrive in 0 at t. N_t is a Poisson process with parameter $\lambda = 2\mu \int_{\mathbb{R}^2} d\rho$. The sizes Z_i of the jumps are independent random variables whose laws are related to $d\rho$:

$$\varepsilon_i = \pm 1 \quad \text{with probability } 1/2,$$

$$(x_i, y_i) \in (A, B) \subset \mathbb{R}^2 \quad \text{with probability } \int_{A \times B} d\rho \Big/ \int_{\mathbb{R}^2} d\rho$$

Moreover, we define the transformation J on the variables Z_i by :

$$J Z_i = (\varepsilon_i x_i , \varepsilon_i y_i , \varepsilon_i) \quad \text{and extend it trivially to } Z(t).$$

15

Thus, we can state :

Proposition I

Given a bounded continuous fonction $\tilde{A}_0(\tilde{q},p)$, the solution of equation (3) which for $t = 0$ reduces to $\tilde{A}_0$ is bounded continuous and can be written as :

$$\tilde{A}(\tilde{q},p,t) = e^{\lambda t}\, E_t\left\{ \hbar^{-N_t}\, e^{i\,\Xi_t}\, e^{2i\pi\left(pX(0)+\pi\hbar\tilde{q}\,JX(0)\right)}\right.$$

$$(5) \qquad \cdot \exp\left[\frac{i}{\hbar}\sum_{\eta=\pm 1}\eta\int_0^t \hbar\left(p-\pi\hbar\,JY(\tau)+\eta\pi\hbar\left(\tilde{q}-Y(\tau)\right)\right)d\tau\right]$$

$$\left. \cdot \exp\left[-2i\pi^2\hbar\int_0^t\left(JY(\tau)dX(\tau)+Y(\tau)dJX(\tau)\right)\right]\tilde{A}_0\left(\tilde{q}-Y(0),\,p-\pi\hbar\,JY(0)\right)\right\}$$

where Ξ_t is a functional of $Z(t)$ given, when $N_t = n$, by

$$\Xi_t(z_i, 1\leq i\leq n) = \sum_{i=1}^{n}\left(\frac{\pi}{2}\varepsilon_i + \varphi(y_i,x_i) + 2\pi^2\hbar\,\varepsilon_i y_i x_i\right)$$

2. <u>Case</u> $h(p) = \dfrac{p^2}{2m}$: <u>a stochastic process in phase space and its classical limit</u>.

<u>2.1</u> In the case of $h(p) = \dfrac{p^2}{2m}$ the above Fourier transform can be inverted explicitely and the result is an expression of $A(q,p,t)$ in terms of a genuine phase space stochastic process.

Proposition II

Given a bounded continuous function A_0, the solution of (3) which for $t = 0$ reduces to A_0 can be written as :

$$A(q,p,t) = e^{\lambda t}\, E_t\left\{ \hbar^{-N_t}\, e^{i\Psi_t}\, \exp\left[-\frac{2i}{\hbar}\int_0^t q(\tau)dJp(\tau)\right]\right. \cdot$$

$$(6) \qquad \left. \cdot \exp\left[2i\pi\int_0^t\left(p(\tau)dX(\tau)-Jx(\tau)dJp(\tau)\right)\right]A_0\left(q(0)+\pi\hbar\,JX(0),\,-p(0)\right)\right\}$$

where :

- $\quad \Psi_t(z_i, 1\leq i\leq n) = \sum_{i=1}^{n}\left(\frac{\pi}{2}\varepsilon_i + \varphi(y_i,x_i)\right)$

- $\quad p(\tau),q(\tau)$ are classically related paths given, when $N_t = n$, by

$$P(\tau) \equiv -p + \pi\hbar \sum_{j=i}^{m} \varepsilon_j y_j = -p + \pi\hbar \, JY(\tau)$$

$$q(\tau) \equiv q - \frac{1}{m} \int_{\tau}^{t} P(\sigma)\, d\sigma$$

$$= q + (p/m)(t-\tau) - (\pi\hbar/m) \sum_{j=i}^{m} \varepsilon_j y_j (t_j - t_{i-1})$$

for $t_{i-1} \leq \tau < t_i$.

This formula exhibits a "quasi" classical action $\int q\, dJp$.
However the residual terms $\int p\, dX - JX\, dJp$ have no classical
counterparts, since they come from instantaneous jumps x_i in configuration space. These terms disappear when the potential does not depend on p.

2.2 The classical limit.

This is most easily performed on the double Fourier trans-
form $\tilde{a}(\tilde{q},\tilde{p},t)$ of $A(q,p,t)$, which can be written after some rearrange-
ment as :

$$\tilde{a}(\tilde{q},\tilde{p},t) = \sum_{m=0}^{\infty} \left(\frac{\mu}{\hbar}\right)^m \int_0^t dt_m \cdots \int_0^{t_2} dt_1 \sum_{\varepsilon_j = \pm 1} \prod_{j=1}^{m} \left[d\rho(y_j, x_j) \right]$$

$$(7) \quad \cdot \exp\left\{ i \sum_{j=1}^{m} \left(\frac{\pi}{2}\varepsilon_j + \varphi(y_j, x_j) + 2\pi^2\hbar\,\varepsilon_j\,\Omega_j \right) \right\} .$$

$$\cdot \tilde{a}_0\left(\tilde{q} - \sum_{j=1}^{m} y_j \,,\, \tilde{p} - \frac{\tilde{q}}{m}t + \frac{1}{m}\sum_{j=1}^{m} y_j (t_j - t_0) \,,\, 0 \right),$$

$$t_0 = 0, \; t_{m+1} = t,$$

$$\Omega_j = -y_j \left[\tilde{p} + \frac{\tilde{q}}{m}(t_j - t) - \frac{1}{m}\sum_{i=j+1}^{m}(t_j - t_i)y_i - \sum_{i=j+1}^{m} x_i \right]$$
$$+ x_j \tilde{q} - x_j \sum_{i=j+1}^{m} y_i \; .$$

Noticing that

$$\sum_{\varepsilon_j = \pm 1} \prod_{j=1}^{m} \varepsilon_j \, e^{2i\pi^2\hbar\,\varepsilon_j\,\Omega_j} = (2i)^m \prod_{j=1}^{m} \sin(2\pi^2\hbar\,\Omega_j)$$

we obtain :

$$|\tilde{a}(\tilde{q},\tilde{p},t)| \leq \sum_{m=0}^{\infty} (4\pi^2\mu)^m \int_0^t dt_m \cdots \int_0^{t_2} dt_1 \int \prod_{j=1}^{m} \left(d\rho(y_j,x_j)|\Omega_j| \right) |\tilde{a}_0|$$

Using now the estimates, in a manner analogous to $\begin{bmatrix} 4 \end{bmatrix}$:

$$\prod_{j=1}^{m} |\Omega_j| \leq \frac{1}{m!} \left(\sum_{j=1}^{m} |\Omega_j| \right)^m$$

$$\left(\sum_{j=1}^{m} |\Omega_j| \right)^m < m! \; \exp\left\{ \sum_{j=1}^{m} |\Omega_j| \right\}$$

$$\sum_{j=1}^{m} |\Omega_j| \leq \left(|\tilde{p}| + \frac{|\tilde{q}|}{m} t + \frac{1}{m} \sum_{j=1}^{m} |y_j| + \sum_{j=1}^{m} |x_j| \right) \left(|\tilde{q}| + \sum_{j=1}^{m} |y_j| \right)$$

we get finally :

$$(8) \qquad |\tilde{a}(\tilde{q},\tilde{p},t)| \leq \sum_{m=0}^{\infty} \left(\frac{4\pi^2 t}{T} \right)^m \left[\int \exp\left\{ \sqrt{\tfrac{K}{m}}\,(T+t)|y| + \sqrt{m\mu}\,|x| \right\} d\rho(y,x) \right]^m$$

$$\cdot \exp\left\{ \sqrt{\tfrac{K}{m}}\,(T+t)|\tilde{q}| + \sqrt{m\mu}\,|\tilde{p}| \right\} |a_0|$$

where T is some time such that

$$T < \frac{4\pi^2 K m c}{\eta (1 + 4\pi^2 K)}$$

Under conditions ensuring the convergence on the right hand side of
(8), we obtain the absolute convergence of series (7), uniformly
in $\hbar$. Under the same conditions and using similar estimates, we get
the same property for all $\hbar$ derivatives of $\tilde{a}(\tilde{q},\tilde{p},t)$.
The precise theorem reads :

Proposition III

Let $A_0(q,p)$ be a classical observable with Fourier transform
$\tilde{a}_0(\tilde{q},\tilde{p})$ in $C(\mathbb{R}^2)$. If there exist constants ε , η and K such that
the potential (cf (2)) verifies

$$\int d\rho(y,x)\, e^{\varepsilon|y| + \eta|x|} < K ,$$

then $\tilde{a}(\tilde{q},\tilde{p},t)$ given by (7) is C^∞ in $\hbar = 0$ provided that

$$\mu < \eta^2/m \;\;, \;\; t < m\varepsilon / (1 + 4\pi^2 K)\eta$$

Moreover, the limit $\hbar = 0$, can again be written as an expectation
value :

$$\tilde{a}(\tilde{q},\tilde{p},t) = e^{\lambda t}\, E_t\left\{(-4\pi^2\mu)^{N_t}\, e^{i\Phi_t}\, \Omega_t\, \tilde{a}_o\big(\tilde{q}(o),\tilde{p}(o) - \chi(o)\big)\right\}$$

where

$$\tilde{q}(\tau) \equiv \tilde{q} - \gamma(\tau)$$

$$\tilde{p}(\tau) = \tilde{p} + \frac{1}{m}\int_t^\tau \tilde{q}(\sigma)\,d\sigma$$

$$\Phi_t(z_i,1\le i\le n) = \sum_{\delta=1}^{n}\varphi(y_\delta,x_\delta)\ ,\quad \Omega_t(z_i,1\le i\le n) = \prod_{\delta=1}^{n}\Omega_\delta$$

Conclusion

Thus, we have shown that the probabilistic interpretation of Feynman integral ([1] , [2]) can be applied to describe the evolution of some classes of observables in a quantum phase space. Formula (5) can be regarded as equivalent to a Feynman integral for this problem. We could have imagined that the stochastic character was specific to the quantum nature of the objects under consideration. However, as it results from the last paragraph, a similar description is also valid for a classical system with a sufficiently regular potential.

References

[1]   V.P. Maslov, A.M. Chebotarev. VINITI, Itogui Nauki, Vol 15 (1978) 5 ; translated in : Journal of Soviet Mathematics 13 (1980) 315.

[2]   Ph. Combe, R. Hoeg-Krohn, R. Rodriguez, M. Sirugue and M. Sirugue-Collin : Feynman path integrals with piecewise classical paths, to appear in J. Math. Phys.

[3]   See, for example, J. Pool, J. Math. Phys. 7 (1966) 66.

[4]   S. Albeverio, R. Hoegh-Krohn, Inventiones Math. 40 (1977) 59.

TRANSFORMATION OF WIENER INTEGRALS AND THE
DESINGULARIZATION OF COULOMB PROBLEM

Ph. Blanchard
Fakultät für Physik
Universität Bielefeld
D-4800 Bielefeld

§1. INTRODUCTION

The Feynman Path Integral, if not the only one, is a privileged tool to
quantize classical systems. Especially it gives a good understanding of the relationship between classical and quantum systems and also though the Euclidean formalism
it allows to use the techniques and results of classical statistical mechanics. It
is obvious that it would be likely to perform change of variables within Path integrals in order to exhibit connections between a priori different systems, and actually there has been studies in this direction. To give here rigorous results we shall
restrict ourselves to zero dimensional space viz to ordinary quantum mechanics.
Even in this case, one encounters interesting phenomena which show the complexity
of the problem, but also some of its advantage since it allows in a way we shall
make precise later to desingularize some problems, and in particular the Coulomb
problem. From a historical point of view, the existence of a correspondence between
the Coulomb problem and the harmonic oscillator problem has been remarked in the
early stage of quantum mechanics, and has been used in different contexts. This correspondence is suggested heuristically by the relationship which exists between the
Hermite functions giving the eigenvalues of the harmonic oscillator problem which
give an orthonormal basis for L^2 ($\mathbb{R}$, $e^{-x^2}dx$) and the Laguerre functions which play
a similar role for the Coulomb problems and the space L^2 ($\mathbb{R}_+$, $e^{-x}dx$).

This is confirmed for instance by the result of D. Fivel [1] who has shown that
a eigenvalue ψ_1 of the radial Schrödinger equation with a spherically symmetric potential of the type:

$$V_1(r) = \frac{g}{r} + G\frac{u(r)}{r} \tag{1.1}$$

and with energy eigenvalue E, is transformed into a solution of another radial
Schrödinger equation ψ_2 associated with the spherically symmetric potential

$$V_2(y) = \frac{1}{2} Ky^2 + Gu(y^2) \tag{1.2}$$

with energy eigenvalue ε, where one has made the following substitutions:

$$r = y^2, \ \psi_1(r) = y^{1/2}\psi_2(y) \tag{1.3}$$

Within such a transformation the respective roles of the coupling constant and energy are exchanged, indeed

$$K = - 8E, \ \varepsilon = - 4g \tag{1.4}$$

At the classical level this correspondence exists also and has been noted in 1965 by P.Kustaanheimo and E.Stiefel [2]. They have shown that there exists a mapping of $\mathbb{R}^3$ into $\mathbb{R}^4$ such that, if furthermore one makes a change of time

$$t \to \tau = \int \frac{dt}{r}, \tag{1.5}$$

the equations of the Kepler problem in $\mathbb{R}^3$ transform into linear equations with constant coefficients in $\mathbb{R}^4$, which are perfectly regular at the center of motion. They have used this regularity to study perturbations of the Kepler problem.

In 1979 I.H.Duru and H.Kleinert [3] have used the same transformation to show that the Green's function of the hydrogen atom in $\mathbb{R}^3$ can be expressed in a formal way using the Feynman Path Integral in terms of the Green's function of an harmonic oscillator in $\mathbb{R}^4$.

Here we want to discuss the same correspondence at the level of stochastic differential equations associated with the Euclidean version of the problem, i.e. with the corresponding imaginary-time Schrödinger equation which can be solved by the Wiener integral. To achieve that program we use of course the classical results of the Ito stochastic calculus as well as the probably less known results about stochastic change of time in stochastic differential equations, but also a classical theorem from A.Hurwitz [4] on the composition of quadratic forms

§2. HURWITZ'S THEOREM ABOUT COMPOSITION OF QUADRATIC FORMS

Historically the problem of composition of quadratic forms ask the following: For whats values of N does there exist a formula

$$(x_1^2 + x_2^2 + \ldots + x_N^2)(y_1^2 + \ldots + y_N^2) = z_1^2 + z_2^2 + \ldots + z_N^2, \tag{2.1}$$

or

$$\| x \|^2 \cdot \| y \|^2 = \| z \|^2, \tag{2.2}$$

where the z_i are homogeneous bilinear forms in x and y?

A famous result of A.Hurwitz states that this is only possible if N = 1,2,4 or 8. The case N = 1 is trivial and we will discuss the other cases briefly.

Such a decomposition proves to be useful for number theory, in the sense that it shows that if two numbers decompose as a sum of N squares, then their product also decomposes as a sum of N squares.

For N = 2 the corresponding identity

$$(x_1^2 + x_2^2)\,(y_1^2 + y_2^2) = (x_1 y_1 - x_2 y_2)^2 + (x_1 y_2 + x_2 y_1)^2, \tag{2.3}$$

is known as the Fibonacci formula and the corresponding decomposition for N = 4 has been established by Euler. The identity discovered by Euler in 1743 is not an obvious application of complex numbers and its discovery was a substantial achievement. Its verification, however, is trivial. It illustrates J.E.Littlewood's famous comment that any identity is trivial if written down by somebody else. To prove the Fermat's four-square theorem it suffices to show that any prime can be written as a sum of four squares. In 1770 Lagrange succeeded in construction a proof and in 1773 Euler (then 66 years old) gave a simpler proof-success after 43 years.

The assumption about z leads to write:

$$z = B(x)\, y, \tag{2.4}$$

where B(x) is an NxN matrix whose entries are linear homogeneous in x. More precisely:

$$B(x) = \sum_{i=1}^{N} B_i\, x_i, \tag{2.5}$$

where the B_i's are N NxN matrices with real entries. The condition $\| z \|^2 = \| x \|^2 \| y \|^2$ implies that:

$${}^t B(x)\, B(x) = \| x \|^2, \tag{2.6}$$

where ${}^t B(x)$ denotes the transpose of the matrix B(x). Consequently the B_i's satisfy the relations:

$${}^t B_i B_i = 1_N$$

$$\tag{2.7}$$

$${}^t B_j B_k + {}^t B_k B_j = 0 , \; j \neq k,$$

from what follows that N has to be even. Furthermore, and that is the Hurwitz's remark, at least 2^{N-2} of the following matrices

$$1, B_i, B_{i_1} B_{i_2}, \ldots, B_1 B_2 \ldots B_N,$$

are linearly independent. Consequently, one has the obvious inequality

$$2^{N-2} \leq N^2 \tag{2.8}$$

which is saturated for N = 8. Directly one can show that the case N = 6 is excluded. The solutions for N = 2,4,8 are given explicitly by the following formulas:

$$
N = 2 \qquad B(x) = \begin{vmatrix} x_1 & -x_2 \\ x_2 & x_1 \end{vmatrix}
$$

$$
N = 4 \qquad B(x) = \begin{vmatrix} x_1 & -x_2 & -x_3 & -x_4 \\ x_2 & x_1 & -x_4 & -x_3 \\ x_3 & x_4 & x_1 & x_2 \\ x_4 & -x_3 & x_2 & -x_1 \end{vmatrix}
\tag{2.9}
$$

$$
N = 8 \qquad B(x) = \begin{vmatrix}
x_1 & x_2 & x_3 & x_4 & x_5 & x_6 & x_7 & x_8 \\
-x_2 & x_1 & x_4 & -x_3 & x_6 & -x_5 & x_8 & -x_7 \\
-x_3 & -x_4 & x_1 & x_3 & x_7 & -x_8 & -x_5 & x_6 \\
-x_4 & x_3 & -x_2 & x_1 & -x_8 & -x_7 & x_6 & x_5 \\
-x_5 & -x_6 & -x_7 & x_8 & x_1 & x_2 & x_3 & -x_4 \\
-x_6 & x_5 & x_8 & x_7 & -x_2 & x_1 & -x_4 & -x_3 \\
-x_7 & -x_8 & x_5 & -x_6 & -x_3 & x_4 & x_1 & x_2 \\
-x_8 & x_7 & -x_6 & -x_5 & x_4 & x_3 & -x_2 & x_1
\end{vmatrix}
\tag{2.10}
$$

For N = 4 the matrix B(x) is connected with the rule of multiplication of quaternions and for N = 8 with the multiplication of octonions. Let us formulate the following:

<u>Theorem:</u> (Hurwitz [4]) If there exists a formula such that for all x,y

$$(x_1^2 + x_2^2 + \ldots + x_N^2)(y_1^2 + y_2^2 + \ldots + y_N^2) = z_1^2 + z_2^2 + \ldots + z_N^2,$$

where the z_1 are bilinear homogeneous functions of x and y then N = 1, 2, 4, 8.

Let us remark that

$$x \to B(x) \, x,$$

is for $N = 2$ a mapping from $\mathbb{R}^2 \to \mathbb{R}^2$
$N = 4$ a mapping from $\mathbb{R}^4 \to \mathbb{R}^3$
$N = 8$ a mapping from $\mathbb{R}^8 \to \mathbb{R}$

<u>Remark</u>: For $N = 2$ the matrix $B(x)$ has the following geometrical interpretation. Let $z = y_1 + iy_2$ be the plane of the trajectory of a moving particle and introduce a parameter plane $w = x_1 + ix_2$ mapped onto the physical plane conformally by the transformation $z = w^2$ ($y_1 = x_1^2 - x_2^2$, $y_2 = 2x_1x_2$). A conical section centered at the origin of the w-plane is transformed into a conical section of the z-plane having one focus at the origin. For this reason the transformation is a very practical method for the discussion of the Kepler problem.

§3. SOME RESULTS ABOUT STOCHASTIC CALCULUS

This section is devoted to the "formula of change of variables", the prototype of which is due to Ito and to a formula of change of time.

Let us briefly recall some well known results. For a more complete treatment, see e.g. [5] [6] and [7].

Suppose we consider an imaginary-time equation of the Schrödinger type:

$$\frac{\partial u(x,t)}{\partial t} + \frac{1}{2} \sum_{i,j,k} b_{ik}\, b_{jk}\, \frac{\partial^2 u(x,t)}{\partial x_i \partial x_j} + V(x)\, u\,(x,t) = 0 \tag{3.1}$$

(For the Schrödinger equation $b = \frac{1}{\sqrt{m}}$).
Assuming furthermore the existence of a strictly positive solution u_0 of the previous equation and putting

$$u(x,t) = u_0\,(x,t)\, \varphi\,(x,t)$$

then φ satisfies the following parabolic equation:

$$\frac{\partial \varphi(x,t)}{\partial t} + \frac{1}{2} \sum b_{ik}\, b_{jk}\, \frac{\partial^2 \varphi(x,t)}{\partial x_i \partial x_j}$$
$$+ \sum \beta_i(x,t)\, \frac{\partial \varphi(x,t)}{\partial x_i} = 0 \tag{3.3}$$

The solution of the previous equation in term of the initial condition $\varphi_0(x)$ is given in term of a stochastic process $\xi_t(s)$ as:

$$\varphi(x,t) = E_x[\varphi_0(\xi_t(s))] \tag{3.4}$$

where the stochastic process ξ_t satisfies the stochastic differential equation:

$$d\xi_t^i(s) = \beta_i(\xi_t(s),s)ds + \Sigma\ b_{ik}\ dW_k(s) \tag{3.5}$$

where W_k is the standard Wiener Process and

$$\xi_t(t) = x. \tag{3.6}$$

The drift term β_i in equation (3.3) allows to calculate the potential up to a constant viz the energy of the eigenfunction more precisely in the case where $b_{ij} = \frac{1}{\sqrt{m}}\ \delta_{ij}$

$$V(x) - E_0 = \frac{1}{2m}\ [\beta \cdot \beta + div\ \beta] \tag{3.7}$$

Formula (3.4) actually expresses the solution of the Schrödinger equation (with imaginary-time) as an integral over the paths $\omega(s)$ of the stochastic process ξ_t and it is possible to perform changes of variable within this integral according to the fundamental Ito's lemma, which is the corner stone of stochastic differential calculus.

Ito's Lemma

Let $\xi_i(t)$, ..., $\xi_N(t)$ be stochastic processes, solutions of the stochastic differential equations:

$$d\xi_k(t) = \beta_k(\xi)\ dt + \Sigma\ b_{kj}\ dW_j \tag{3.8}$$

let u be a C^2 function, then

$$u(\xi_1(t),\ \dots\ ,\ \xi_N(t)) \tag{3.9}$$

is a stochastic process solution of the stochastic differential equation

$$d\eta(t) = A(\eta)\ dt + \sum_j\ B_j dW_j \tag{3.10}$$

where

$$A = \sum_k \frac{\partial u}{\partial x_k}\ \beta_k + \frac{1}{2} \Sigma\ \frac{\partial^2 u}{\partial x_i\ \partial x_k}\ \sum_j\ b_{ij}\ b_{kj} \tag{3.11}$$

$$B_j = \Sigma_k\ \frac{\partial u}{\partial x_k}\ b_{kj} \tag{3.12}$$

We now consider the question of time-substitutions in stochastic differential equation whose coefficents do not depend explicitly on time.

<u>Theorem</u> (Gihman - Skorokhod) [5]

Let $\xi(t)$ be a diffusion process solution of the stochastic differential equation:

$$d\xi(t) = \beta\,(\xi)\,dt + b\,dW$$

Let us define a family τ_t of random variables by

$$t = \int_0^{\tau_t} ds\ g(\xi(s)).\tag{3.13}$$

For each t the variable τ_t is a Markov time. Provided that

$$\int_0^\infty g(\xi(s))ds = \infty$$

with probability one, then $\eta(t) = \xi(\tau_t)$ is a stochastic process which satisfies the following stochastic differential equation:

$$d\eta(t) = \frac{\beta(\eta(t))}{g(\eta(t))}\,dt + \frac{b(\eta(t))}{\sqrt{g(\eta(t))}}\,dW\tag{3.14}$$

We shall gather all these results to transform a problem with a singularity of the Coulomb type at the origin into a harmonic oscillator like problem

§4. TRANSFORMATION OF THE WIENER INTEGRAL

Let us consider the imaginary-time Schrödinger equation in $\mathbb{R}^N$ (N = 1,2,4,8)

$$\frac{\partial u(x,t)}{\partial t} = -\frac{1}{2m}\Delta u(x,t) + V(||x||^2)u(x,t),\tag{4.1}$$

where we assume that the potential V is sperically symmetric, and moreover goes to infinity in such a way that its spectrum is positive and purely discrete. Under these conditions the ground state is strictly positive and spherically symmetric. Consequently the stochastic differential equation associated with (4.1) has the form:

$$d\xi_i(s) = -F(||\xi(s)||^2)\,\xi_i(s)\,ds + \frac{1}{\sqrt{m}}\,dW_i(s)\tag{4.2}$$

where the W_i are independent Wiener Processes.

Using the matrices B_i of the Hurwitz's theorem, one can define a new stochastic process $\Theta\,(t)$

$$\Theta(s) = B(\xi(s))\,\xi\,(s)\tag{4.3}$$

then by Ito's Lemma:

$$d\Theta(s) = (-2F(||\Theta(s)||)\Theta_i(s) + 2c_i)ds$$
$$+ \frac{2}{\sqrt{m}} \sum_{k,l} (B_k)_{il}\,\xi_k(s)\,dW_l\tag{4.4}$$

moreover $c_i = \sum_k (B_k)_{ii}$. For N = 2 and 4 the c_i's are zero but for N = 8 C_1 = 16. Solutions of stochastic differential equations of this form have been called "distorted Brownian motion".

Now if we introduce the Markov time τ_t

$$t = \int_0^{\tau_t} ds \, ||\Theta(s)|| \tag{4.5}$$

and a new stochastic process η_t:

$$\eta(t) = \Theta(\tau_t) \tag{4.6}$$

it satisfies the stochastic differential equation

$$d\eta_i(t) = \left(-2\frac{F(||\eta(t)||)}{||\eta(t)||} \eta_i(t) + \frac{2c_i}{||\eta(t)||}\right) dt \tag{4.7}$$

$$+ \frac{2}{\sqrt{m}} \frac{1}{||\eta(t)||^{1/2}} \sum_{k,l} (B_k)_{1i}\xi_k(\tau_t)dW_1$$

The last factor in the previous expression looks a bit complicated, however it can be simplified by the following lemma:

<u>Lemma</u>: Let $\xi_i(t)$ be stochastic processes whose differentials are given by:

$$d\xi_i(s) = \frac{1}{m^{1/2}} \frac{2}{||\eta(s)||^{1/2}} \sum_{k,l} (B_k)_{1i} \, \xi_k(\tau_s) \, dW_1$$

They are independent Wiener processes.

Proof is a direct application of the previous general results, indeed let h be a smooth function, then

$$H(x,t) = E_x[h(\xi(s))], \tag{4.8}$$

satisfies the partial differential equation:

$$\frac{\partial H(x,t)}{\partial t} = -\frac{2}{m} \sum_i \frac{\partial^2}{\partial x_i^2} H(x,t) \tag{4.9}$$

With the initial condition:

$$\lim_{t \to 0} H(x,t) = h(x)$$

Let us take the function

$$h(x) = \exp(i\lambda.x)$$

The solution of 4.9 is given by

$$H(x,t) = \exp\left(- \frac{2t}{m} ||\lambda||^2 + i\lambda.x\right)$$

which is precisely the characteristic function of a Wiener process. We can summarize our results in the following

<u>Theorem:</u>

If $\eta(t)$ is a stochastic process in $\mathbb{R}^N$ ($N = 1,2,4,8$) whose stochastic differential equation is

$$d\eta_i(t) = - F(||\eta(t)||^2)\, \eta_i(t)\, dt + \frac{1}{\sqrt{m}}\, dW_i$$

one can associate a stochastic process in $\mathbb{R}^n$ ($n = 1,2,3,1$) through the relation

$$\Theta(t) = B(\eta(\tau_t))\eta\,(\tau_t)$$

where the B's are the Hurwitz's matrices satisfying

$$^tB(x)\ B\ (x) = \Sigma x_i^2 = ||x||^2$$

and the Markov time τ_t is defined by

$$t = \int_0^{\tau_t} ds\,||\eta(s)||.$$

The random process Θ in $\mathbb{R}^n$ satisfies the stochastic differential equations

$$d\Theta_i(s) = \left(\frac{-2F(||\Theta(s)||)}{||\Theta(s)||}\, \Theta_i(s) + \frac{2\ c_i}{||\Theta(s)||}\right)\, ds + \frac{2}{m^{1/2}}\, dW_i$$

where the W_i are independent Wiener processes.

One can see easily that if we start from a potential V of the form

$$V(x) = \frac{1}{2m}\, [(F(||x||^2)^2 + F'(||x||^2))||x||^2 + NF(||x||^2)]$$

the stochastic process Θ solves an (imaginary time) Schrödinger equation associated with a potential W given by

$$W(x) = \frac{2}{m}\, [F(||x||)^2 + F'(||x||) + (n-1)\, \frac{F(||x||)}{||x||}\,]$$

As a special case the pure harmonic oscillator problem in four dimensions $F(x^2) \equiv 1$ is transformed into the Coulomb problem in 3 dimensions. As in classical mechanics (cf.[2]) one can go from the Coulomb problem to the harmonic oscillator problem by using the inverse of the transformation (2.9). However, this transformation is singular but we can choose an inverse by imposing the auxiliary condition.

$$x_4 \, dx_1 - x_3 \, dx_2 + x_2 dx_3 - x_1 \, dx_4 = 0$$

and the two problems are then completely equivalent.

REFERENCES

[1]   D. Fivel
       Phys. Rev. 142, 1219-1235 (1966).

[2]   P.Kustaanheimo, E.Stiefel
       J.Reine Angew. Math. 218, 204-219 (1965).

[3]   I.H.Duru, H.Kleinert
       Phys. Lett. B 84, 185-188 (1979).

[4]   A. Hurwitz
       Math. Werke II, 565-571. See also T.Y.Lam,
       The Algebraic Theory of Quadratic Forms, Benjamin, New York, 1973.

[5]   I.I.Gihman, A.V.Skorohod
       Stochastic Differential Equations, Springer, 1972.

[6]   I.I.Gihman, A.V.Skorohod
       Introduction à la Théorie des Processus Aléatoires,
       Editions Mir (1980).

[7]   Ph.Blanchard, M.Sirugue
       J.Math.Phys. 22 1372-1376 (1981)

<u>DYNAMICAL SYSTEMS WITH FEW DEGREES OF FREEDOM</u>

P. COLLET

Centre de Physique Théorique de l'Ecole Polytechnique

Plateau de Palaiseau - 91128 Palaiseau - Cedex - France

"Groupe de Recherche du C.N.R.S. n° 48"

I - <u>Introduction</u>

The recent years have witnessed new developments in the theory of
dynamical systems with few degrees of freedom. One of the reason for
these developments (but not the only one) is the possibility of
visualizing complicated mathematical objects using a computer.
Strange Julia sets were known to exist fifty years ago, but they were
not studied in details, mainly because there was no global representation of them. Simultaneous to these computer analysis of the problems,
an important mathematical development took place. Two major ideas came
out from these studies : *simple dynamical systems can have chaotic behaviors, and chaos is frequent in physical systems* (or at least in models of physical systems). A corollary of the above remark is that there is no need for infinitely many degrees of freedom in order to have
a chaotic behavior. There is even more, chaos in systems with infinitely many degrees of freedom can be "driven" by a finite dimensional
subsystem ([T.F.],[M.P]). This is the whole spirit of the truncations
of the various hydrodynamic equations. We shall come back to this question later on. Such a behavior seems to occur near the onset of turbulence in some experiments. But, at the present time, the nature of fully developped turbulence is not at all settled.

Using a Galerkin approximation, the evolution equations of a physical system (usually field equations) are reduced to a finite system of coupled non linear differential equations,

$$\frac{d\vec{x}}{dt} = \vec{F}(\vec{x}) \ , \ \vec{x} \in \mathbb{R}^n \ .$$

One of the most important example is the Rayleigh-Benard experiment in which a container of fluid is heated from below. Truncating the Boussinesq equations and keeping only two Fourier modes of the temperature field, and one Fourier mode of the velocity potential, one gets the Lorenz system [Lo.1],[La],[Ru.1].

$$\frac{dx}{dt} = \sigma x + \sigma y$$

$$\frac{dy}{dt} = -xz + rx - y$$

$$\frac{dz}{dt} = xy - bz$$

where b is a geometrical constant, σ the Prandtl number, and r the Rayleigh number. Of course we have obtained this system after a drastic approximation of the initial equations. It is not clear that we have captured any interesting physical behavior. This question will be briefly considered later on, and we shall assume that our truncation is still an interesting model. Another type of dynamical systems appears frequently. In those systems, the time evolution is discrete. The successive states of the system are described by a vector $\vec{X}_q$ in $\mathbb{R}^n$, and

$$\vec{X}_{q+1} = T(\vec{X}_q)$$

where T is a map from $\mathbb{R}^n$ to $\mathbb{R}^n$. The case of discrete time is sometimes simpler, and one can often reduce a continuous time system to a discrete one using for example a Poincaré map or an invariant foliation. From now on, we shall discuss most of the time the discrete case. It is easy in general to derive the corresponding statements for the continuous time case. The most popular examples of maps are the logistic map $x \to \mu x(1-x)$ in one dimension, and the Henon map in two dimensions [He]

$$\begin{pmatrix} x' \\ y' \end{pmatrix} = \begin{pmatrix} 1+y - ax^2 \\ bx \end{pmatrix} \, .$$

Despite their apparent simplicity, these two maps, as well as the Lorenz system can exhibit very complicated time behavior. We shall be mainly interested in the dissipative case, which is defined by

$$\sup_x \left| \det DT_x \right| < 1 \, ,$$

and we shall assume that T is not linear.

A natural question comes up immediatly : given a simple dynamical system, is it possible to decide whether it has chaotic behaviors or not? We are far from any satisfactory answer to this question in the general case, although the notion of hyperbolicity described briefly in paragraph IV can be considered as a partial answer. We shall now review some interesting notions which help clarifying the above question. They have emerged from the mathematical development of the theory, although, as we shall see, they have an important physical content. We first describe the notion of attractor. Transient regimes are more complicated and less interesting than asymptotic regimes. The attractors are the subsets of phase space which describe the long time behavior of the system. A possible definition is : Ω is an attractor for T, if there exists a neighborhood U of Ω such that $T U \subset U$, and $\bigcap_n T^n U = \Omega$ where $T^n = T \circ T \circ \ldots \circ T$ n times. The physical meaning of this definition is that an orbit which penetrates in U will be asymptotic to Ω . This definition is probably too restrictive, see [Ru.3]. A simple example is given by the map $f : x \to 2x(1-x)$ for $x \in [o,1]$. $x = \frac{1}{2}$ is the only attractor for f in $[o,1]$, and one can take $U =]o,1[$. This is an example of the simplest kind of attractor : a stable fixed point. Let T be a c^1 map of $\mathbb{R}^n$, we shall say that M_o is a stable fixed point for T

32

If $\qquad T(M_O) = M_O$, and $SP(DT_{M_O}) \subset \{z \in C, |z| < 1\}$.

It is easy to verify that if M is near enough to M_O, then $T^n M \to M_O$. M_O is an attractor (one can show that DT_{M_O} is a contraction in some norm equivalent to the Euclidean norm).

In the general case, it is useful to add an irreducibility condition to the definition of an attractor. For example, one can impose that Ω should contain a dense orbit. Note that an attractor is not necessarily connected and that a map T can have more than one attractor (the Henon map has infinitely many attractors for some values of a and b [Ne]). If T has more than one attractor, one can ask which attractor is relevant for a given initial condition. The concept of basin of attraction is associated with the notion. The basin $B(\Omega)$ of the attractor Ω is the set of points whose orbit is asymptotic to Ω, i.e.

$$B(\Omega) = \{M, d(T^n(M),\Omega) \to O \text{ if } n \to +\infty\},$$

where d is the usual distance in R^n . Even a stable fixed point can have a basin of attraction with a complicated structure (for example, the boundary of the basin can have a Hausdorff dimension greater than 1 [Ma]).

Another important physical input is the parameter dependence. During an experiment, one tries to record the changes in the behavior of the system which are due to the modifications of some physically relevant parameter. One of the most well known parameters is the Reynolds number. Therefore, it is important to study one parameter families of transformations. We shall denote them by T_μ, μ being the parameter. It turns out that it is in some sense easier to study one parameter families than single maps. There has been recently some interesting results for the case of two parameters ([Gu],[Ch]), however the problem is much more difficult. When the parameter changes, many things can happen. The simplest phenomenon is a change in the topologocal structure of the attractor. This is called a bifurcation. A simple example is given by the family of maps f_μ of $[o,1]$: $x \to f_\mu(x) = \mu x(1-x)$. If $1 < \mu < 3$, f_μ has a stable fixed point, and this is the only attractor. If μ is slightly bigger than 3, the only attractor of f_μ is a stable period 2, i.e. a couple of points $\{x_\mu^{(1)}, x_\mu^{(2)}\}$ such that

$$f_\mu(x_\mu^{(1)}) = x_\mu^{(2)}, f_\mu(x_\mu^{(2)}) = x_\mu^{(1)} \text{ and } |(f_\mu \circ f_\mu)'(x_\mu^{(1)})| = |(f_\mu \circ f_\mu)'(x_\mu^{(2)})| < 1$$

$\mu = 3$ is the bifurcation point. Note that for μ greater than 3, f_μ still has a fixed point, but this fixed point is not stable anymore. The derivation of f_μ at this point has a modulus greater than one, and this is sometimes called a repulsor. Another important notion is the stability under small perturbations. A physical system is usually not isolated, it interacts with the exterior world. This creates a lot of uncontrolled perturbations which are assumed to be small (for example the thermal fluctuations). Besides, we are using models which have been obtained by truncating the real equations. The true models are considered as perturbations of the simplified ones. A related idea is that we want to study "typical cases", i.e. cases which occur in general. Special situations (for example when additional symmetries are present) have to be studied separately. A possible mathematical version of "typical case" is the notion of genericity. This means typical in the topological sense : a property is generic if it is true on a residual set. This notion has led to many interesting developments [Ar] and we shall use it frequently. However, when looking at the turbulent regime, this definition is not adequate, and must be replaced by some measure theoric concept [Ja, C.E.1].

A typical experiment (or computer simulation) starts with a value of μ for which the system is in a state as simple as possible. Usually this state is described by a stable fixed point. As μ increases, bifurcations take place, the behavior becomes more and more complicated, and eventually a chaotic state is reached. A succession of bifurcations from a stable fixed point to a turbulent state is called a route to turbulence or a scenario [E].

II - <u>Generic bifurcations of a stable fixed point. Routes to turbulence</u>

Let M_μ be a fixed point of $T_\mu (T_\mu (M_\mu) = M_\mu)$. From the stability condition, we have a bifurcation from a stable fixed point if $Sp\ DT_\mu|M_\mu$ intersects the unit circle for some critical μ_c. There are three generic possibilities for this intersection. Each of them is the first milestone in a route to turbulence. Note that $Sp\ DT_\mu|M_\mu$ is invariant by complex conjugation since $DT_\mu|M_\mu$ is a real matrix. We shall now examine these three important bifurcations. We shall assume that for $\mu < \mu_c$, $Sp\ DT_\mu|M_\mu$ is inside the unit disk.

<u>Case I</u> - $1 \in Sp\ DT_{\mu_c}|M_{\mu_c}$: Saddle node bifurcation.

Generically 1 is a simple eigenvalue. When μ crosses μ_c, the stable fixed point desappears, and a complicated turbulent behavior can occur. This turbulent behavior is usually intermittent [P.M] : turbulent bursts of random durations are separated by laminar phases of duration of order $|\mu-\mu_c|^{1/2}$.

<u>Case II</u> - $e^{\pm i\theta} \in Sp\ DT_{\mu_e}|M_{\mu_e}$ $\theta \neq 0\ mod(\pi)$, Hopf bifurcation.

Generically one can assume that only one pair of complex conjugate eigenvalues is of modulus one. For μ greater than, but near μ_c, the map T_μ has an invariant closed curve. This bifurcation is the first step in the Ruelle-Takens route to turbulence. We shall now describe this route for flows. A stable fixed point bifurcates to a stable closed curve, which then bifurcates to a stable 2 torus. The next event would be a bifurcation to a 3 torus. However, in every neighborhood of this vector field, one can find a vector field with a strange attractor [R.T.][R.T.H.]. In other words, there are small perturbations which produce a strange attractor.

Case III - $1 \in \mathrm{Sp}\ DT_{\mu_c}|_{M_{\mu_c}}$, Pitchfork bifurcation.

For $\mu > \mu_c$, but near μ_c , the map T_μ has a stable period 2. That is to say there is a pair of points $M_\mu^{(1)}$, $M_\mu^{(2)}$ such that

$$T_\mu(M_\mu^{(1)}) = M_\mu^{(2)}, \quad T_\mu(M_\mu^{(2)}) = M_\mu^{(1)}, \quad \text{and } \mathrm{Sp}\ DT_\mu^2{}_{M(1)} = \mathrm{Sp}\ DT_\mu^2|_{M_\mu^{(2)}} \quad \text{is}$$

inside the unit disk. This bifurcation is also known as the period doubling route to turbulence is a succession of infinitely many period doubling bifurcations.

We have now described the three generic bifurcations of a stable fixed point together with some well known routes to turbulence. However there are many other routes. One can encounter for example combinations of the above routes, inverse bifurcations, hysteresis, bifurcations of higher dimensional objects, and so on. A large variety of routes has already been observed experimentally [Li],[Go].

III - The period doubling route to turbulence

This route, which was discovered by Feigenbaum [Fe] and Coullet and Tressere [Te], is interesting because it shows some universal phenomenon which should make it easy to identify. However, although we have now many numerical and experimental examples, this route does not seem to be more frequent than any other route [Go]. It is however favored if the system has only one excited mode since no Hopf bifurcation can take place. We now describe some properties of this route, and refer to [C.E2] for proofs in some cases, and references. The first property is that there is an open set U of C^ω one parameter families of maps where this route occurs. This can be considered as an answer to the stability question. For noisy perturbations, see [C.F.H.]. Let μ_n be the value of the parameter for which a stable period 2^n bifurcates to a stable period

2^{n+1} for a one parameter family in U.

Then
$$\lim_{n \to +\infty} \frac{\mu_{n+2} - \mu_{n+1}}{\mu_{n+1} - \mu_n} = \delta$$

and this number δ is the same for any family in U. The numerical value of δ is 4.6692.. It is possible to make some quatitative predictions about the behavior of the system for $\mu > \mu_c = \lim_n \mu_{n+1}$. For example one can give an approximate value of μ where a stable period 3 can occur [C.E.1].However, as we have mentionned before, this estimate can be destroyed by the branching of another route. Some universal properties of the power spectrum have also been proven. See [C.E.T.] and references there. Similar phenomenon occur in the conservative case but with a different value of δ . See [C.H.E.] and references there.

IV - <u>Some remarks about the turbulent behaviors of simple systems</u>.

A natural question at this point is : what is at the end of a route to turbulence ? the answer is not known and at this moment we can only make few remarks about the complexity of the turbulent state. We shall denote by μ_e the value of μ at the end at the route (for example the accumulation point of period doubling bifurcations). It is not true that every μ greater than μ_e gives rise to a turbulent behavior. Consider for example the family $f_\mu(x) = \mu x(1-x)$ for $x \in [o,1]$. In the segment $]\mu_e,4]$, there is a dense set of values of μ for which f_μ has a stable periodic orbit [D.H.]. However, the set of values of μ for which f_μ has an invariant measure absolutely continuous with respect to the Lebesgue measure is of positive Lebesgue measure [Ja]. Therefore, if the parameter μ is choosen at random in $]\mu_e,4]$ with respect to the Lebesgue measure, there is a large probability (of the order of 95 % [Lo.2]) to pick a μ giving rise to a turbulent state. This kind of randomness can be imposed to the system by external constraints (for example thermal fluctuations), and we have here an example where the topological notion of genericity is not adequate. Even less is known in higher dimensions. We shall mention however the Newhouse phenomenon [Ne], which for the Henon map implies the existence of infinitely many stable periodic points for an infinite set of values of the parameter. We shall now briefly discuss the question of the "quality" of chaos. This is essentially the subject of ergodic theory. An interesting notion of strong chaos has been introduced by Ruelle [Ru.2]: sensitive dependence on initial conditions. From the relation $T^n(x) - T^n(y) = DT^n_z(x-y)$, we conclude that it is interesting to look at the behavior of DT^n_z when $n \to \infty$. The logarithms of the eigenvalues of $((DT^n_z)^t DT^n_z)^{\frac{1}{2n}}$ are called the Lyapunov exponents. Although the positivity of some Lyapunov exponents is a good criterion for chaos (this is also called hyperbolicity), it turns out to be very difficult to verify this condition either from a numerical or from a mathematical point of view. The picture of an attractor with a hyperbolic structure is as follows. The directions corresponding to positive Lyapunov exponents are inside the attractor , while the transversal directions, which correspond to negative Lyapunov exponents, are contracting directions. For more details, see [Bo].

REFERENCES

The interested reader is urged to consult [Ar] and [Ec].

[Ar]   V. Arnold. Chapitres supplémentaires de la théorie des équations différentielles. Editions MIR, Moscou, 1980.

[Bo]   R. Bowen. Equilibrium states and the ergodic theory of Anosov diffeomorphisms. Lecture notes in Mathematics, Vol.470,Springer Verlag, Berlin, Heidelberg, New York, 1975.

[C.E.1] P. Collet, J.P. Eckmann. Comm. Math. Phys. 73, 115 (1981).

[C.E.2] P. Collet, J.P. Eckmann. Iterated maps on the interval as dynamical systems. Birkhaüser, Basel, Boston, 1980.

[C.E.K.] P. Collet, J.P. Eckmann, H. Koch. Physica $\underline{30}$, 457 (1981).

[C.E.T.] P. Collet, J.P. Eckmann, L. Thomas. Commun. Math. Phys. $\underline{81}$, 261 (1981).

[C.F.H.] J.P. Crutchfield, J.D. Farmer, B.A. Huberman. Preprint, University of Santa Cruz, 1980.

[Ch] A. Chenciner. In Proceedings of Les Houches Summer School 1981. R.H. Helleman and G. Iooss editors. To appear.

[C.T.] P. Coullet, C. Tressere. Comptes Rendus Acad. Sciences $\underline{287}$, 577 (1978).

[D.H.] Douady, Hubbard. To appear.

[Ec] J.P. Eckmann. Rev. Mod. Phys. $\underline{53}$, 643 (1981).

[Fe] M. Feigenbaum. J. Stat.Phys. $\underline{19}$, 25 (1978); $\underline{21}$, 669 (1979).

[Go] J. Gollub. In Proceedings of Les Houches Summer School 1981. R.H. Helleman and G. Iooss editors. To appear.

[Gu] Guckenheimer. Preprint, University of Santa Cruz 1981.

[He] M. Henon. Commun. Math. Phys. $\underline{50}$, 69 (1976).

[Ja] M.V. Jakobson. Commun. Math. Phys. $\underline{81}$, 39 (1981).

[La] O.E. Lanford. In statistical Mechanics and Dynamical Systems. D. Ruelle editor. Duke university, Durham, 1977.

[Li] A. Libchaber. In Proceedings of Les Houches Summer School 1981. R.H. Helleman and G. Iooss editors. To appear.

[Lo.1] E.H. Lorenz. J. Atm. Sciences $\underline{20}$, 130 (1963).

[Lo.2] E.H. Lorenz. In global analysis . M.Ormela and J. Marsden editors Lecture Notes in Mathematics Vol. 755. Springer-Verlag, Berlin, Heidelberg, New York,1979.

[Ma] B.B. Mandelbrot. In Nonlinear Dynamics. R.G. Helleman editor. Annals of the New York Academy of Sciences. Vol. 357 (1980).

[MP] J. Mallet-Paret. J. Diff. Equ. $\underline{22}$, 331 (1976).

[Ne] S.E. Newhouse. Publ.Math.I.H.E.S. $\underline{50}$, 101 (1979).

[P.M.] Y. Pomeau, P. Manneville. Physica $\underline{01}$, 219, (1980).

[R.T.] D. Ruelle, F. Takens. Commun. Math. Phys. $\underline{20}$, 167 (1971).

[R.T.N.]D. Ruelle, F. Takens, S. Newhouse. Commun. Math. Phys. $\underline{64}$, 35
        (1978).

[Ru.1] D. Ruelle. Statistical Mechanics and dynamical systems.
        Duke University, Durham, 1977.

[Ru.2] D. Ruelle. Ann. New York Acad. Sci. $\underline{316}$, 408 (1978).

[Ru.3] D. Ruelle. Commun. Math. Phys. $\underline{82}$, 137 (1981).

[T.F.] T. Temam, C. Foias. J. Math. Pures et Appl. $\underline{58}$, 339 (1979).

LAPLACE EXPANSIONS OF CONDITIONAL WIENER INTEGRALS

AND APPLICATIONS TO QUANTUM PHYSICS

IAN DAVIES   AUBREY TRUMAN

Mathematics Department, Heriot-Watt University
Riccarton, Currie, Edinburgh EH14 4AS

1. Introduction

This paper is an expository account of rigorous results on the Laplace expansions
of conditional Wiener integrals, their proofs, and some of their applications to quantum mechanics[1]. This study was largely inspired by the earlier pioneering work on
Wiener integrals by Michael Schilder[2], one of Donsker's students, and by Barry Simon's
excellent book on functional integration[3]. Most of the ideas in this subject can be
traced back to Monroe Donsker[4] and we are grateful to him and to Barry Simon for pointing us in the direction of Schilder's work.

Basically our results show how Schilder's rigorous work on the Laplace expansions
of Wiener integrals can be extended to conditional Wiener integrals with functional
integrands having a finite number of non-degenerate global maxima. Recently abstract
results of this kind have been published by Ellis and Rosen[5], but, as we explain below,
our results cannot easily be deduced from theirs.

After explaining our Laplace theorem we give two applications to Schrödinger
operators. The first application is to a generalised Mehler kernel formula for
Schrödinger operators which was implicit in earlier work of Cécile DeWitt[6] and
Maurice Mizrahi[7] (for related results see Ref. 8). Here we only give a generalised
Mehler kernel formula for quantum mechanical Hamiltonians, $H(\hbar) = -2^{-1}\hbar^2 \dfrac{d^2}{dx^2} + V(x)$,
acting on $L^2(\mathbb{R})$ (c.f. Example 1). However, our methods also yield results for quantum
mechanical Hamiltonians, $H(\hbar) = -2^{-1}\hbar^2\Delta_x + V(\underline{x})$, acting on $L^2(\mathbb{R}^n)$, even for non-
separable systems, and in the case $n = 3$ for Zeeman effect Hamiltonians,
$H(\hbar) = 2^{-1}(i\hbar\underline{\nabla}_x + 2^{-1}\underline{B} \wedge \underline{x})^2 + V(\underline{x})$. We give only a simple example of our results for
Zeeman effect Hamiltonians in this paper (c.f. Example 2).

The second application is to the remarkable formula of Bender and Wu for the
X^{2N} anharmonic oscillator[9]. Following formal ideas of Lipatov[10] and Brezin[11] et al,
and Simon's[12] more rigorous work, we obtain Bender-Wu type formulae for the large
order behaviour of the Rayleigh-Schrödinger perturbation coefficients for the ground
state energy of the X^{2N} anharmonic oscillator from 'instanton' contributions to the
Laplace expansion of a conditional Wiener integral. Our 'instantons' are somewhat
different from those considered by previous authors, but remarkably the Laplace expansion still gives correctly the rapidly varying factors in the large order behaviour.
Our expansion offers some possibility of determining the large order behaviour to

higher orders, but this is only at the expense of evaluating some very involved conditional Wiener integrals. This does not seem practicable at present.

2. Preliminaries and the Laplace Theorem

In what follows $C_0[0,T]$ denotes the Banach space of continuous functions $z:[0,T] \to \mathbb{R}$, with $z(0) = z(T) = 0$, and sup norm

$$\|z\| = \sup_{\tau \in [0,T]} |z(\tau)|.$$

$C_0[0,T]$ supports the conditional Wiener measure $\mu_{0,0;0,T}$ with covariance

$$\int z(s)z(t)d\mu_{0,0;0,T}(z) = (2\pi T)^{-\frac{1}{2}}s(1 - t/T),$$

for $0 \leqslant s < t \leqslant T$, with zero mean. For the associated probability measure $\mu_{0,0;0,T}\{C_0[0,T]\}^{-1}\mu_{0,0;0,T} = (2\pi T)^{\frac{1}{2}}\mu_{0,0;0,T}$, we write

$$(2\pi T)^{\frac{1}{2}} \int_{C_0[0,T]} F(z)d\mu_{0,0;0,T}(z) = \mathbb{E}_z^T\{F(z)\}$$

and, for measurable sets A with characteristic function χ_A,

$$\mathbb{E}_z^T\{\chi_A(z)\} = \mathbb{E}_z^T\{A\}.$$

$C_0^*[0,T]$ is the reproducing kernel Sobolev space associated with $C_0[0,T]$; $z \in C_0^*[0,T]$ if z is absolutely continuous with derivative $\dot{z} \in L^2[0,T]$, $\int_0^T [\dot{z}(\tau)]^2 d\tau < \infty$. Defining the inner product (z,w) by

$$(z,w) = \int_0^T \dot{z}(\tau)\dot{w}(\tau)d\tau,$$

for the kernel $\rho(\sigma,\tau) = \sigma(1 - \tau/T), \sigma < \tau,$ we have the reproducing property

$$(z(\cdot), \rho(\sigma,\cdot)) = z(\sigma),$$

$\sigma \in [0,T]$, for all $z \in C_0^*[0,T]$. This is useful in evaluating conditional Wiener integrals.

We now come to the statement of our Laplace theorem.

Theorem 1

Let $F(z)$ be a real-valued continuous functional defined on $C_0[0,T]$ and suppose that the functional $\left\{F(z) - 2^{-1}\int_0^T [\dot{z}(\tau)]^2 d\tau\right\}$ has a finite number of distinct global maxima $x_i \in C_0^*[0,T]$, with

$$\left\{ F(x_i) - 2^{-1} \int_0^T [\dot{x}_i(\tau)]^2 d\tau \right\} = b \ ,$$

for $i = 1,2,\ldots,n$. If F satisfies conditions 1 - 6 below, then

$$\exp\{-b\lambda^{-2}\} \, \mathbb{E}_z^T\{\exp\{\lambda^{-2}F(\lambda z)\}\} = \Gamma_0 + \lambda\Gamma_1 + \lambda^2\Gamma_2 + \lambda^3\Gamma_3 + \ldots + \lambda^{m-3}\Gamma_{m-3} + O(\lambda^{m-2}),$$

as $\lambda \to 0$, where the Γ_j are conditional Wiener integrals depending only on the functional F and its Frechet derivatives evaluated at x_i, $i = 1,2,\ldots,n$.

Conditions 1 - 6

(1) $F(z)$ is measurable.

(2) $F(z) \leqslant (b+L_1) + L_2\|z\|^2$, $\mu_{0,0;0,T}$ a.e., L_1 and L_2 being positive numbers, with $L_2 < \min\{\gamma/2T,\ 1/4T\}$, γ being the constant in Lemma 1 below.

(3) $F(z)$ is continuous for $\|z\| \leqslant \max\left\{ \dfrac{(L_1 + 1)^{\frac{1}{2}}}{|L_2 - 1/2T|^{\frac{1}{2}}} ,\ \dfrac{(2T(L_1 + 1))^{\frac{1}{2}}}{\gamma^{\frac{1}{2}}} \right\}$ and upper semi-

continuous on $C_0[0,T]$. (Here we do not preclude the possibility that, for some z_0, $F(z_0) = -\infty$, as long as $F(z) \to -\infty$ as $z \to z_0$.)

(4) $F(z)$ has $m \geqslant 3$ continuous Frechet derivatives $D^j F$ $(j \leqslant m)$ in a ball of radius $\delta > 0$, centred at x_i in $C_0[0,T]$, for $i = 1,2,\ldots,n$. Further we assume that $D^j F(x_i + \eta)z^j = O(\|z\|^j)$, if $\|\eta\| < \delta$, for $i = 1,2,3,\ldots,n$, for $j = 1,2,\ldots,m$.

(5)   For some $\varepsilon > 0$, for $\|\eta\| < \delta$, $\mathbb{E}_z^T\{\exp\{(1+\varepsilon)D^2F(x_i + \eta)z^2/2\}\}$ is uniformly bounded for $i = 1,2,\ldots,n$.

(6) $\dot{x}_i(\cdot)$ is of bounded absolute variation on $[0,T]$, for $i = 1,2,\ldots,n$.

Remarks

(i)   The conditional Wiener integrals Γ_j can be represented by Feynman diagrams.

(ii)   To some extent the larger the value of the constant γ in conditions (2) and (3) above the better is the corresponding theorem.

(iii) The bound in condition (2) is quadratic in the sup-norm. The results of Ellis and Rosen[5] required similar quadratic bounds but in L^2-norms. These bounds are more restrictive than the Schilder-type bound above. As they stand these Ellis and Rosen bounds are, in fact, too restrictive for our purposes. Following the methods of Pincus, Simon[12] has established results similar to those of Ellis and Rosen, but these only give information about Γ_0. Simon gives a very clever argument to avoid the difficulty with the L^2-norms for an important class of F's.

(iv) When $\varepsilon = 0$ and $\eta = 0$ condition (5) is merely equivalent to the fact that the conditional Wiener integral for Γ_0 is finite.

In the next section we given an outline proof of the above theorem in the case when there is a unique nondegenerate global maximum. In the case when there are several global maximisers the proof is slightly **more involved**[13], but **the** basic idea is the same.

3. Outline Proof of Laplace Theorem

The proof follows very closely Schilder's original paper. Here we only discuss three of the basic estimates required in the proof. We require the reflection principle for Brownian motion for one of our estimates.

Reflection Principle

Let $x(s)$ be Brownian motion starting at 0. Let $\tau(a)$ be the first hitting time of a, so that

$$\tau(a) = \inf\{s \geqslant 0 \,|\, x(s) = a\}.$$

Now let B be any open interval, $B \subset (-\infty, 0)$ and denote by W_0 the expectation with respect to Wiener measure $d\mu^W(x)$. Then, for any fixed $t > 0$,

$$W_0\{x(t) \,\varepsilon\, (a + B),\ \tau(a) < t\} = W_0\{x(t) \,\varepsilon\, (a - B), \tau(a) < t\}$$

i.e. if the Brownian path is known to have got to a before time t, at time t, it is as likely to be to the left of a as it is to the right of a.

The proof of the reflection principle uses the strict Markov property (see Refs 14 and 15).

We now set $B = (-a + A)$, where $A \subset (-\infty, a)$, some $a > 0$, giving in above, with $t = T$,

$$W_0\{x(T) \,\varepsilon\, A, \tau(a) < T\} = W_0\{x(T) \,\varepsilon\, (2a - A), \tau(a) < T\}.$$

But $x(T) \,\varepsilon\, (2a - A) \Rightarrow \tau(a) < T$, since $(2a - A) \subset (a, \infty)$ and the sample paths are continuous. Hence we have shown that

$$W_0\{\tau(a) < T, x(T) \,\varepsilon\, A\} = W_0\{x(T) \,\varepsilon\, (2a - A)\},$$

or

$$W_0\{\tau(a) < T \,|\, x(T) \,\varepsilon\, A\} = \int_{(2a-A)} e^{-b^2/2T} db \left/ \int_A e^{-b^2/2T} db \right. .$$

Choosing $A = (-\delta, \delta)$ and letting $\delta \to 0$, gives for the conditional process z

44

$$\mathbb{E}_z^T\left\{\sup_{0\leqslant s\leqslant T} z(s) > a\right\} = e^{-2a^2/T} \ .$$

In exactly the same way, we show that

$$\mathbb{E}_z^T\left\{\sup_{0\leqslant s\leqslant T} - z(s) > a\right\} = e^{-2a^2/T} .$$

The last two identities finally yield

$$e^{-2a^2/T} < \mathbb{E}_z^T\left\{\sup_{0\leqslant s\leqslant T} |z(s)| > a\right\} < 2e^{-2a^2/T} .$$

This proves Lemma 1.

Lemma 1

For some fixed constants $C, \gamma > 0$

$$\mathbb{E}_z^T\{\|z\| > a\} < C \exp\{-\gamma a^2/T\}.$$

The best possible value of γ is 2.

Initially we proved a version of Lemma 1 by using a modification of the proof of the Kolmogorov Lemma. This gave a much inferior value for γ. We are grateful to Peter Baxendale for suggesting that the reflection principle could be used to obtain the best possible value for γ.

For $z \in C_0[0,T]$ denote by $z_n(\cdot)$ its polygonalisation defined by

$$z_n(s) = z\left(\frac{jT}{n}\right) + \left(s - \frac{jT}{n}\right)\left[z\left(\frac{(j+1)T}{n}\right) - z\left(\frac{jT}{n}\right)\right]\frac{n}{T} \ ,$$

$jT/n \leqslant s \leqslant (j+1)T/n$, $j = 0,1,2,\ldots,n-1$. The argument above can now be extended to yield, after a little calculation, Lemma 2.

Lemma 2

For each $\delta > 0$, $n = 1,2,\ldots,$ for a constant $D > 0$,

$$\mathbb{E}_z^T\{\|z - z_n(\cdot)\| \geqslant \delta\} \leqslant D \ \exp\left\{\frac{-m\delta^2}{8T}\right\} \ .$$

The next lemma shows that the dominant contribution to the Laplace integral comes from a neighbourhood of points z with $\lambda z \sim x_1$, the unique global maximiser of $\left\{F(z) - 2^{-1}\int_0^T [\dot{z}(\tau)]^2 d\tau\right\}$.

<u>Lemma 3</u>

Let F satisfy the conditions in Theorem 1 (with $n = 1$) and let $\delta > 0$. Then, for sufficiently small λ,

$$I(\lambda) = \mathbb{E}_z^T\left\{\left[1 - \chi\left(\frac{\delta}{\lambda}, \frac{x_1}{\lambda}, z\right)\right] \exp\{\lambda^{-2}F(\lambda z)\}\right\} = 0(\exp(\alpha\lambda^{-2})),$$

for some constant $\alpha < 0$, $\chi\left(\frac{\delta}{\lambda}, \frac{x_1}{\lambda}, z\right)$ being the characteristic function of

$$\left\{z \,\middle|\, \|z - \frac{x_1}{\lambda}\| < \frac{\delta}{\lambda}\right\}.$$

<u>Proof</u>

Following Schilder[2] we write $I(\lambda) = I_2(\lambda) + I_3(\lambda) + I_4(\lambda)$, where

$$I_2(\lambda) = \mathbb{E}_z^T\left\{\left[1 - \chi\left(\frac{\delta}{\lambda}, \frac{x_1}{\lambda}, z\right)\right]\left[1 - H\left(\frac{\eta}{\lambda}, n, z\right)\right] \exp(\lambda^{-2}F(\lambda z))\right\},$$

$$I_3(\lambda) = \mathbb{E}_z^T\left\{\left[1 - \chi\left(\frac{\delta}{\lambda}, \frac{x_1}{\lambda}, z\right)\right] H\left(\frac{\eta}{\lambda}, n, z\right) \chi\left(\frac{a}{\lambda}, 0, z\right) \exp(\lambda^{-2}F(\lambda z))\right\},$$

$$I_4(\lambda) = \mathbb{E}_z^T\left\{\left[1 - \chi\left(\frac{\delta}{\lambda}, \frac{x_1}{\lambda}, z\right)\right] H\left(\frac{\eta}{\lambda}, n, z\right)\left[1 - \chi\left(\frac{a}{\lambda}, 0, z\right)\right] \exp(\lambda^{-2}F(\lambda z))\right\}$$

$H\left(\frac{\eta}{\lambda}, n, z\right)$ being the characteristic function of $\left\{z \in C_0[0,T] \,\middle|\, \|z - z_n(\cdot)\| \leq \frac{\eta}{\lambda}\right\}$.

From the Cauchy-Schwarz inequality

$$I_2(\lambda) \leq \mathbb{E}_z^T\left\{\left[1 - H\left(\frac{\eta}{\lambda}, n, z\right)\right]\right\}^{\frac{1}{2}} \mathbb{E}_z^T\left\{\exp(2\lambda^{-2}F(\lambda z))\right\}^{\frac{1}{2}}.$$

The known bounds on F and Lemma 2 enable us to choose η so that $I_2(\lambda) = 0(\exp - \lambda^{-2})$.   Similarly the known bounds on F and Lemma 1 enable a to be chosen so that $I_4(\lambda) = 0(\exp - \lambda^{-2})$. The difficult term to handle is $I_3(\lambda)$ for which we refer to our preprint. $\qquad\qquad\square$

The remainder of the proof of the theorem is not difficult but slightly tedious. It depends upon Taylor's theorem for functionals on Banach space and the use of the bounds in condition 4. $\qquad\qquad\square$

4. <u>Application to a Generalised Mehler Kernel Formula</u>

The next theorem is a basic ingredient in our results.

Theorem 2

Let the potential $V \in C^\infty(\mathbb{R})$ be bounded below. Set
$A(z) = 2^{-1} \int_0^T [\dot{z}(\tau)]^2 d\tau + \int_0^T V(z(\tau)) d\tau$. Then $A(z)$ attains its global minimum at at

least one path $X_{min} \in \mathcal{A}(x,y,T) = \{$absolutely continuous $X:[0,T] \to \mathbb{R}$, with

$X(0) = x$, $X(T) = y\}$. X_{min} is smooth and satisfies

$$\ddot{X}_{min}(\tau) = V'(X_{min}(\tau)), \qquad \tau \in [0,T].$$

Moreover, if the self-adjoint quantum mechanical Hamiltonian
$H(\hbar) = -2^{-1}\hbar^2 \dfrac{d^2}{dx^2} + V(x)$ and $\lambda = \hbar^{\frac{1}{2}}$,

$$\exp\{-TH(\hbar)/\hbar\}(x,y) = (2\pi T\hbar)^{-\frac{1}{2}} \exp\{-A(X_{min})/\hbar\} \, \mathbb{E}_z^T\{\exp\{\lambda^{-2}F(\lambda z)\}\},$$

where $F(z) = -\int_0^T \{V[X_{min}(\tau)+z(\tau)] - V[X_{min}(\tau)] - z(\tau)V'[X_{min}(\tau)]\}d\tau$, so that

$$\left\{F(z) - 2^{-1}\int_0^T [\dot{z}(\tau)]^2 d\tau\right\} = A(X_{min}) - A(X_{min}+z).$$

Proof

The first part of the theorem follows from standard results in the direct methods
of the calculus of variations[16]. The second part of the theorem follows by using the
Cameron-Martin[17] formula for translations in the Feynman-Kac formula. $\quad\Box$

In order for the functional F above to satisfy condition (5) the time T has
to be so small as to ensure that X_{min} the global minimiser of $A(z)$ over
$\mathcal{A}(x,y,T)$ is unique. The next lemma helps to explain how this arises.

Lemma 4

When $V \in C^\infty(\mathbb{R})$ is real-valued, bounded below together with its second deriva-
tive so that, for some constant β, $V'' \geq -|\beta|$, the global minimiser of
$A(z) = 2^{-1}\int_0^T [\dot{z}(\tau)]^2 + \int_0^T V(z(\tau))d\tau$ over $\mathcal{A}(x,y,T)$ is unique if $T < \pi/|\beta|^{\frac{1}{2}}$.

Proof

If there are two global minimisers $X_1, X_2 \in \mathcal{A}(x,y,T)$ they satisfy the Euler-
Lagrange equation above and are smooth. Hence, if $h(\tau) = X_1(\tau) - X_2(\tau)$,
$h(0) = h(T) = 0$ and, for some point $\xi(\tau)$ intermediate to $X_1(\tau)$ and $X_2(\tau)$,

$$\ddot{h}(\tau) = h(\tau)V''(\xi(\tau)).$$

Multiply both sides of above equation by $h(\tau)$ and integrate by parts to give

$$\int_0^T [\dot{h}(\tau)]^2 d\tau = -\int_0^T h^2(\tau)V''(\xi(\tau))d\tau.$$

But the Rayleigh-Ritz quotient gives $\int_0^T [h'(\tau)]^2 d\tau \geq \frac{\pi^2}{T^2} \int_0^T h^2(\tau)d\tau$, $\frac{\pi^2}{T^2}$ being the least

eigenvalue λ of the equation $-\ddot{h}(\tau) = \lambda h(\tau)$, $h(0) = h(T) = 0$. By hypothesis $-V'' \leq |\beta|$ and so

$$\frac{\pi^2}{T^2} \int_0^T h^2(\tau)d\tau \leq |\beta| \int_0^T h^2(\tau)d\tau.$$

Hence, for $T < \pi/|\beta|^{\frac{1}{2}}$, $\int_0^T h^2(\tau)d\tau = 0$ and result follows. □

Remark

The above result is a best possible result in the sense that, for $V(x) = -|\beta|x^2/2, |x| < \delta; V(x) = -|\beta|\delta^2/2$, otherwise, A is minimised over $\oint \left[0,0, \frac{\pi}{|\beta|^{\frac{1}{2}}}\right]$ by $X(\tau) = \alpha \sin(\tau|\beta|^{\frac{1}{2}})$, for any $\alpha, -\delta < \alpha < \delta$, $\tau \in [0,T]$.

(V is not smooth).

The next result explains why we refer to our application as a Mehler kernel formula.

Theorem 3

Let the self-adjoint quantum mechanical Hamiltonian $H(\hbar) = \left[\frac{-\hbar^2}{2} \Delta_1 + V\right]$, where $V \in C^\infty(\mathbb{R})$ and is bounded below together with its second derivative. Then, setting $A(X_{min}) = A(x,y,T)$, for sufficiently small finite time $T > 0$,

$$\exp\{-TH(\hbar)/\hbar\}(x,y) = (2\pi\hbar)^{-\frac{1}{2}} \exp\{-A(x,y,T)/\hbar\}\left\{\frac{\partial^2 A}{\partial x \partial y}(x,y,T)\right\}^{\frac{1}{2}}$$

$$\times (1 + \hbar K_1 + O(\hbar^2)),$$

where

$$K_1 = -\frac{1}{8}\int_0^T V^{iv}(X_{min}(\tau))G^2(\tau,\tau)d\tau + \frac{1}{24}\int_0^T\int_0^T V'''(X_{min}(\tau))V'''(X_{min}(\sigma))$$

$$\times [3G(\tau,\iota)G(\iota,\sigma)G(\sigma,\sigma) + 2G^3(\tau,\sigma)]d\tau d\sigma,$$

48

$G(\tau,\sigma)$ (Feynman-Green's function) being the Green's function of the Sturm-Liouville differential operator $\left[\dfrac{d^2}{d\sigma^2} - V''(X_{min}(\sigma))\right]$, with $G(0,\tau) = G(T,\tau) = 0$.

Proof

This follows from a tedious calculation and the identity:

$$(2\pi T)^{-\frac{1}{2}} \mathbb{E}_z^T\left\{\exp\left\{\sum_{i=1}^{n} \alpha_i z(t_i)\right\}\exp\left\{-\frac{1}{2}\int_0^T V''(X_{min}(\sigma))z^2(\sigma)d\sigma\right\}\right\}$$

$$= (2\pi)^{-\frac{1}{2}}\left\{\frac{\partial^2 A}{\partial x \partial y}(x,y,T)\right\}^{\frac{1}{2}} \exp\left\{\frac{1}{2}\sum_{i,j=1}^{n}\alpha_i\alpha_j G(t_i,t_j)\right\} .$$

For the proof of this identity we refer to the preprint[18]. $\square$

Observe that if V is strictly convex $V'' > 0$ the above result is valid for all finite time $T > 0$. When V satisfies $V'' \geqslant -|\beta|$ the last theorem is only valid for $T < \min\left\{\left(\dfrac{\gamma}{|\beta|}\right)^{\frac{1}{2}}, \dfrac{1}{\sqrt{2}|\beta|^{\frac{1}{2}}}\right\}$. This is one reason why we require the best possible value of γ. When $V'' \equiv 0$, above result reduces to the classical Mehler kernel formula exactly.

We now give some simple examples of our results. Observe that to calculate $\exp\{-TH(\hbar)/\hbar\}(x,y)$ upto arbitrarily high orders in powers of $\hbar$ all that is required is the classical path X_{min} for A over $\mathcal{A}(x,y,T)$. We give X_{min} here in two cases.

Example 1

Let V be the X^4 anharmonic oscillator so that, for $A,B > 0$, $V(x) = 2^{-1}(A^2 x^2 + 2^{-1}B^2 x^4)$, $x \in \mathbb{R}$. For given $x < 0$ and $y > 0$ and $T > 0$, define $E_0 = E_0(x,y,T)$ by $T = \int_x^y (A^2 u^2 + 2^{-1}B^2 u^4 + 2E_0)^{-\frac{1}{2}}du$, where we assume

$T > \int_x^y (A^2 u^2 + 2^{-1}B^2 u^4 + 2^{-1}A^4 B^{-2})^{-\frac{1}{2}}du$. Then the minimising path X_{min} of A over

$\mathcal{A}(x,y,T)$ is X_{min}, where

$$X_{min}(t) = \alpha\, \text{tn}(\beta(t - t_0),k), \qquad t \in [0,T],$$

for $\alpha^2 = 2(1 - k^2)A^2/B^2(2 - k^2)$, $\beta^2 = A^2/(2 - k^2)$, k being defined by $E_0 = A^4(1 - k^2)/B^2(2 - k^2)^2$, tn being a Jacobi function.

We emphasise that our results also carry over to higher than one dimension, where they are valid even for non-separable systems. Moreover in three dimensions similar results are valid for Zeeman effect Hamiltonians, $H(\hbar) = 2^{-1}(i\hbar\nabla x + 2^{-1} \underset{\sim}{B} \wedge \underset{\sim}{x})^2 + V(\underset{\sim}{x})$, $\underset{\sim}{B}$ being the constant magnetic field vector (See Ref 1(c)). Here for some potentials $\exp\{-itH(\hbar)/\hbar\}(\underset{\sim}{x},\underset{\sim}{y})$ is given in terms of a classical path X_{min} in much the same way as in Theorems $\underset{\sim}{2}$ and $\underset{\sim}{3}$, except that in this case

$$A(z) = 2^{-1}\int_0^T \underset{\sim}{\dot{z}}^2(\tau)d\tau - \int_0^T V(\underset{\sim}{z}(\tau))d\tau + 2^{-1}\int_0^T (\underset{\sim}{B} \wedge \underset{\sim}{z}(\tau)).\underset{\sim}{\dot{z}}(\tau)d\tau.$$

Example 2

For cartesian coordinates (x_1,x_2,x_3) let $V(x_1,x_2,x_3) = a_1^2 x_1^2 + a_2^2 x_2^2 + a_3^2 x_3^2$, where $a_1,a_2,a_3 \in \mathbb{R}$ and let the constant magnetic field vector $\underset{\sim}{B} = (B_1,B_2,B_3)$. When two of B_1,B_2,B_3 are zero, the above Zeeman system is separable and the corresponding X_{min} is easily determined. If two or more of B_1,B_2,B_3 are nonzero the system is non-separable. In this case we consider the cubic in λ

$$(\lambda - a_1^2)(\lambda - a_2^2)(\lambda - a_3^2) - \lambda^2 \underset{\sim}{B}^2 + \lambda(B_1^2 a_1^2 + B_2^2 a_2^2 + B_3^2 a_3^2) = 0.$$

For real B_1,B_2,B_3, this cubic in λ has three positive roots. Setting $\alpha = \pm\sqrt{\lambda}$ gives six values $\alpha_j(B)$, satisfying $\det(\alpha^2 \delta_{ij} + i\alpha \, \varepsilon_{ijk} B_k - A^2_{ij}) = 0$, with $A^2_{ij} = a_i^2 \delta_{ij}$, for $i,j = 1,2,3$ (no summation). Let $\underset{\sim}{c}_j(B)$ be the corresponding eigenvectors suitably normalised. Then, for small $|B|$, by the implicit function theorem, for fixed $\underset{\sim}{X},\underset{\sim}{Y} \in \mathbb{R}^3$ and $T > 0$, $\exists$ a unique $(\gamma_1(B),\gamma_2(B),\ldots,\gamma_6(B)) \in \mathbb{C}^6$, satisfying $\overset{6}{\underset{1}{\Sigma}} \gamma_j(B)\underset{\sim}{c}_j(B) = \underset{\sim}{X}$, $\overset{6}{\underset{1}{\Sigma}} \gamma_j(B)\underset{\sim}{c}_j(B)\exp(i\alpha_j T) = \underset{\sim}{Y}$. We write $\gamma_j(B) = \gamma_j(B)(x,y,T)$, $j = 1,2,\ldots,6$. In this case the first term in the power series expansion in $\hbar$ is the only contribution which is nonzero. Here, for small $|\underset{\sim}{B}|$,

$$X_{min}(t) = \sum_{j=1}^6 \gamma_j(B)(x,y,T)\underset{\sim}{c}_j(B)\exp(i\alpha_j t), \quad t \in [0,T].$$

Further details are given in Refs (1(a)) and (1(c)).

5. <u>Application to the Bender-Wu Formula for the X^{2N} Anharmonic Oscillator</u>

Let $H(\beta) = 2^{-1}\left[\dfrac{-d^2}{dx^2} + x^2\right] + \beta x^4$, for $\beta > 0$, denote the self-adjoint quantum mechanical x^4-anharmonic oscillator. Then, as is well known, the Rayleigh Schrödinger perturbation series for the ground state energy $E_0(\beta)$

$$E_0(\beta) = \sum_{n=0}^{\infty} E_n \beta^n$$

is divergent. However, the perturbation series is Borel summable and, using known analyticity properties of $E_0(\beta)$ and Watson transforms, the correct value of $E_0(\beta)$

is recoverable from the above divergent series[19]. One therefore needs to know the large order behaviour of E_n as $n \to \infty$.

This large order behaviour of E_n was established numerically by Bender and Wu[9], in what to my mind is one of the most remarkable formulae of theoretical physics, as

$$E_n \sim -\frac{1}{\pi}\sqrt{\frac{6}{\pi}}(-3)^n \Gamma\left(n+\frac{1}{2}\right)\left[1 - \frac{95}{72n} - \frac{20099}{10368n^2} + O\left(\frac{1}{n^3}\right)\right] ,$$

as $n \to \infty$. Formal functional integral derivations of this result were given by Lipatov[10], Brezin[11] et al (see also Refs 20 - 22). All these authors use the identity:

$$E_0(\beta) = \lim_{T \to \infty} -T^{-1} \ln \, \text{tr}\{\exp\{-TH(\beta)\}\}$$

and express $\dfrac{d^n}{d\beta^n} \, \text{tr}\{\exp\{-TH(\beta)\}\}\Big|_{\beta=0}$ as a functional integral. In these treatments, formally commuting the T and n limits, the large order behaviour of E_n is expressed in terms of

$$\left\{\lim_{n \to \infty} \frac{d^n}{d\beta^n} \, \text{tr}\{\exp\{-TH(\beta)\}\}\right\} .$$

The latter quantity is given by the Laplace expansion of a functional integral with integrand having a manifold of maxima. Because of this the methods require involved changes of integration variables known as collective coordinate methods[23]. Moreover, because of the manifold problem, the above argument is extremely difficult to make rigorous[24]. The only published rigorous results to date in this direction are the works of Simon[12] and Spencer[25]. Below we detail one possible way of avoiding the manifold problem. Formally commuting limits, in much the same spirit as above, we obtain the numerically correct rapidly varying factors in the large order behaviour of E_n for the x^{2N} anharmonic oscillator $(N \geqslant 2)$ from our Laplace theorem. This is a consequence of the theorem below.

<u>Theorem 4</u>

Let $H(\beta) = \left[2^{-1}\left(-\dfrac{d^2}{dx^2} + x^2\right) + \beta x^{2N}\right]$ be the quantum mechanical anharmonic oscill-

ator Hamiltonian, for $\beta > 0$, $N \geqslant 2$, with eigenvalues $E_n(\beta)$, arranged in ascending order and corresponding orthonormal eigenfunctions $\phi_n(\cdot)$, $n = 0,1,2,\ldots$. Define $g_n(T)$ by

$$g(T,\beta) = \sum_0^\infty e^{-TE_n(\beta)} |\phi_n(0)|^2 = \sum_0^\infty g_n(T)\beta^n .$$

Then, for sufficiently large fixed T, as $n \to \infty$,

$$g_n(T) \longrightarrow \frac{\gamma(T)n^{\beta(T)}}{n!} e^{-n\alpha(T)} \, n^{Nn}(-1)^n\left[1 + O\!\left(\frac{1}{n}\right)\right] ,$$

where $\beta(T) \equiv 0$, $\gamma(T) = \left[-\pi E \frac{\partial T}{\partial E}\right]^{-\frac{1}{2}}(N-1)^{\frac{1}{2}}$, E being the 'instanton' energy for 'instanton' Y_0 with period $2T$, Y_0 satisfying

$$\ddot{Y}_0(s) = Y_0(s) - 2NY_0^{2N-1}(s), \qquad s \in [0,T], \ Y_0(0) = Y_0(T) = 0,$$

$$\alpha(T) = N + (N-1)\ln\left\{\int_0^T Y_0^{2N}(s)ds\right\}, \quad \text{so that}$$

$$\alpha(T) \longrightarrow N + (N-1)\ln\left\{\frac{2^{1/N-1}\Gamma^2\left[\frac{N}{N-1}\right]}{(N-1)\Gamma\left[\frac{2N}{N-1}\right]}\right\} ,$$

as $T \to \infty$.

Proof

Again we limit ourselves to giving only an outline proof. Firstly one proves that for $T > 0$

$$\sum_{n=0}^{\infty} e^{-E_n(\beta)T} |\phi_n(0)|^2 = \int d\mu_{0,0;0,T}(z)\exp\left\{-2^{-1}\int_0^T z^2(s)ds - \beta\int_0^T z^{2N}(s)ds\right\} .$$

Following Simon[12a] from this one can show that

$$\frac{(-1)^n g_n(T)n!}{n^{Nn}} = \mathbb{E}_z^T\{\exp\{nF(n^{-\frac{1}{2}}z)\}\} ,$$

where $F(z) = -2^{-1}\int_0^T z^2(s)ds + \ln\left[\int_0^T z^{2N}(s)ds\right]$, $F(0) = -\infty$. The burden of proof then is to show that F satisfies conditions 1 - 6 and, in particular, to show that, for large T,

$$G(z) = \left\{2^{-1}\int_0^T [\dot{z}(s)]^2 ds + 2^{-1}\int_0^T z^2(s)ds - \ln\left[\int_0^T z^{2N}(s)ds\right]\right\}$$

has exactly two global minimisers in $C_0^*[0,T]$, $X(s) = \pm Y_0(s)\left[\int_0^T Y_0^{2N}(t)dt\right]^{-\frac{1}{2}}$, $s \in [0,T]$, the corresponding minimum being $\alpha(T)$. This depends upon two basic lemmas.

Lemma 5

The functional $G(z) = 2^{-1}\int_0^T [\dot{z}(s)]^2 ds + 2^{-1}\int_0^T z^2(s)ds - \ln\left[\int_0^T z^{2N}(s)ds\right]$ $(N \geqslant 2)$

attains its global minimum $\alpha(T)$ at at least one path $X \in C_0^*[0,T]$, X satisfies the Euler-Lagrange equation

$$-\ddot{X}(t) + X(t) - 2NX^{2N-1}(t)/\int_0^T X^{2N}(s)ds, \quad t \in [0,T].$$

After some calculation this reduces the problem to finding $\min\limits_{n=1,2,\ldots} \left\{2^{-1}nS[\frac{2T}{n}]\right\}$, where

$$S[T'] = 2\sqrt{2}\int_0^{y_1} y^{2N}[E(T') - V(y)]^{-\frac{1}{2}}dy ,$$

$V(y) = -2^{-1}y^2 + y^{2N}$ and $y_1 > 0$ is such that $V(y_1) = E(T')$; $E(T')$ being the energy of the solution Y of $-\ddot{Y}(t) + Y(t) - 2NY^{2N-1}(t) = 0$, $Y(0) = Y(T'/2) = 0$, with period T'. To complete the argument we use the following lemma:

Lemma 6

Let $S[T']$ be a real-valued continuous function, for $T' \in (0,\infty)$, with $S[T'] \to S[\infty]$ as $T' \to \infty$. Then, if $S[T'] > 2^{-1}S[\infty] > 0$, for $T' \in (0,\infty)$, we can deduce that, for sufficiently large T,

$$\min\limits_{n=1,2,\ldots} \left\{2^{-1}nS[\frac{2T}{n}]\right\} = 2^{-1} S[2T] .$$

The last lemma and the fact that in our case S satisfies the stated hypotheses completes the proof of Theorem 4. We refer the reader to our preprint for further details (See Ref 1(b)). $\qquad\square$

The connection with the Bender-Wu formula comes about because

$$E_0(\beta) = \sum_{n=0}^{\infty} E_n \beta^n = \lim_{T\to\infty} -T^{-1}\ln g(T,\beta).$$

Formally commuting the T and n limits, gives as $n \to \infty$

$$E_n \to \lim_{T\to\infty}\left\{-T^{-1}g_0(T)^{-1}\lim_{n\to\infty} g_n(T)\right\} ,$$

which from Stirling's formula yields

$$E_n \rightarrow [n(N-1)]! \left\{ -\frac{1}{2} \left[\frac{\Gamma(2N/N-1)}{\Gamma^2(N/N-1)} \right]^{N-1} \right\}^n \lim_{T\to\infty} \left[\frac{-T^{-1}\gamma(T)n^{\beta(T)-1}}{2\pi g_0(T)(N-1)^{\frac{1}{2}}} \right] .$$

In the above formula, in contradistinction to previous treatments, we have explicit values for $\gamma(T)$ and $\beta(T)$. To get complete agreement for the leading behaviour with the numerical Bender-Wu formula would require that $\beta(T) \rightarrow \frac{1}{2}$ as $T \rightarrow \infty$. In fact our calculation gives $\beta(T) \equiv 0$. Remarkably, though, the method gives correctly the first two rapidly varying factors (c.f. Brezin[11] et al). The source of the disagreement here presumably is commuting the T and n limits above. This suggests that apart from an overall multiplicative factor the method might yield correctly the higher order behaviours. Unfortunately this is not easily checked, because of the complexity of the coefficients of the higher order terms. The method does lead to the analogue for the x^{2N} anharmonic oscillator of Simon's Theorem 18.3.

Corollary

Define $g_n(T)$ by

$$\sum_{n=0}^{\infty} e^{-TE_n(\beta)} |\phi_n(0)|^2 = \sum_{n=0}^{\infty} g_n(T)\beta^n ,$$

where $E_n(\beta)$ are the eigenvalues, ϕ_n the corresponding eigenfunctions of

$$H(\beta) = \left[-2^{-1}\left(\frac{d^2}{dx^2} - x^2 \right) + \beta x^{2N} \right], \quad \beta > 0. \quad \text{Then, as} \quad n \rightarrow \infty ,$$

$$\left[\frac{(-1)^n n! \, g_n(T)}{n^{Nn}} \right]^{1/n} \rightarrow e^{-\alpha(T)} ,$$

where, as $T \rightarrow \infty$,

$$\alpha(T) \rightarrow N + (N-1)\ln\left\{ 2^{1/N-1}\Gamma^2\left(\frac{N}{N-1}\right) \Big/ (N-1)\Gamma\left(\frac{2N}{N-1}\right) \right\} .$$

Barry Simon's theorem[12a] is the corresponding result for $\sum_0^{\infty} e^{-TE_n(\beta)}$ for the x^4 anharmonic oscillator.

6. Acknowledgement

It is a pleasure to thank David Elworthy, Peter Baxendale, Ken Brown, Michel Sirugue and Madeleine Sirugue-Collin for helpful conversations. One of us

(A.T.) is grateful to CNRS-CPT for a research grant (ATP no. 055) which assisted in the completion of some of this work. The other (I.D.) is grateful to the SRC for a research studentship.

References

[1a]   I.M. DAVIES and A. TRUMAN, 'Laplace asymptotic expansions of conditional Wiener integrals and generalised Mehler kernel formulas', accepted for publication by J. Math. Phys.

[1b]   I.M. DAVIES and A. TRUMAN, 'On the Laplace asymptotic expansion of conditional Wiener integrals and the Bender-Wu formula for X^{2N} anharmonic oscillator, accepted for publication by J. Math. Phys.

[1c]   I.M. DAVIES and A. TRUMAN, 'Laplace asymptotic expansions of conditional Wiener integrals and generalised Mehler formulas for Hamiltonians on $\mathbb{R}^n$, to be submitted to J. Phys. A.

[2]   M. SCHILDER, Trans. Amer. Math. Soc., 125, 63-85 (1965).

[3]   B. SIMON, 'Functional Integration and Quantum Physics', (Academic Press, New York 1979).

[4]   M.D. DONSKER and S.R.S. VARADHAN, Phys. Rep. 77, 3, 235-37 (1981) and references cited therein.

[5a]   R.S. ELLIS and J.R. ROSEN, Bull. Amer. Math. Soc., 3, 1, 705-9 (1980).

[5b]   R.S. ELLIS and J.R. ROSEN, 'Asymptotic analysis of Gaussian integrals, I: Isolated minimal points', to appear in Trans. Amer. Math. Soc.

[5c]   R.S. ELLIS and J.R. ROSEN, Commun. Math. Phys. 82, 153-81 (1981).

[6]   C. DeWitt-Morette, Ann. Phys. (N.Y.) 97, 367-99 (1976).

[7]   M. MIZRAHI, J. Math. Phys. 20, 844-55, (1979).

[8a]   S. ALBEVERIO and R. HOEGH-KROHN, Inv. Math. 40, 59-106 (1977).

[8b]   S. ALBEVERIO, P. BLANCHARD and R. HOEGH-KROHN, 'The Trace formula for the Schrödinger operators', Preprint Bielefeld 1980.

[9a]   C. BENDER and T.T. WU, Phys. Rev. 184, 1231-60 (1969).

[9b]   C. BENDER and T.T. WU, Phys. Rev. Lett. 27, 7, 461-5 (1971).

[9c]   C. BENDER and T.T. WU, Phys. Rev. D7, 6, 1620-36 (1973).

[10]   L.N. LIPATOV, J. E. T. P. Lett. 25, 2, 104-7 (1977).

[11]   E. BREZIN et al, Phys. Rev. D15, 6, 1544-57, 1558-64 (1977).

[12a] See Ref. 3, Theorem 18.3 and Chapter 18 in general.

[12b] E. HARREL and B. SIMON, Duke Math. J. 47, 845-902 (1980).

[13]   See Ref. 1(b) and 1(c).

[14] K. ITO and H.P. MCKEAN, 'Diffusion Processes and their sample paths', (Springer-Verlag, Berlin, New York 1965).

[15] D. WILLIAMS, 'Diffusions,Markov Processes and martingales Vol. 1: Foundations', (Wiley 1979).

[16] N.I. AKHIEZER, 'The Calculus of Variations', (Blaisdell, New York, London 1962). See Chapter 4.

[17] H.H. KUO, Lecture Notes in Mathematics $\underline{463}$, (Springer-Verlag, Berlin, Heidelberg, New York 1975). See page 113.

[18a] See Ref 1(a).

[18b] A.TRUMAN, 'The polygonal path formulation of the Feynman Path integral', Lecture Notes in Physics $\underline{106}$, 73-102 (1979).

[19] B. SIMON, 'Large Orders and Summability of Eigenvalue Perturbation Theory: A Mathematical Overview', to appear in Int. J. Quant. Chem., Proceedings of 1981 Sanibel workshop.

[20] J.C. COLLINS and D.C. SOPER, Ann. Phys. $\underline{112}$, 209-34 (1978).

[21] G. AUBERSON et al, Il Nuovo Cimento $\underline{48A}$, 1-23 (1978).

[22] V. FIGEROU, Commun. Math. Phys. $\underline{79}$, 401-33 (1981).

[23a] N. BOGOLIUBOV and S. TYABLIKOV (1949) 'N. Bogoliubov's collected papers', (Moscow 1972).

[23b] L.D. FADEEV and V.N. POPOV, Phys. Lett $\underline{25B}$, 29-30 (1969)

[24] We have been informed by Barry Simon that the problem of the commutativity of the limits in T and n has been solved by Steven Breen, a former student of T. Spencer.

[25] T. SPENCER, Commun. Math. Phys. $\underline{74}$, 273-80 (1980).

<u>STOCHASTIC PROCESSES AND FERMI FIELDS</u>

G.F. De Angelis[1]
Centre de Physique Théorique
CNRS
Luminy, Marseille, France

D. de Falco[1,2]
Joseph Henry Laboratories of Physics
Princeton University, Princeton N.J.
08544 U.S.A.

F. Guerra
Istituto Matematico "Guido Castelnuovo"
Università di Roma, Italy

Talk given by D. de Falco at the "Workshop on Stochastic Processes in Quantum Theory and Statistical Physics: Recent Progress and Applications" Marseille, June 29-July 4, 1981.

1. Permanent address:Istituto di Fisica, Università di Salerno,
   I-84100 Salerno, Italy
2. Research supported in part by NSF PHY 78-23952.

We wish to report here some results (1,2) pointing to the possibility of introducing ordinary probabilistic concepts in the analysis of Fermi fields.

The specific system considered is the Fermi oscillator described by hermitian "field" operators q,p satisfying the equal time anticommutation relations

$$q^2 = p^2 = 1 \qquad \{q,p\} = 0$$

and the Heisenberg equations

$$\dot{q} = p \qquad \dot{p} = -q$$

generated by the Hamiltonian

$$H_o = a^* a = \frac{q-ip}{2} \frac{q+ip}{2}$$

On this example we give, in terms of ordinary numerical valued Markov processes, a Feynman-Kac formula for the hamiltonian semigroup, a path integral representation of the configurational Schwinger functions, and a complete characterization of the dynamics in terms of stochastic field equations.

The main interest of these results lies, of course, in the fact that Fermi oscillators are building blocks of the quantum Dirac field, as briefly reviewed below.

In the description of the Dirac field we use, for the sake of simplicity, the real eight component notations of Schwinger (3) .

In this notational scheme, the Dirac matrix algebra is generated by four real symmetric 8 x 8 matices $\alpha_1, \alpha_2, \alpha_3, \alpha_4$ and by three imaginary antisymmatic matrices $\alpha_5, \alpha_6, \alpha_7$ satisfying:

$$\{\alpha_i, \alpha_j\} = \delta_{ij} \qquad i,j = 1, \cdots, 7$$

The Dirac equation reads

$$\dot{\psi} = -\sqrt{-\Delta + m^2} \, K \psi$$

where

$$K = i \frac{\alpha \cdot \nabla + m \alpha_5}{\sqrt{-\Delta + m^2}}$$

is an orthogonal, antisymmetric ($K^2 = -\mathbb{1}$, $K^T = -K$) operator in the linear phase space of initial conditions

$$f(x) = \begin{pmatrix} f_1(x) \\ \vdots \\ f_8(x) \end{pmatrix}$$

equipped with the scalar product

$$\langle f, g \rangle = \int dx \sum_{i=1}^{8} f_i(x) g_i(x)$$

Let $\mathcal{H}_\pm$ be two orthogonal subspaces in phase space such that

$$K\,\mathcal{H}_\pm = \mathcal{H}_\pm$$

(e.g. in Ref. 1 the choice $\mathcal{H}_\pm$ = range $\dfrac{1 \pm \alpha_4 K}{2}$ was shown to be particularly well suited to the inbedding of the Minkowski theory in the Euclidean scheme).

Let $(f_n)_{n \in \mathbb{N}}$ be an orthonormal basis in $\mathcal{H}_-$, $(f_n, K f_n)_n$ a canonical basis in phase space.

Then

$$\psi(t,\underline{x}) = \sum_{n \in \mathbb{N}} \varphi_n(t)\, f_n(\underline{x}) + P_n(t) K f_n(\underline{x})$$

and the Dirac equation becomes:

$$\begin{cases} \dot{\varphi}_i(t) = \sum_j \left\langle f_i,\, \sqrt{-\Delta + m^2}\, f_j \right\rangle P_j(t) \\[2ex] \dot{P}_i(t) = -\sum_j \left\langle f_i,\, \sqrt{-\Delta + m^2}\, f_j \right\rangle \varphi_j(t) \end{cases}$$

while the C.A.R.

$$\left\{ \psi_a(\underline{x}),\, \psi_b(\underline{y}) \right\} = 2\,\delta_{ab}\,\delta(\underline{x} - \underline{y})$$

become

$$\left\{ \varphi_i,\, \varphi_j \right\} = \left\{ P_i,\, P_j \right\} = \delta_{ij}$$

$$\left\{ \varphi_i,\, P_j \right\} = 0$$

In a finite space box Λ we can choose the f_n's to be eigenfunctions of the laplacian (with suitable boundary conditions):

$$\left(-\Delta + m^2 \right) f_n = \omega_n^2\, f_n$$

The equations of motion then become:

$$\begin{cases} \dot{\varphi}_i = \omega_i\, P_i \\[2ex] \dot{P}_i = -\omega_i\, \varphi_i \end{cases}$$

The (inverse) Jordan-Wigner transformation (4)

$$q_{\jmath} = (i\, P_1\, q_1) \cdots \cdots (i\, P_{\jmath-1}\, q_{\jmath-1})\, q_{\jmath}$$

$$P_{\jmath} = (i\, P_1\, q_1) \cdots \cdots (i\, P_{\jmath-1}\, q_{\jmath-1})\, P_{\jmath}$$

preserves the form of the equations of motion

$$\begin{cases} \dot{q}_i = \omega_i\, P_i \\[2mm] \dot{P}_i = -\omega_i\, q_i \end{cases}$$

and the anticommutation relations for equal indices

$$q_i^2 = P_i^2 = 1 \qquad \{q_i, P_i\} = 0$$

but variables corresponding to different degrees of freedom <u>commute</u>, i.e. :

$$[q_i, q_j] = [P_i, P_j] = [q_i, P_j] = 0 \qquad\qquad \text{for}\quad i \neq j$$

After this analysis, the Fermi oscillator appears as one of the independent Jordan-Wigner degrees of freedom of the Dirac field.

The considerations we are going to give for the single Fermi oscillator extend to the full Dirac field in terms of the Jordan-Wigner configurational part of the quantum Dirac field defined by

$$q(t, \underline{x}) = \sum_{n \in I\!N} q_n(t)\, f_n(\underline{x})$$

Thus motivated, we go back to the single Fermi oscillator.

We introduce the Hilbert space $L^2(Z_2, d\sigma)$,where

$$Z_2 = \{-1, 1\} \qquad \text{and} \qquad \int \cdot\, d\sigma = \frac{1}{2} \sum_{\sigma = \pm 1} \cdot$$

on which we realize the relevant operators in the "Q representation":

$$(q\psi)(\sigma) = \sigma\, \psi(\sigma)$$

$$(P\psi)(\sigma) = i\sigma\, \psi(-\sigma)$$

$$(H_0\psi)(\sigma) = \frac{1}{2}\left[\psi(\sigma) - \psi(-\sigma)\right]$$

The ground state wave function turns out to be $\Omega_o(\sigma) = 1$

Our first observation is that the hamiltonian semigroup $\exp{-tH_o}$ is Markovian in this representation : it is sufficient to observe tha H_o annihilates constants and has non positive off-diagonal elements:

$$\langle \delta_{\sigma_1} \cdot , H_o \delta_{\sigma_2} \cdot \rangle = \frac{\delta_{\sigma_1 \sigma_2} - \delta_{\sigma_1 -\sigma_2}}{2}$$

or, more explicitly, to observe that

$$\left(\exp{-tH_o}\, \psi \right)(\sigma) = \int_{Z_2} P_t^o(\sigma, d\sigma')\, \psi(\sigma') \qquad t \geqslant o$$

where the quantities

$$P_t^o(\sigma, d\sigma') = \left(1 + \sigma\sigma'\exp{-t} \right) d\sigma'$$

have the characteristic properties of the transition probabilities for a jump Markov process with values in Z_2.

Call $q_o(t)$ the stochastic process characterized by the previous transition probabilities and by the probability density

$$\rho_o(t,\sigma) = |\Omega_o(\sigma)|^2 = 1$$

Call $\mathbb{P}_o$ the measure on the set Ω of trajectories

$$\omega : t \in [o,+\infty) \longrightarrow \omega(t) \in Z_2$$

determined by the process q_o.

Then the following Feynman-Kac formula holds

$$\langle \varphi, \exp{-tH_o}\, \psi \rangle = \int_\Omega d\mathbb{P} \, \overline{\varphi(q_o(0))}\, \psi(q_o(t)) = \mathbb{E}\left(\overline{\varphi(q_o(0))}\, \psi(q_o(t)) \right)$$

The correlation functions of the process $q_o(t)$ are, in turn, given by:

$$\mathbb{E}\left(q_o(t_1)\cdots q_o(t_n) \right) = \begin{cases} o & \text{for } n \text{ odd} \\ \exp{-(t_{i_2}-t_{i_1})} \cdots \exp{-(t_{i_{2K}}-t_{i_{2K-1}})} & \text{for } n = 2K \end{cases}$$

where $\begin{pmatrix} 1 & \cdots\cdots & 2K \\ i_1 & \cdots\cdots & i_{2K} \end{pmatrix}$ is such that $t_{i_1} \leqslant t_{i_2} \leqslant \cdots t_{i_{2K}}$

Namely:

$$\mathbb{E}\left(q_o(t_1)\cdots\cdots q_o(t_n)\right) =$$

$$=\langle \Omega_o, q \, \exp-(t_2-t_1)H_o \, q \, \exp-(t_3-t_2)H_o q \cdots \exp-(t_n-t_{n-1})H_o \, q \, \Omega_o\rangle$$

The entire quantum mechanical structure is encoded in the ground state process $q_o(t)$: from it one can indeed recostruct the Hilber space as the L^2 space on the states of the process at some fixed time, the operator $q(0)$ as multiplication by the independent variable, the hamiltonian by the definition

$$(\exp-t\,H_o\,\psi)(\sigma) = \mathbb{E}\left(\psi(q_o(t))\,\big|\,q_o(0)=\sigma\right)$$

After which one can set:

$$q(t) = \exp it\,H_o\,q(0)\exp-it\,H_o$$

$$P(t) = \dot{q}(t) = i\left[H_o, q(t)\right]$$

Here we give a characterization of the ground state process $q_o(t)$ and, for the previous considerations, of the full quantum system in terms of purely probabilistic concepts:

$q_o(t)$ is the unique stationary Markov process with values in Z_2 satisfying the dynamical condition

$$\frac{D^+D^- + D^-D^+}{2}\,q_o(t) + q_o(t) = 0 \qquad\qquad 1)$$

(the definition of the mean forward and backward derivatives $D^\pm$ will be recalled in a moment (5)).

A fully probabilistic reconstruction procedure starts from the observation that:

The search for non stationary solutions of 1) leads in a canonical way to consider non stationary solutions of the Schrodinger equation

$$i\frac{d}{dt}\,\psi(t,\sigma) = \frac{1}{2}\left[\psi(t,\sigma) - \psi(t,-\sigma)\right] \qquad\qquad 2)$$

generated by the hamiltonian H_o.

In this context, the mean forward and backward derivatives of the stochastic process $q(t)$ are defined as the processes

$$D^\pm q(t) = P^\pm(t, q(t))$$

where

$$P^{\pm}(t,\sigma) = \pm \lim_{\Delta t \searrow 0} \mathbb{E}\left(\frac{q(t\pm\Delta t)-q(t)}{\Delta t} \,\middle|\, q(t)=\sigma\right)$$

One needs just to recall the definition of the conditional expectation appearing above to realize that $p^{\pm}$ are not independent concepts but are related by

$$\varrho(t,\sigma)P^{+}(t,\sigma) = \varrho(t,-\sigma)P^{-}(t,-\sigma) \qquad 3)$$

where ϱ is the probability density of the process.

A further relation is imposed by the continuity equation, which for a Markov process with values in Z_2 reads:

$$\frac{d}{dt}\,\varrho(t,\sigma) = \sigma\,\varrho(t,\sigma)\,\frac{P^{+}(t,\sigma)+P^{-}(t,\sigma)}{2} \qquad 4)$$

The meaning of the second order stochastic derivatives appearing in 1) is clear once one recalls that for a function F of the process $q(t)$

$$D^{\pm}F(t,q(t)) = \frac{\partial}{\partial t}\,F(t,q(t)) + P^{\pm}(t,q(t))\nabla F$$

(here and in the following, if $f(x)=f_0+xf_1$ is a function on Z_2, we set $\nabla f = f_1$).

The dynamical condition 1) is best expressed in terms of the auxiliary fields

$$P = \frac{P^{+}+P^{-}}{2}$$

$$\delta P = \frac{P^{+}-P^{-}}{2}$$

In these notations 1) reads:

$$\left(\frac{\partial P}{\partial t} + P\nabla P - \delta P\nabla\delta P\right)(t,\sigma) = -\sigma \qquad 1')$$

to be solved under the conditions

$$\sum_{\sigma\in Z_2}\varrho(t,\sigma)\,\delta P(t,\sigma) = 0 \qquad 3')$$

$$\sum_{\sigma\in Z_2}\sigma\,\varrho(t,\sigma)\,P(t,\sigma) = 0 \qquad 3'')$$

equivalent to 3), and 4) reads

$$\frac{d}{dt}\varrho(t,\sigma) = \sigma\, P(t,\sigma)\,\varrho(t,\sigma)$$ 4')

Summarizing, the solution of 1) requires the following steps:

-Start with some initial information about the "position" (say $\varrho(o,x)$) and the "velocity" (say $p^+(0,x)$) of the process;

-Equations 1'),3'),3"),4') then determine $\varrho(t,x)$ and $p^+(t,x)$;

-with the solution found for $p^+(t,x)$, the forward and backward Kolmogorov equations

$$\frac{\partial}{\partial\sigma} P(t,x\mid\mathfrak{z},y) = \frac{y}{2}\left(p^+(\mathfrak{z},y)\, P(t,x\mid\mathfrak{z},y) + p^+(\mathfrak{z},-y)\, P(t,x\mid\mathfrak{z},-y)\right)$$ 5)

$$\frac{\partial}{\partial t} P(t,x\mid\mathfrak{z},y) = \frac{x}{2}\, p^+(t,x)\left(P(t,-x\mid\mathfrak{z},y) - P(t,x\mid\mathfrak{z},y)\right)$$ 6)

determine then the transition probabilities: $P(t,\sigma\mid\mathfrak{z},\sigma')$.

(In all the previous discussion p^- would do as well)

Our first statement is now easily proven: for a stationary solution of 1) it must be $\dot\varrho = 0$
4') then implies p=0, which satisfies 3"). 1') imposes then $\delta p\nabla\delta p = \sigma$
namely $\delta p(t,\sigma) = \pm\sigma$ and $p^+(t,\sigma) = \pm\sigma$ where only the minus sign is acceptable because it must be $\sigma\, p^+(t,\sigma) \leq 0$
3')imposes that $\varrho(t,x)$ is independent of x and therefore, by nomalization

$$\varrho(t,\sigma) = 1$$

5), 6), with $p^+(t,x)=-x$, determine then

$$P(t,\sigma\mid\mathfrak{z},\sigma') = \frac{1+\sigma\sigma'\exp-(\mathfrak{z}-t)}{2}$$

Namely the only stationary solution of 1) <u>is</u> the independently defined ground state process $q_o(t)$.

Next we show how the non linear problem 1) reduces to the linear problem 2)

The change of unknown function

$$\varrho(t,\sigma) = \exp 2\, R(t,\sigma)$$

reduces 4') to

$$\frac{d}{dt} R(t,\sigma) = \frac{\sigma}{2}\, P(t,\sigma)$$

The constraints 3'), 3") are easily parametrized noticing that a condition of the form

$$\sum_{\sigma \in Z_2} F(t,\sigma) \exp 2 R(t,\sigma) = 0$$

implies for the function F the form

$$F(t,\sigma) = \sigma f(t) \exp -2\sigma \nabla R$$

where f depends only on t.

In particular

$$p(t,\sigma) = \alpha(t) \exp -2\sigma \nabla R$$

$$\delta p(t,\sigma) = \beta(t)\sigma \exp -2\sigma \nabla R$$

where the unknown functions α, β are to be determined by 1'), which is equivalent to:

$$\begin{cases} \dfrac{d\alpha}{dt} + \beta^2 \sin h \, 2 \nabla R = 0 \\[2mm] \alpha^2 + \beta^2 = 1 \end{cases}$$

A further constraint is imposed on α and β by the observation that

$$p^+(t,\sigma) = (\alpha + \sigma\beta) \exp -2\sigma \nabla R$$

so that the condition (to be satisfied in particular by the initial data $p^+(0,x)$)

$$\sigma p^+(t,\sigma) \leq 0$$

requires

$$\alpha + \beta \leq 0$$
$$\alpha - \beta \geq 0$$

The constraints on α and β are parametrized by

$$\alpha(t) = \sin 2 S_1(t)$$

$$\beta(t) = \mp \cos 2 S_1(t)$$

where the minus sign corresponds to the choice

$$2 S_1 \in \left[-\tfrac{\pi}{4}, \tfrac{\pi}{4} \right]$$

and the plus sign corresponds to the choice

$$2 S_1 \in \left[\tfrac{3\pi}{4}, \tfrac{5\pi}{4} \right]$$

The unknowns of our problem are now R(t,x), to be determined by

$$\frac{d}{dt}\, R(t,\sigma) = \frac{1}{2}\, \sin 2\sigma\, S_1(t)\, \exp -2\sigma\, \nabla R$$

equivalent to 4'), and

$$\frac{d}{dt}\, S_1(t) = \frac{1}{2}\, \nabla\left(\cos 2\sigma\, S_1(t)\, \exp -2\sigma\, \nabla R\right)$$

equivalent to 1').

Introduce now an auxiliary function $S_o(t)$, defined, within an additive constant by

$$\frac{d}{dt}\, S_o(t) = \frac{1}{2}\left[\cosh 2\nabla R \cos 2 S_1(t) - 1\right]$$

and set $\quad S(t,\sigma) = S_o(t) + \sigma\, S_1(t)$

Our equations become then:

$$\frac{d}{dt}\, R(t,\sigma) = \frac{1}{2}\, \sigma\, \sin 2\nabla S\, \exp -2\sigma\nabla R$$

$$\frac{d}{dt}\, S(t,\sigma) = \frac{1}{2}\left[\cos 2\nabla S\, \exp -2\sigma\, \nabla R - 1\right]$$

or, equivalently, in terms of the unknown function $\psi(t,\sigma) = \exp\left[R + iS\right]$

$$i\,\frac{d}{dt}\, \psi(t,\sigma) = \frac{1}{2}\left[\psi(t,\sigma) - \psi(t,-\sigma)\right]$$

which <u>is</u> the Schrodinger equation for the Fermi oscillator in the "Ω representation".-

References :

1. D. de Falco, F. Guerra: Journal of Mathematical Physics, $\underline{21}$, 1111 (1980)

2. G.F. De Angelis, D. de Falco, F. Guerra: Physical Review $\underline{D23}$, 1747 (1981);

3. J. Schwinger: Physical Review $\underline{115}$, 721 (1959);

4. P. Jordan, E. Wigner: Zeitschrift für Physik $\underline{47}$, 631 (1928);

5. E. Nelson: Dynamical Theories of Brownian Motion. Princeton University Press (1967).

PERIODIC ORBITS OF DYNAMICAL SYSTEMS WITH CHAOTIC BEHAVIOR

S. DE GREGORIO +o

E. SCOPPOLA ++

B. TIROZZI +

+   Istituto di Matematica dell'Università di Roma

o   Istituto di Matematica dell'Università dell'Aquila

++ Presently at the Centre de Physique Théorique, CNRS, Luminy, Marseille France.
   On leave of absence from the Accademia Nazionale dei Lincei, Roma.

The numerical study of systems of ordinary differential equations which present a stochastic behavior, has had an increasing development in connection with the theory of turbulence.

In particular many numerical results were obtained on the Lorenz model and on the five modes model for an incompressible bidimensional fluid ([1],[2]) where the chaotic behavior appears to be connected with the stability properties of certain periodic orbits.

In such a situation it is interesting to have a theory which enables us to determine when an orbit which numerically appears to be periodic is rigorously periodic.

Sinai and Vul have pointed out a criterion to solve this problem, that is they have stated a set of conditions which implies the existence and the unicity of a fixed point of the Poincaré map in a certain neighborhood of the numerical fixed point [3].

We apply in [4] a modified version of the method of Sinai and Vul in order to study closed orbits which appear in the Lorenz model and in the five modes model. In order to show how this method works we introduce some notations : let $X = \{ x_i , i = 1,\dots,d \}$ be a vector in R^d and $(X,Y) = \sum_{i=1}^{d} x_i y_i$ $|X| = (\sum_{i=1}^{d} x_i^2)^{1/2}$. Consider the system of ordinary differential equations:

$$\frac{dx_i}{dt} = f_i (x_1,\dots,x_d) \qquad \text{or briefly:} \qquad \dot{X} = F(X) \qquad \text{where}$$

$F = f_i$, $i = 1,\dots,d$ is defined by

$$f_i(X) = (G^i,X) + (B^i X,X) + C^i \qquad i = 1,\dots,d$$

and
- G^i is a constant vector
- B^i is a constant dxd matrix
- C^i is a constant.

We denote with S_t the one parameter group of shift along the trajectories generated by the differential equations, $X^0(t) = S_t X^0$ is the solution with initial condition X^0.

Let Γ be a fixed hyperplane (in the applications $\Gamma = \{ X | x_j = a \}$, $a \in R$, $j = \in \{1,\dots,d\}$ fixed) and let $X_0 \in \Gamma$ be a point such that for some $T > 0$, $X^0(T) \in \Gamma$ and in a sufficiently small neighborhood of X^0, contained in Γ, is defined the Poincaré map P induced by the flow on the hyperplane Γ: $P(X^0) = S_T X^0 = \bar{X} \in \Gamma$.

If we define $Y = X-X^0$, $Y^0 = \bar{X}-X^0$, $\varepsilon = |Y^0|$ and $Q(Y) = P(X)-X^0$ we can expand the Poincaré map in a neighborhood of the point X^0 :

$$Q(Y) = Y^0 + LY + k(Y)$$

where L is the derivative of the Poincaré map evaluated at $X = X^0$, and $K(Y)$ is the non linear part. We can now state the main criterion of the method.

<u>CRITERION</u>

If $K(Y)$ satisfies the following condition :

there exist positive constants ρ_0 and k_0 such that for any $\rho < \rho_0$ and any $Y^1 = X^1 - X^0$, $Y^2 = X^2 - X^0$; $X^1, X^2 \in \Gamma$, $|Y^1| < \rho$, $|Y^2| < \rho$,

$$|K(Y^1) - K(Y^2)| \leq k_0 \, \rho \, |Y^1 - Y^2|$$

and if for some $\rho_1 < \rho_0$

$$\| (L-E)^{-1} \| \left(\frac{\varepsilon}{\rho_1} + k_0 \rho_1 \right) < 1$$

where $\varepsilon = |Y^0|$ and E is the identity matrix then there exists a unique fixed point of the Poincaré map in the ρ_1-neighborhood of the point X^0.

The proof of this criterion is an application of the Newton method and can be found in [3]. We want to figure out now how one can apply the above criterion. Suppose we have chosen a numerical method of integration (for instance the method of successive approximations with step Δ:

$$X^{(0)}(t) = X^0$$
$$\vdots$$
$$X^{(i)}(t) = X^0 + \int_0^t F(X^{(i-1)}(S))ds$$
$$i \leq k \qquad\qquad 0 \leq t \leq \Delta)$$

in such a way that at each step the error of integration is of the order of a small fixed constant α , which is related to the computer precision, (this means in the example k sufficiently large and Δ sufficiently small).

We obtain thus a numerical trajectory starting at $X^0 \in \Gamma$:

$$X_{(0)} = X^0 \, , \, X_{(1)} \, , \, X_{(2)} \, \ldots, \, X_{(m)}, \, \ldots$$

If we look for the intersection of this numerical trajectory with the hyperplane Γ , by successive bisections of the integration step Δ we can find a point $X_{(n)}$ of intersection of the trajectory with Γ and a period with a precision of the order of α , that is $|(X_{(n)})_j - a| \leq \alpha$. Suppose now we are dealing with a numerical periodic orbit, that is $|X_{(n)} - X^0|$ of the order of α , then the constant ε appearing in the criterion can be evaluated by

$$\varepsilon \equiv |\bar{X} - X_0| \leq |\bar{X} - S_{n\Delta} X^0| + |S_{n\Delta} X^0 - X_{(n)}| + |X_{(n)} - X^0|$$

70

Since through the previous remark the first and the third terms in the r.h.s. are of the order α , the greatest contribute to ε is the quantity $\left| S_{n\Delta} X^0 - X_{(n)} \right|$ that is the distance between the numerical trajectory and the solution $S_{n\Delta} X^0$.

To study this quantity we use the following results : let $F'(X)$ be the matrix with elements $(F'(X))_{ij} = \dfrac{\partial f_i(X)}{\partial x_j}$ and consider the linearized equation:

$$\frac{dZ}{dt} = F'(X^0(t))Z \quad .$$

Let $\mathcal{L}(s,t)$ be the fundamental matrix of the solutions of this differential equation, that is

$$Z(t) = \mathcal{L}(s,t) Z(s) \quad .$$

We define $C_1 = \sup_{0 \leqslant s,t \leqslant T} \| \mathcal{L}(s,t) \|$ where the norm of the matrix $\mathcal{L}$ is given by $\| \mathcal{L} \| = (\text{max eigenvalue of } \mathcal{L}\mathcal{L}^*)^{1/2}$

and let C_2 be a constant such that

$$\sum_{i,j,\kappa} \left| \frac{\partial^2 f_i}{\partial x_j \partial x_\kappa} Y_j Z_\kappa \right| \leqslant C_2 |Y| |Z|$$

for any $Y, Z \in \mathbb{R}^d$.

We shall use the symbol F_i'' for the matrix $\dfrac{\partial^2 f_i}{\partial x_j \partial x_\kappa}$, $j,k = 1,\ldots,d$.

LEMMA 1

For any fixed $T > 0$ if $\rho_0 = (C_1^2 C_2 T)^{-1}$, $|X - X^0| < \rho_0$, then for $t \in [0,T]$

$$\left| S_t X - S_t X^0 \right| \leqslant 2 C_1 |X - X^0| \quad .$$

<u>Proof</u> : If $Y(t) = X(t) - X^0(t)$,

$$\frac{d}{dt} Y(t) = F(X(t))) - F(X^0(t)) = F'(X^0(t))Y(t) + \frac{1}{2}(F''Y(t),Y(t)).$$

We can write :

$$Y(t) = \mathcal{L}(0,t)Y(0) + \int_0^t ds \, \mathcal{L}(s,t) \frac{1}{2} (F''Y(s),Y(s)),$$

and then

$$\left| Y(t) \right| \leqslant C_1 |Y(0)| + \frac{1}{2} C_1 C_2 \int_0^t |Y(s)|^2 \, ds$$

$$|Y(t)| \leq \frac{c_1 \, |Y(o)|}{1 - \frac{1}{2} c_1^2 c_2 t |Y(o)|} \leq 2C_1 \, |Y(0)|$$

With analogous computations we can prove the following more interesting result on the evolution of two points near X^o :

LEMMA 2

If $\quad |X^1 - X^o| < \rho_o \quad , \quad |X^2 - X^o| < \rho_o \quad$ with ρ_o as in Lemma 1, then

$$|S_t X^1 - S_t X^2| \leq 8 \, C_1 \, |X^1 - X^2| \qquad \forall \, t \, \epsilon \, [0,T]$$

Let us write $\quad |X_{(n)} - S_{n\Delta} X^o| \quad$ in the following way :

$$|X_{(n)} - S_{n\Delta} X^o| \; = \; \sum_{i=0}^{k-1} \; (S_{i\Delta} X_{(k-i)} - S_{i\Delta} (S_\Delta X_{(k-i-1)}))$$

we obtain

$$|X_{(n)} - S_{n\Delta} X^o| \leq 8 \, C_1 \, n \, \alpha$$

if $\quad \alpha \leq (8 m c_1^3 c_2 T)^{-1} \quad$ where we used this conditions on α in order to apply Lemma 2.
This estimate and the lemmas hold also in the case of an unstable orbit. We observe by the way that all the method can be applied leaving out of account the stability properties of the orbit.

As far as the estimate of the non linear part of the Poincaré map is concerned, we can explicitely compute the constant k_o.
In fact, let $\quad X^1, X^2 \, \epsilon \, \Gamma \quad$ be such that

$$|Y^1| \; = \; |X^1 - X^o| < \rho \quad , \qquad |Y^2| \; = \; |X^2 - X^o| < \rho$$

and let T_1, T_2, T be such that $\quad S_{T_1} X^1 \, \epsilon \, \Gamma \; , \; S_{T_2} X^2 \, \epsilon \, \Gamma \; , \; S_T X^o \, \epsilon \, \Gamma \; .$

We have to estimate the quantity $X^1(T_1) - X^2(T_2) - L(X^1 - X^2) \; .$

If we write the difference $X^1(T_1) - X^2(T_2)$ in terms of the matrix $\mathcal{L}$ as in the lemma, using the identity $(X^1(T_1))_j - (X^2(T_2))_j = 0$ and Lemma 2 we can find :

$$|K(Y^1) - K(Y^2)| \leq k_o \, \rho \, |Y^1 - Y^2|$$

where

$$k_0 = 16 \, c_1^2 \left(1 + \frac{c_4}{c_3}\right) \left[c_1 c_2 T + \frac{c_5}{c_3} \left(2 + \frac{c_4}{c_3}\right)\right]$$

with

$$c_3 = \inf_{X \in U_\rho(X^0)} |f_j(X)|$$

$$c_4 = \max_{X \in U_\rho(X^0)} \max_i |f_i(X)|$$

$$c_5 = \max_{X \in U_\rho(X^0)} \| F'(X) \|$$

and

$$U_\rho(X^0) = \left\{ X \ / \ \sum_{i \neq j}^d (x_i - x_j^0)^2 < \rho^2 \ , \ |x_j - a| < \rho \right\}$$

We also obtain that the linear part L of the Poincaré map is given in terms of the matrix $\mathcal{L}$ by

$$L_{ik} = \mathcal{L}_{ik}(0,T) - \frac{f_i(\bar{X})}{f_s(\bar{X})} \, \mathcal{L}_{jk}(0,T)$$

So the study of the linear part L and of the constant C_1 is reduced to the evaluation of the matrix $\mathcal{L}(s,t)$ which must be studied with the computer.

In fact this matrix can be evaluated by introducing a new matrix $\overline{\mathcal{L}}$ defined on the numerical trajectory $X^0, X_{(1)}, \ldots, X_{(k)}, \ldots$, by

$$\overline{\mathcal{L}}(m\Delta, n\Delta) = \prod_{i=m}^{n-1} (E + \Delta F'(X_{(i)})) \quad \text{where } m < n \text{ and } E \text{ is the identity matrix.}$$

All the approximations so introduced can be estimated [4].

The last problem is the study of the constant α , that is the evaluation of the necessary computer's precision.

This is a crucial problem ; in fact the applicability of the criterion depends on the smallness of α : each periodic orbit can be tested by the criterion if we can choose a constant α and then ε sufficiently small, by taking a suitable constant ρ_1 . In order to study the constant α we need the intervals arithmeticthat is we represent the numbers on the computer with intervals and then all the computations are made with intervals arithmetic.

APPLICATIONS

This method has been applied to prove the existence of the following closed orbits.

(1) Lorenz model [1]

$$\begin{cases} \dot{x} = -\sigma x + \sigma y \\ \dot{y} = -y + (\sigma+z)x \\ \dot{z} = -bz + xy - r \end{cases}$$

$$\sigma = 10$$

$$b = 8/3$$

$$r = 294.133333 \quad .$$

The Poincaré map is defined on the plane $z = -11$ and the numerical fixed point considered is

$$X^0 = (-9.3249753511062, \; 7.70489278613672, \; -11).$$

We find $\rho_1 = .5 \; 10^{-9}$, $C_1 = 2.6 \; 10^2$

(2) Lorenz model

$$\begin{cases} \dot{x} = a_1 x + b_1 yz + b_1 xz \\ \dot{y} = a_2 y - b_1 yz - b_1 xz \\ \dot{z} = - a_3 z + (x+y)(b_2 x + b_3 y) \end{cases}$$

$$a_1 = 9.700378782$$

$$a_2 = - 16.700378782$$

$$a_3 = 2.666666667$$

$$b_1 = - .227266206$$

$$b_2 = 2.616729797$$

$$b_3 = - 1.783396463 \quad .$$

(These values of the parameters correspond to the values $r = 28$, $\sigma = 6$, $b = 8/3$ of the usual form of the Lorenz model). The Poincaré map is defined on the plane $z = 27$ and the numerical fixed point is

$$X^0 = (3.5007872047249 \; ; \; 3.3303317970426 \; ; \; 27).$$

and $\rho_1 = .33 . 10^{-6}$ $C_1 = 7$

This orbit has also been studied by Sinai and Vul [3].

(3) <u>Five modes model</u> [2]

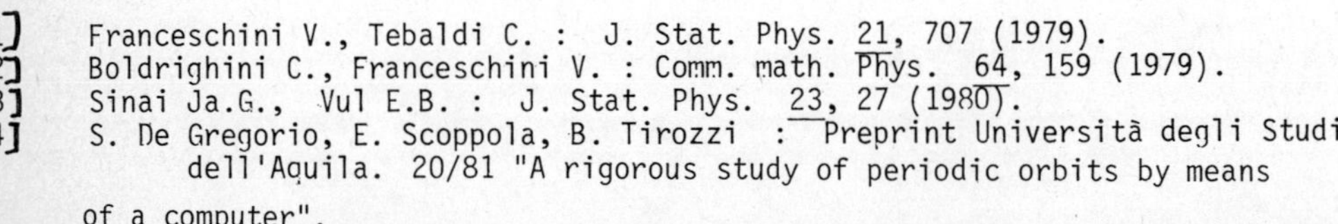

$$\begin{cases} \dot{x} = -2x + 4yz + 4uv \\ \dot{y} = -9y + 3xz \\ \dot{z} = -5z - 7xy + r \\ \dot{u} = -5u - xv \\ \dot{v} = -v - 3xu \end{cases}$$

$$r = 25 \quad .$$

The Poincaré map is defined on the plane $z = 3$ and the coordinates of the numerical fixed point are

$$x_0 = .4666202$$
$$y_0 = .6421500$$
$$z_0 = 3$$
$$u_0 = .6876595$$
$$v_0 = -2.979946$$

and $\rho = 10^{-7}$, $c_1 = 10$

REFERENCES
==========

[1] Franceschini V., Tebaldi C. : J. Stat. Phys. 21, 707 (1979).
[2] Boldrighini C., Franceschini V. : Comm. math. Phys. 64, 159 (1979).
[3] Sinai Ja.G., Vul E.B. : J. Stat. Phys. 23, 27 (1980).
[4] S. De Gregorio, E. Scoppola, B. Tirozzi : Preprint Università degli Studi
       dell'Aquila. 20/81 "A rigorous study of periodic orbits by means
of a computer".

The Van Hove limit in Classical and Quantum Mechanics

G. F. Dell'Antonio

Istituto Matematico G. Castelnuovo

Univ. di Roma

1 Introduction

The problem of describing the motion of a particle (classical
or quantal) in a random medium presents itself very naturally, in
view of the "uncontrollable" fluctuations of the media through which
particle pass in many phenomena of interest in physics.
Such madia could be plasmas, inhomogeneous magnetic and electric
fields, ... To consider such media "random" is the best possible
assumptions if only statistical information is available.
On small intervals of time, one would expect that the behaviour of
the particle be itself random; however, on a large time-scale, and
if the local interaction with the medium is sufficiently small, one
would expect, mainly from the law of large numbers, that a more regular
regime is attained. In particular, if the medium is "stationary"
(essentially, homogeneous) one expects that the velocity distribution
be rather accurately described by a Markov process on the large time
scale.
One is then led naturally to study the Van-Hove limit (small coupling,
long time) for the momentum distribution, both classically and
quantum-mechanically.
This analysis can be also useful in the study of the equilibrium
(Gibbs) distribution, if one considers the motion of a tagged particle
in an isolated system composed of infinitely many particles. One can
describe the "remaining" particles through Markov partitions, and
recover, under suitable assumptions,the setting we consider here. It
is therefore interesting to notice that the Markov processes one
obtains in the Van Hove limit have the microcanonical as unique irreducible invariant measure.
In describing the Van Hove limit, we must require that the average
effect of the interaction be zero, otherwise the scale of time involved
is not large enough to led to microscopic fluctuations. Also, the
mean-square deviation must be small, roughly as the inverse of the
time scale, otherwise the overall effect would increase without bound.

All these intuitive ideas have been known for many years to theoretical phisicists and applied mathematicians, and several attempts to formalize them have been made in the past. The "reasonable" results have therefore been known, although no proof was available; see e.g. [1] and in particular the interesting review paper by N. Van Kampen. Several methods of approach have been suggested, among others the method of Bourret, and led to acceptable results. In the Quantum-Mechanical context, the germinal papers are [2] ; in these papers Van Hove correctly describes the leading contributions to the perturbation series for correlation functions, and argues why the remaining terms should vanish in the limit considered.

While an actual proof is not given, his method has inspired all later research on the Quantum-Mechanical case. Some further results and some proofs were given in [3] , where motion in a random environment is considered. Later, in a very interesting paper [4] , H. Spohn has provided further rigorous results, proving convergence of dynamics for extensive observables for short times (on the long time scale) under rather stringent conditions on the random potential.

From a mechanical point of view, proving that the expected behaviour actually occurs requires implementing the idea that thereis a "small" correlation between the interactions experienced by the particle at comparatively distant times, so as to allow the application of some form of the central limit theorem, as described e.g. in [5] . The difficulty lies of course in controlling those events for which this is not true. One may try to modify the original law of motion so as to avoid all dangerous events, and then make use of the properties of the limit process to show that such modification was actually superflous.

This is precisely what is done in rigorous treatments of the motion of a particle in a random homogeneous field of force. In this case, the limit process is a diffusion, and it is well known that diffusions, much as Brownian motion, are non recurrent when $d \geq 3$, where d is the number of space dimensions; since dangerous events are essentially recurrent trajectories, one expects to be able to give a proof only if $d \geq 3$. Of course, some technical assuption about the weakness of

the correlations _in_ _space_ for the random force field will be required;
a part from this, specific properties of the random field should not
be relevant.

If one wants to obtain a simple result about the limit invariant
distribution, a Hamiltonian should exist, and therefore we will take
the random force field to be conservative, so that there will be a
random potential.

This program can be implemented in the classical case, see e.g. [6] .
Sufficient control over the measure of dangerous trajectories allows
to bring the problem to a stage in which it can be attacked by techniques such as the ones developed in [5] . The main technical tools in
proving convergence are martingale estimates, which allow among other
things to use Prokorov's compactness condition. As a consequence,
convergence is best possible, i.e. weak convergence of processes.

Recall that, for a classical particle in a random force field, with
realizations $\omega \in \Omega$, the velocity becomes a process (for fixed initial
velocity $\underline{v}$ and position $\underline{x}$, the velocity at time t is a function of
ω). It is _not_ a Markov process since the velocity at time t depends,
for fixed $t_o < t$, on ω, $v(t_o)$, $x(t_o)$. Call $\underline{v}(\omega, \underline{x}_o, \underline{v}_o, t)$ this process,
when the force field is $\lambda F(x,\omega)$

One proves [6] that $\underline{v}$ converges in distribution, when $t = \tau \lambda^{-2}$, $\lambda \to 0$,
to a diffusion process, whose generator depends only on the two-point
correlation of the force field.

We shall give the explicit formulae in $\S 2$ and $\S 5$, where we discuss
also the "classical limit". For more details, we refer to [6] .

Here we want to discuss briefly the Quantum Mechanical case. We shall
outline the main ideas and technical tools which enter in the proof,
and then give a short discussion of the classical limit. For details,
we refer to [7] .

The approach which works in the classical case does not carry over
to Quantum Mechanics, for a basic reason: there is now no position-
velocity process, since position and momentum are not simultaneously
observable. Therefore it is not obvious how to proceed to discard
"trajectories which come back close to the initial point".

A second obvious difficulty comes from the equation of motion.

While classical equations are "local", and have local solutions, the Schroedinger equation does not.

Indeed, if $f(\underline{x})$ has compact support in R^d (d=space dimension), and $t \mapsto f(\underline{x},t)$ is a solution with $f(\underline{x},0)=f(\underline{x})$, then for <u>all</u> $t_o > 0$ the function $f(\underline{x},t_o)$ does <u>not</u> have compact support. Of course, the amount (in the L^2 sense) of $f(x,t)$ outside some compact set can be very small, especially if $f(p)$ is different from zero essentially only in a small neighborhood of some $p^o \in R^d$. Therefore <u>some</u> approximate form of localization still holds.

In principle, using a suitable class of test functions f, a time lattice of sufficiently large scale (compared with λ^{-1}) and concentrating on the motion of the "center of mass" of the distribution $f(\underline{x},t)$, one may be able to replicate the steps of the Classical Mechanics proof, and perhaps obtain a stronger version of convergence than the one given here.

We have not succeeded in doing so, and so our proof will be of totally different nature (although there are intriguing relations between our estimates and the ones necessary to control the motion of the baricenter of $f(\underline{x},t)$ as compared to its spreading).

In particular, we will be unable to use martingale inequalities (again, some of the estimates will however "look like" martingale estimates), and will have no Prohorov's lemma, mainly due to lack of processes to which one could apply our estimates.

As a consequence, our results for Quantum Mechanics are weaker than those one obtains for Classical Mechanics, and in particular we only prove convergence of all finite-dimensional distributions to those of a Markov process (not a diffusion, when $\hbar > 0$).

As we shall explain in $\S 2$ and $\S 3$, there are two main steps in our proof. The convergence result proved in the first step are <u>uniform</u> in $\hbar \geq 0$; this implies in particular, as we shall see in $\S 5$, that one has convergence, strongly in semi-group, when $\hbar \to 0$, of the limit Quantum process to the limit Classical diffusion.

Convergences in the second step are uniform in $M \geq \hbar \geq \varepsilon$ for all $\varepsilon > 0$, $M < \infty$. If one considers expectations in suitable coherent states, one has also uniformity in $\varepsilon \geq \hbar \geq 0$, where ε depends on the state. Therefore the methods outlined here should also provide an alternative proof of the Classical Mechanics case, although in a weaker form and for potential fields. We shall not pirsue here this point.

Before beginning, in the next $\S$, an outline of relevant definitions and proofs, we want to comment briefly on another point.

Both in the Quantum and in the Classical (potential) case the limit process is reduced by each sphere $S_a \equiv \{ \underline{P} \mid |\underline{p}| = a \}$, where $\underline{P}$ is the momentum of the particle. In the Quantum case, on each sphere the process is Poisson; when $\hbar \to 0$, the probability of having finite jumps goes to zero, and the Poisson process converges to a diffusion, which is precisely the Classical Mechanics result.

There is an intriguing connection between the fact that the limit process is Poisson and the description of quantum mechanical processes in terms of Poisson processes, recently stressed by a Marseille group [8] . Of course, such results hold only for a special class of potentials; but one should recall that there can be several concrete realizations of a random field, and moreover the specific form of the random potential plays a secondary role, and only properties of the correlation functions are relevant.

There are indeed some indications that one could use this representation of Quantum Mechanical correlation functions to improve on the results outlined here.

It would seem that the jumps of the limit Poisson process correspond to <u>pairs</u> of jumps of the limit Poisson process through which correlations are calculated, and that the integrals in [8] are only sensitive, when $\lambda \to 0$, to the "tiny" subset of configurations for which the sums of the $(2n-1)^{th}$ and $(2n)^{th}$ jumps are again Poisson distributed, somewhat in the spirit of the analysis of Donsker and Varedhan , [9] .

We shall not dwell on this point here, but we would like to remark the following: the basic intuition that it is only successive interactions which are correlated, when the time scale is sufficiently large, and in fact that such correlations occur on a much smaller time-scale, (so as to be "instantaneous") is not only at the basis of the pioneering work by Van Hove [2] , and of the formal approaches [1] (the method of cumulants is an attempt to make this intuition into a mathematical proof) but it is also the starting point in the analysis of [4] , [5] , [6] .

Finally, before closing the introduction, I am pleased to record here my appreciation and thanks to the organizers for an inspiring Conference, and to Ph. Combe for many stimulating discussions.

2 <u>Motivations</u> <u>and</u> <u>expected</u> <u>results</u>

We shall describe briefly in this   the motivations and expected results, and introduce the new time-scale. Emphasis will be placed on the similarities between the Quantum and Classical cases. In this section, many arguments are heuristic, and their role is only to make the expected results plausible. Still, it should be noticed that the proof which will be outlined in the following $\S$ are just a rigorous way of looking at the content of the present section.

<u>A</u>. <u>Classical Mechanics</u>   (see [6])

Let $V(x)$ be a random field, so that $V(x,\omega)$ is, for each $\omega \in \Omega$ (Ω a probability space) a potential which produces a force acting on the classical particle at the point $\underline{x} \in R^d$.
We shall treat here only conservative random forces, bacause we want to stress the analogies with Quantum Mechanics. One could extend without difficulty the treatment to cover non-conservative random force fields.
If $V(x,\omega)$ is a potential, Newton's equation for a particle of mass one reads

$$\frac{d^2 x}{dt^2} = \lambda \nabla V \qquad (2.1)$$

$\lambda (>0)$ is a "coupling costant".
If we require, e.g. , that $V(x,\omega)$ be of class C^2 for all ω, then (2.1) has a unique maximal solution (for given initial data $x_o, u_o \equiv \dot{x}(0)$) With $u(t) \equiv \frac{dx}{dt}$, the process ($x(t), u(t)$) is deterministic, in particular a Markov process. The process $u(t)$ is not, by itself, a Markov process.
We scale time according to

$$t = \lambda^{-2} \tau \tag{2.2}$$

and consider the limit $\lambda \to 0$.

That (2.2) is the correct scaling can be seen in many heuristic ways; we refer for this to [1] , but remark that the formal analisis given below can also be looked at as a heuristic justification of (2.2) . After scaling, the equations take the form, with

$$x^\lambda(\tau) \equiv x(\tau/\lambda^2) \quad ; \quad u^\lambda(\tau) = \frac{dx}{dt}(\tau/\lambda^2)$$

$$\frac{dx^\lambda}{d\tau} = \frac{1}{\lambda^2} u^\lambda \quad ; \quad \frac{du^\lambda}{d\tau} = \frac{1}{\lambda} \nabla V(x^\lambda) \tag{2.3}$$

The initial data are $u^\lambda(0) = u_0$, $x^\lambda(0) = x_0$.

We expect to obtain, when $\lambda \to 0$, a simple description of time-evolution (with τ as time-parameter) for the velocity distribution.

We consider therefore the Liouville equation associated to (2.3). Using the notation

$$\phi^\lambda(t, x_0, u_0, \omega) \equiv u^\lambda(t, x_0, u_0, \omega)$$

we consider

$$F^\lambda(\tau, x, u, \omega) \equiv f(\phi^\lambda(\tau\lambda^{-2}, x, u, \omega)) \tag{2.4}$$

where $f(\varsigma)$, $\varsigma \in R^d$, is a sufficiently smooth function.

The function F satisfies then the following Liouville equation

$$\frac{\partial F^\lambda}{\partial \tau} = \frac{1}{\lambda^2} (u \cdot \nabla_x) F^\lambda + \frac{1}{\lambda} (\nabla V) \cdot \nabla_u F^\lambda \tag{2.5}$$

$$F^\lambda(0, x, u, \omega) = f(u) \qquad \forall \; \omega, x, \lambda$$

We want to argue that , when averaged over Ω with the given measure μ, the distribution F satisfies in the limit $\lambda \to 0$ a diffusion equation, which we shall describe explicitely in terms of the two-point correlation of the field V.

Write formally

$$F^{\lambda} = F_0 + \lambda F_1 + \lambda^2 F_2 + \dots \qquad (2.6)$$

Substituting (2.6) in (2.5) and equating terms of equal order in λ one obtains

a) $\quad \nabla_x F_0 = 0 \qquad i.e. \qquad F_0 = F_0(u)$

b) $\quad u \cdot \nabla_x F_1 + \nabla_x V \cdot \nabla_u F_0 = 0$

  i.e. , at least weakly

$$F_1 = \int_0^{\infty} \nabla_x V(x + u s) \, \nabla_u F_0(u, \tau) \, ds$$

c) $\quad u \cdot \nabla_x F_2 + \nabla_x V \cdot \nabla_u \int_0^{\infty} ds \, (\nabla_x V(x + u s) \, \nabla_u F_0) - \dfrac{\partial F_0}{\partial \tau} = 0 \qquad (2.7)$

Integrate now c) over Ω . Since V is a stationary field, $E(\nabla_x(uF_2)) = 0$ and therefore, setting

$$E(F_0)(u) \equiv \int_{\Omega} F_0(u, \omega) \, P(d\omega) \equiv \bar{F}_0(u)$$

$$\frac{\partial \bar{F}_0}{\partial \tau} = E\left(\int_0^{\infty} ds \, \nabla_x V \cdot \nabla_u (\nabla_x V(x + u s) \cdot \nabla_u F_0) \right) \qquad (2.8)$$

This is not yet an equation for $\bar{F}$. One can however argue as follows. From the definition (2.4) of F, one sees that $F_0(u, \omega)$ only depends on the velocity at time $\tau \lambda^{-2}$ of a particle which at time zero has position x and velocity u , in the limit $\lambda \to 0$. One can also interpret a) above as pointing to the fact that $F(u, \omega)$ depends on the configuration of the potential only at points which are λ^{-1} distant from x . If the correlation of the potential field decreases sufficiently fast with distance, one expects not only that the integral in (2.8) be convergent but also that, in its dependence of ω , $F_0(u)$ be uncorrelated from the potentials appearing in (2.8).

One should then replace (2.8) by

$$\frac{\partial \bar{F}_0}{\partial \tau}(u, \tau) = (\mathcal{L} \bar{F}_0)(u, \tau) \qquad (2.9)$$

where the operator $\mathcal{L}$, acting on C^2-functions, is defined by

$$\mathcal{L} \equiv \sum_{i,j=1}^{d} \frac{\partial}{\partial u_i} A_{ij}(u) \frac{\partial}{\partial u_j} \tag{2.10}$$

$$A_{ij}(u) = \int_0^{\infty} E(\nabla_i V(x) \nabla_j V(x+us)) \, ds \tag{2.11}$$

In fact, if the diffusion generated by (2.10) is non-recurrent (i.e. if $d \geq 3$), one would expect that the approximation made in replacing (2.8) by (2.9) be a rather accurate one (when λ is small) for a large fraction Ω_λ of the configurations in Ω, the measure of Ω_λ converging to one when $\lambda \to 0$.

One would therefore expect not only that $E(F(x,u))$ converge for all x to $\overline{F}_o(u)$ satisfying (2.9), but that for all x the process $\phi(x,u,t\lambda^{-2},\omega)$ converge weakly to the diffusion generated by $\mathcal{L}$.

The integral in (2.11) will converge if the two-point correlation function

$$E(V(x)V(y)) \cong G(x-y)$$

vanishes sufficiently fast when $|x-y| \to \infty$.

One can then give to (2.11) the alternative form

$$A_{ij}(u) = \int_{R^d} \delta(Ku) K_i K_j \, G(K) \, d^3K \tag{2.12}$$

From this expression it is easy to verify that

$$\sum_j A_{ij}(u) u_j = \sum_i A_{ij}(u) u_i = 0$$

and therefore that $\mathcal{L}$ is reduced by each sphere $S_c \equiv \{u \mid |u| = c\}$, $c > 0$, in the sense that, if $f(u) = f_1(|u|) g(\hat{u})$ then

$$(\mathcal{L} f)(u) = f_1(|u|)(\mathcal{L}^{|u|} g)(\hat{u}) \tag{2.10}'$$

where $\mathcal{L}^{|u|}$ is an operator acting on C^2 functions on S_1.

It is also easy to verify that, for every $c > 0$, the operator $\mathcal{L}^c$ is self-adjoint on $L^2(S_1, d\hat{\Omega})$, with $d\hat{\Omega}$ the uniform measure on S_1, and has negative spectrum.

Under reasonable assumptions on $\widetilde{G}(K)$, each operator $\mathcal{L}^c$ is, for $c > 0$, strictly elliptic, its resolvent is compact, zero is a simple eigenvalue.

The functions constant on S_1 span the corresponding one-dimensional space.

Therefore, the diffusion induced on S_1 by each $\mathcal{L}^c$, has the homogeneou distribution as unique equilibrium state.

The operator $\mathcal{L}$ defined in (2.10) induces correspondingly a (degenerate) diffusion on R^d .

B. Quantum Mechanics

We perform also in this case the same heuristic analisis as before The evolution equation is now

$$i\hbar\, \frac{\partial\psi}{\partial t} = \left(-\hbar^2\Delta + \lambda V\right)\psi \equiv \left(H_o + \lambda V\right)\psi \tag{2.13}$$

where $\psi \in L^2(R^d) \cap C^2$, $\Delta = \sum_1^d \frac{\partial^2}{\partial x_i^2}$

Solving (2.13) provides a one-parameter group of unitary transformatior at least if for each ω the function $V(x,\omega)$ is sufficiently regular and well-behaved at ∞ , so that $-\hbar^2\Delta + \lambda V$ can be extanded to a self-adjoir operator.

We shall come back to this (non-trivial) question in $\S$ 3, but remark here that the behaviour of V <u>at</u> <u>large</u> <u>distances</u> can be an obstruction to the solvability of (2.13) in $[0,T]$ for any $T > 0$.

As in the classical case, we would expect that the limit $\lambda \to 0$, $t = \tau\lambda^{-2}$ give meaningful results only for a subclass of observables, corresponding to the functions of velocity in the classical case.

By observable we mean here any bounded symmetric operator on $L^3(R^d)$. Corresponding to functions of velocity one has here the observables which commute with translations in space. This set, which will be denoted by $\mathcal{Q}_o$, can be put in correspondence in a natural way with bounded real functions of the momentum operator.

The evolution of an observable is given, at least formally, by

$$A \to A^\lambda(t) \equiv \exp(i(H_o + \lambda V)t)\, A\, \exp(-i(H_o + \lambda V)t) \tag{2.14}$$

For fixed ω, $A^\lambda(t,\omega)$ will <u>not</u> be an element of $\mathcal{a}_o$ for $t > 0$.
On the other hand, if V is stationary, $E(A^\lambda(t)) \in \mathcal{a}_o$ for all t, λ .
We expect that $E(A^\lambda(\tau \lambda^{-2}))$ be the quantity that should be studied in
the limit $\lambda \to 0$.
One can write (2.14) as

$$i \frac{dA^\lambda}{dt} = \left[H_o + \lambda V , A^\lambda \right] \tag{2.15}$$

Scaling through $t = \lambda^{-2} \tau$, one gets

$$i \frac{dA^\lambda}{dt} = \frac{1}{\lambda^2} \left[H_o, A^\lambda \right] + \frac{1}{\lambda} \left[V^\lambda, A^\lambda \right] \tag{2.16}$$

Again we expand formally

$$A^\lambda(\tau \lambda^{-2}) = A_o(\tau) + \lambda A_1(\tau) + \lambda^2 A_2(\tau) + \dots$$
$$A_i(0) = 0 \qquad i \geq 1$$

Substitution in (2.23) and comparison of terms of same order in λ gives

a) $\left[H_o , A_o(\tau) \right] = 0$

This <u>suggest</u> that A be translation invariant

b) $\left[H_o , A_1(\tau) \right] + \left[V , A_o(\tau) \right] = 0$

Defining $V(\sigma) = \exp(iH_o\sigma) \, V \, \exp(-iH_o\sigma)$, one has

$$A_1(\tau) = i \int_0^\infty \left[V(\sigma), A_o(\tau) \right] d\sigma$$

Also here, the integral should be taken in a weak sense.

c) $i \frac{d A_o}{d\tau} = \left[H_o , A_2 \right] + i \left[V , \int_0^\infty \left[V(\sigma) , A_o(\tau) \, d\sigma \right] \right] \tag{2.17}$

Integrating (2.17) over Ω space, and using that

$$E(H_o A_2) = E(A_2 H_o)$$

due to the fact that V is stationary, one has, with $\overline{A}_o \equiv E(A_o)$

$$\tag{2.18}$$

As in the classical case, one would like to argue that, due to the time-scale chosen, $A(\tau)$ in (2.18) can be taken to be statistically independent from V and $V(\sigma)$.

The argument is here less transparent than in the classical case, due to the lack of description in terms of classical trajectories (the use of Poisson processes, as described in [8], could be of use here). There are now two competing mechanisms: the "spreading of the wave packet" would tend to bring in long-distance correlations, while the large separation for (most) trajectories of the baricenters would tend to decouple $A^{\lambda}(\tau \lambda^{-2})$, for small λ, from $V(\sigma)$ and V.

The best way to look at this is to take matrix elements of (2.18) between coherent states, with a Gaussian spread in momentum. If the potential is the Fourier trasform of a measure, one can develop, using the Weyl operators, an analisis similar to the one used for the classical case. If the potential is more general, the argument becomes less convincing, even at a heuristic level.

On the basis of these heuristic arguments, one substitutes (2.18) with

$$\frac{d\bar{A}_0}{d\tau} = \mathcal{L}_{\hbar} \, \bar{A}_0(\tau) \tag{2.19}$$

where $\quad (\mathcal{L}_{\hbar} A)(p) = \frac{1}{\hbar^2} \int \delta(p \cdot \kappa + \frac{\hbar \kappa^2}{2}) \, \tilde{G}(\kappa) \left(A(p + \hbar \kappa) - A(p) \right) d^3\kappa \tag{2.20}$

and we have used the representation of $\mathcal{Q}_0$ as bounded functions of momentum. As seen from (2.20) one should expect to be able to obtain (2.19) only for the subclass of $\mathcal{Q}_0$ generated by the constants and the continuous functions vanishing at ∞.

We shall call $\mathcal{Q}$ this subalgebra of $\mathcal{Q}_0$.

From now on, and until §5, we shall take $\hbar = 1$.

Then $\mathcal{L}_1$ can be given the form

$$(\mathcal{L}_1 A)(p) = \int \delta(p'^2 - p^2) \, \tilde{G}(p' - p) \left(A(p') - A(p) \right) d^d p$$

As we shall see, the heuristic argument which leads to (2.19) can be turned into a proof, if $d \geq 3$. In our proof, this condition is necessary to control the dominance of center-of-mass motion over spreading. It is not unlikely that it can be interpreted as a non-recurrence condition on the Poisson process generated by $\mathcal{L}_{\hbar}$, but we have not checked this point.

It is easy to verify that $\mathcal{L}_{\hbar}$ is a generator of a Markov contraction semigroup on $\mathcal{A}$, and, just as in the classical case, the semi-group it generates is reduced by each sphere $S_c \equiv \{ p \mid |p| = c \}$, $c > 0$. Its reduction $\mathcal{L}_{\hbar}^c$ generates a Poisson process on S_1 . Under rather mild conditions on $\widetilde{G}(p)$, each of these processes has a unique invariant measure, i.e. the uniform distribution on S_1 . We remark that, while convergence of $E(A^\lambda(\tau \lambda^{-2}))$ to $\overline{A}$ parallels the corresponding result in the Classical Mechanics context, it is here difficult to argue which stronger form of convergence one should expect. The point is of course that in Quantum Mechanics time-evolution in presence of a random potential does not lead to a Markov process, due to lack of commutativity (recall that $A^\lambda(t) \notin \mathcal{A}_0$ if $A \in \mathcal{A}_0$). One can prove that all finite-dimensional distributions of elements of $\mathcal{A}$ converge to the corresponding distributions of the process generated by $\mathcal{L}_{\hbar}$. The proof is given in [7] , and is outlined in § 5. However, a stronger form of convergence is an open problem.

To conclude these heuristic considerations, we remark that $\mathcal{L}_{\hbar}^c$, defined through (2.20), depend explicitely on $\hbar$. It is then natural to inquire what is their limit when $\hbar \to 0$ (classical limit).

We treat the problem in [7] , and give some details in § 5 of the present exposition.

It is shown that $\mathcal{L}_{\hbar}^c$ admits a unique extension to a bounded self-adjoint operator on $L^2(S_1)$, denoted again by $\mathcal{L}_{\hbar}^c$.

When $\hbar \to 0$, the sequence $\mathcal{L}_{\hbar}^c$ converges, in the strong resolvent sense, to a self-adjoint operator $\mathcal{L}^c$ which is the unique self-adjoint extension of $\mathcal{L}^c$, as defined through (2.10) , (2.10)' on (a dense domain in) $L^2(S_1)$. In fact, for each $c > 0$, strong convergence as semigroups can be proved in $L^p(S_1)$, $1 \le p < \infty$, but we shall not give the details here.

3 Evolution of Quantum Mechanical observables in a random potential

If $V(x)$ is sufficiently regular, so that $-\frac{1}{2}\hbar^2\Delta + \lambda V$ has a self-adjoint extension, the solution of the Schroedinger equation is given by

$$\psi(x,t) = \left(\exp(iHt\cdot\hbar^{-1})\,\psi\right)(x) \tag{3.1}$$

where $H = -\frac{\hbar^2}{2}\Delta + \lambda V(x) \equiv H_o + \lambda V$ (we have set the mass of the particle equal to one).

Correspondingly, the evolution of observables is

$$A(t) = \exp(iHt/\hbar)\,A\,\exp(-iHt/\hbar) \tag{3.2}$$

Up to § 5 we shall choose units in which $\hbar = 1$.

When $V(x)$ is a random potential, it is not obvious that (3.1) and (3.2) make sense, since, for most ω, $V(x,\omega)$ will not satisfy the properties of local regularity and especially regular behaviour at $|x| \to \infty$, needed to prove that $-\Delta + V$ is self-adjoint.

We recall that a random potential field is a linear map from a space Γ of functions (usually, S , or D) to random variables on a probability space $(\Omega, \mathcal{F}, \mu)$.

With $f \in \Gamma$, one often writes

$$V(f) = \int f(x)\,V(x)\,d^d x \tag{3.3}$$

In favourable cases, one can indeed construct a map

$$R^d \ni x \to V(x,\omega)$$

where $V(x,\omega)$ is measurable on $R^d \times \Omega$ with respect to the product measure, and for almost all ω is a continuous function of x.

We assume that $V(f)$ is stationary in space. When $V(x)$ exists as a random variable, this implies

$$E(\ V(x_1)\ \ldots\ V(x_n)\ ) = E(\ V(x_1+a)\ \ldots\ V(x_n+a)\ ) \qquad (3.4)$$

Otherwise (3.4) holds in a weak sense.

When (3.4) holds, one can construct a d-parameter group of measure-preserving transformations T_a in Ω, $a \in R^d$, such that

$$V(x+a,\omega) = V(x,T_a(\omega))$$

Without essential loss in generality, one can assume that T_a acts ergodically on Ω.

All these assumptions are satisfied if e.g. $V(f)$ is a Gaussian random field with mean zero and covariance operator G (defined by $E(V(f)V(g)) = (f,Gg)$, with scalar product in $L^2(R^d)$) which can be represented by a kernel $G(x-y)$ with the property $G \in L_1(R^d)$.
For technical reasons, we shall also assume $\widetilde{G} \in L_1$.

Even when $V(x,\omega)$ is for all ω continuous in x, its behaviour at $|x| \to \infty$ may be such as to prevent application of the standard criteria for self-adjointness. This is the case, e.g., of the example given above.

In this case, (3.1) and (3.2) must be given a meaning through a limit procedure.

This is done in [7] for the gaussian case, but, as outlined there, the techniques used extend without difficulty to more general situations.

We limit ourselves here to outline the procedure and give the results. One begins by studying the case in which the field $V(x,\omega)$ is substituded by the more regular one

$$V_\varepsilon(x,\omega) = V(x,\omega)\exp(-\varepsilon^2 V^2(x,\omega)) \qquad (3.5)$$

(As remarked in [7], the potential $V_\varepsilon'(x,\omega) = V(x,\omega)\exp(\varepsilon V(x,\omega))$ would be more natural, since it is only the negative part of V which requires a priori bounds to prove self adjointness of $-\Delta + V$, once boundedness of V over compact sets in R^d is known. In fact, the use of V_ε' instead of V_ε is necessary to treat more general situations. For the Gaussian case, (3.5) is more convenient, and leads to less involved calculations).

By construction, $V(x,\omega)$ is jointly measurable in $\underline{x},\omega$ if this is true for V, and moreover for almost all $t > 0$ and $\omega \in \Omega$ is bounded uniformly in x.

Correspondingly, for all $\lambda \in R$, $t \in R^+$, one can define for a.a. ω the unitary continuous group

$$\exp i(-\Delta + \lambda V_\varepsilon)t$$

In [7] it is proved

<u>Lemma 1</u> For each $t \in R^+$, $\lambda > 0$ there exists a subset $\bar{\Omega}_{t,\lambda} \subset \Omega$ $\mu(\bar{\Omega}_{t,\lambda}) = 1$, and a sequence $\varepsilon_{(n)}^{t,\lambda}$, $\varepsilon_{(n)}^{t,\lambda} \to 0$ when $n \to \infty$, such that, if $\omega \in \bar{\Omega}_{t,\lambda}$

$$\text{strong limit}_{n \to \infty} \ (\exp i (-\Delta + \lambda V_\varepsilon) t)(\omega) \equiv U_\lambda(t)(\omega) \quad \text{exists.}$$

For each t, λ and all $\omega \in \bar{\Omega}_{t,\lambda}$, $U_\lambda(t)$ is unitary. Moreover $U_\lambda(t)$ is μ-measurable. $\rfloor$

The proof goes by proving convergence of

$$\exp\left(i (-\Delta + \lambda V_\varepsilon) t \right) \psi_n \tag{3.6}$$

when $\varepsilon \downarrow 0$ along subsequences, on a denumerable basis in $L^2(R^d)$; one uses the fact that the unit sphere in $L^2(R^d)$ is strong sequentially closed, and that pointwise limits of measurable functions are themselves measurable.

It should be stressed that, in general, $\bar{\Omega}_{t,\lambda}$ will depend on t, λ . In particular, for a.a. ω , continuity in t and regularity in λ may not follow. Therefore there will be no operator $-\Delta + V(x,)$; this will require some care in interchanging integration on Ω and formal manipulations with the evolution equations.

One should of course compare this situation with the much simpler one for the classical case, where local regularity of $V(x,\omega)$ and some control over the growth at $|x| \to \infty$ suffice to provide, for almost all ω , a unique maximal smooth solution to the equations of motion.

From Lemma 1 one has immediately

<u>Corollary</u> For all values of t, λ there is a set $\Omega_{t,\lambda}$, of μ-measure one, such that, if $\omega \in \Omega_{t,\lambda}$

$$A \longmapsto A_\lambda(t)(\omega) \equiv U_\lambda(t) \, A \, U_\lambda^{-1}(t) \tag{3.7}$$

provides a unitary equivalence on the observables, which represents the evolution up to time t ⏚

Again, we remark that in general (i.e. for μ-almost all ω), $t \mapsto A(t)(\omega)$ is not continuous.

We close this § with four remarks

a) For fixed $\omega \in \Omega$, $\lambda \in R$, if $A_\lambda(t)(\omega)$ is defined for $t=t_1$, $t=t_2$, $t=t_1+t_2$, one has

$$U_\lambda(t_1+t_2) \;=\; U_\lambda(t_1) \, U_\lambda(t_2) \;=\; U_\lambda(t_2) \, U_\lambda(t_1) \tag{3.8}$$

as easily seen from strong convergence.

It is then not difficult to prove that, for fixed $\lambda \in R$ and a.a. ω , $U_\lambda(t)(\omega)$ is defined for $t \in K(\lambda,\omega)$, where $K(\lambda,\omega)$ is a discrete subgroup of R , which can be taken to be dense in R . Moreover, $t \to U_\lambda(t)$ is a representation of this subgroup.

b) Since, for each t, λ , $A_\lambda(t)(\omega)$ is measurable, and of course bounded in norm since $\| A_\lambda(t)(\omega) \| = \| A \|$, one can define

$$E(\, A_\lambda(t)\,) \equiv \int \mu \, (d\omega) \, A\,(t)(\omega) \tag{3.9}$$

the integral being defined in the strong sense.

For fixed λ , $E(A_\lambda(t))$ exists for all t, and it is proved in [7] that $E(A_\lambda(t))$ is differentiable in t.

We call (3.9) the "averaged dynamics" and shall seek its limit when $t = \tau \lambda^{-2}$, $\lambda \to 0$.

In the same way, one can define

$$E(\, A_\lambda^{(1)}(t_1) \, \ldots \, A_\lambda^{(n)}(t_n) \,) \tag{3.10}$$

where $A^{(k)} \in B\,(\, L^2(R^d)\,)$, $k = 1 \ldots n$, and t

We shall call (3.10) an n-point distribution function, and seek its
limit when $t_i = \tau_i \lambda_i^{-2}$, $\lambda \to 0$.

c) It is shown in [7] , using the integral equation satisfied by
$\exp(-iH_0 t)\exp(iH_o t + iV_\xi t)$ (essentially Duhamel's formula) and its
solution by iteration (Dyson series), that

$$E(A_\lambda(t)) = \sum_{n=1}^{\infty} i^n \lambda^n \int_0^t dt_1 \cdots \int_0^{t_{n-1}} dt_n \, E\left\{ [V(t_n), \cdots [V(t_1), A]\cdots] \right\} \quad (3.11)$$

where $V(t) \equiv U_\lambda(t)VU_\lambda^{-1}(t)$ and $[A_1, A_2] \equiv A_1 A_2 - A_2 A_1$.
The series (3.11) is norm-convergent if V is a Gaussian random field,
with covariance restricted as above, and also if $V(x)$ is a random
field with continuous realizations and having moments of all orders
satisfying suitable bounds.
Expression (3.11) justifies the otherwise formal use of the Dyson
series, done routinely in the physics literature to study the motion
of a Quantum particle in a random field. One should remark that one
must not, in (3.11), interchange integration over the t 's with
integration over   .

d) Let a_o be the (Von Neumann) algebra of essentially bounded
functions of momentum, and a the (norm-closed) subalgebra of a_o
generated by the identity in a_o and the functions which are continuous
and vanish at infinity.
If $A \in a_o$, $E(A_\lambda(t)) \in a_o$ for all λ , t, due to the stationary character
of the field V . One can also verify (see [7]) that $A \in a$ implies
$E(A_\lambda(t)) \in a$.
We shall prove convergence, when $\lambda \to 0$, of the finite-dimensional
distributions of elements of a . The use of a rather than a_o is
purely technical, and connected with the properties of the semi-group
generated by $\mathcal{L}_\xi^c$.
Similar statements can presumably be obtained for a larger class of
observables, and would be in some sense a statement about the asymptotic behaviour of their "translation-invariant" part.
We shall not pursue further this point here.

$\underline{4}$ Convergence of averaged dynamics when $\lambda \to 0$

We have remarked in §2 that heuristic arguments suggest that $E(A_\lambda(\tau \lambda^{-2}))$,as defined through (3.11) ,converge to $\bar{A}_0(\tau)$,solution of (2.19) ,where $\mathcal{L}_t$ is defined in (2.20) .

We want to outline in this § and in the next how this can be proved, when $A \in \mathcal{A}$.

The strategy will be essentially to write a differential equation for the difference of the two quantities,and then use Duhamel's formula and some a-priori estimates,plus a version of Gronwall's lemma,to prove that $\forall \tau > 0, \lim\limits_{\lambda \to 0} E(A_\lambda(\tau \lambda^{-2}) - \bar{A}_0(\tau)) = 0$,in the topology of $\mathcal{A}$,uniformly in τ over compact sets of R^+_0.

In fact,it proves useful to do the proof in two steps,introducing an intermediate function $\hat{A}_\lambda(\tau)$,with values in $\mathcal{A}$.This function is defined as the unique solution in $\mathcal{A}$ of the integral equation

$$\hat{A}_\lambda(\tau) = A + \lambda^2 \int_0^{\tau \lambda^{-2}} d\sigma \int_0^\sigma d\nu E([V(\sigma), [V(\nu), \hat{A}_\lambda(\nu)]]) \tag{4.1}$$

with $A \in \mathcal{A}$.

Remark that $\hat{A}_\lambda(\tau)$ is $\underline{\text{not}}$ a function of ω .

One then proves (see [7])

<u>Theorem 1</u> Let $G(x-y) = E(V(x),V(y))$,V a Gaussian random field.Assume that $G \in L_1$, $\tilde{G} \in L_1$,and that the space-dimension is ≥ 3.Then

A)For all $\tau > 0$,and uniformly on compacts
$$|\hat{A}_\lambda(\tau) - \bar{A}_0(\tau)| \underset{\lambda \to 0}{\to} 0 \quad , \quad \text{where } |\cdot| \text{ is the } \mathcal{A}\text{-norm.}$$

B)For all $\tau > 0$,uniformly on compacts
$$|E(A_\lambda(\tau \lambda^{-2}) - \hat{A}_\lambda(\tau)| \underset{\lambda \to 0}{\to} 0$$

Remark: the proof of B) in [7] is given explicitely under the assumption that V is a Gaussian random field.

It is hovever pointed out there,and some details will be taken up again here,that the proof can be extended ,with comparatively minor modifications,to the case when all moments of V exist and satisfy suitable bounds.The proof of A) involves of course only the properties of the two-point correlation function of V.

We now give some indication on the proof of A) and B). Details and further comments can be found in [7] .

The proof of A) is given in [7] through a direct method,using the properties of the Laplace transforms of the quantities involved.

We give here some formal manipulations and indicate how to construct an alternative proof.We write all formulae for d = 3.

One can write (4.1) in the form ($\hbar$ =1)

$$\hat{A}_\lambda(\tau)(p) = A(p) - (2\pi)^{-3}\lambda^2 \int_0^{\tau\lambda^{-2}} d\sigma_1 \int_{-\sigma_1}^{\sigma_1} d\sigma_2 \int d^3 p' \, \exp\left(\left(\tfrac{i}{2}(p'^2 - p^2)\sigma_2\right)\right)\cdot$$
$$\cdot \tilde{G}(p'-p)\left[\hat{A}_\lambda((\sigma_1 - |\sigma_2|)\lambda^2)(p') - \hat{A}_\lambda((\sigma_1 - |\sigma_2|)\lambda^2)(p)\right] \qquad (4.2)$$

or ,in differential form

$$\frac{d\hat{A}_\lambda(\tau)}{d\tau} = -(2\pi)^{-3}\int_{-\tau\lambda^{-2}}^{\tau\lambda^{-2}} d\sigma \int d^3 p' \, \exp\left(\tfrac{i}{2}(p'^2 - p^2)\right)\tilde{G}(p'-p)\left[\hat{A}_\lambda(\tau - |\sigma|\lambda^2)(p') - \hat{A}_\lambda(\tau - |\sigma|\lambda^2)(p)\right] \qquad (4.3)$$

If $\hat{A}_\lambda(\tau)(p)$ remains continuous,when $\lambda \to 0$,and vanishing at ∞,one could expect that "most" of the contribution will come,when $\lambda \to 0$,from $|\sigma| < \tau\lambda$ $\alpha < 2$.Therefore,in the limit $\lambda \to 0$,(4.3) should become

$$\frac{d\hat{A}(\tau)}{d\tau} = -\int d^3 p' \, \delta(p'^2 - p^2) \, \tilde{G}(p'-p)\left[\hat{A}(\tau,p') - A(\tau,p)\right] \qquad (4.4)$$

where $\hat{A}_\lambda(\tau) \to \hat{A}(\tau)$.

This would prove A) because,as proved in [7] (see also [5]) $\mathcal{L}$ generates on $\mathcal{Q}$ a contraction semi-group,and therefore the solution of(4.4) with initial data $A \in \mathcal{Q}$ is unique.

A proof along these lines can be constructed either from (4.2) (by i-

teration and estimates to prove convergence uniformly in $\lambda \geq 0$) or from
(4.3) (by writing the equation for $\hat{A}_\lambda(\tau)$ and $\bar{A}_o(\tau)$ and then using the
estimates to apply Gronwall's lemma) using the following facts,which
can be easily verified

1) $(B_\lambda A)(p) = -(2\pi)^{-3} \lambda^2 \int\limits_0^{\tau\lambda^{-2}} d\sigma_1 \int\limits_{-\sigma_1}^{\sigma_1} d\nu \int d^3p' \, \exp\left(\frac{i}{2}(p'^2-p^2)\nu\right)\widetilde{G}(p'-p) \quad A \quad (p')$

$A \in \mathcal{a} \cap L^2(R^3)$ defines an increasing (with λ^{-1}) family of symmetric
bounded positive operators on $L^2(R^3)$,which converge in strong re-
solvent sense to $-\tau\mathcal{L}_1(A)$,where

$$(\mathcal{L}_1 A)(p) = \int d^3p' \, \delta(p'^2-p^2) \, \widetilde{G}(p'-p) \, A(p')$$

2) $(D_\lambda A)(p) = (2\pi)^{-3} \int\limits_0^{\tau\lambda^{-2}} d\sigma_1 \int\limits_{-\sigma_1}^{\sigma_1} d\nu \int d^3p' \, \exp\left(\frac{i}{2}(p'^2-p^2)\nu\right)\widetilde{G}(p'-p))A(p) \equiv D_\lambda(p)A(p)$

define a family of uniformly bounded multiplication operators $D_\lambda(p)$
which converge,uniformly over compacts in R^3,when $\lambda \to 0$,to the operator

$$(DA)(p) = \tau D(p)A(p) \;, \quad D(p) = \int d^3p \, \delta(p'^2-p^2) \, G(p'-p)$$

3) $\mathcal{L}A = \mathcal{L}_1 A - DA$
4) $\hat{A}_\lambda(\tau)(p)$ converges in $L^2(R^3)$,when $\lambda \to 0$,uniformly in τ over
compacts,to $\bar{A}_o(\tau)(p)$,if $A(p)$ has compact support.
5) Using also the fact that $\mathcal{L}$ generates a contraction on C ,the
$\hat{A}_\lambda(\tau)(p)$ have a common modulus of continuity,for $\tau \leq T$,when $\lambda < \lambda_o(T)$.
Therefore,convergence takes actually place in the supnorm(which is
the norm of $\mathcal{a}$),uniformly for τ over compacts.
We shall now concentrate on the proof of B).
Let K_λ,an operator on $C(0,T,\mathcal{a})$,be defined by

$$(K_\lambda A)(\tau) = \int\limits_0^{\tau\lambda^{-2}} d\sigma \int\limits_0^{\lambda^2\sigma} d\nu \, E\left(\left[V(\sigma), \left[V(\nu \lambda^{-2}), A(\nu) \right] \right] \right) \tag{4.5}$$

and Z_λ be defined on bounded operators on $L^2(R^3)$ by

$$Z_\lambda \cdot A = [V_\lambda, [V_\lambda, A]] - E[V_\lambda, [V_\lambda, A]] \qquad (4.6)$$

where

$$[V_\lambda, B](s) = \lambda^{-1} \int_0^{s\lambda^{-2}} [\hat{V}(\sigma, \lambda), B(\sigma)]\, d\sigma$$
$$\hat{V}(\sigma, \lambda) = \exp(iH_o \sigma \lambda^{-2})\, V \exp(-iH_o \sigma \lambda^{-2})$$

Let
$$W_n^\lambda \cdot A \equiv E(K_\lambda \cdot Z_\lambda^n A)$$
$$W_\lambda = \sum_{n=1}^\infty W_{(n)}^\lambda$$
$$U_\lambda = \sum_n K_\lambda^n$$

Then one easily verifies, for $\lambda > 0$, that

$$E(A_\lambda) - \hat{A}_\lambda = (1 - U_\lambda W_\lambda)^{-1} U_\lambda W_\lambda \hat{A}_\lambda \qquad (4.7)$$

One easily verifies that there exists $\lambda_0 > 0$ such that U_λ is uniformly bounded when $\lambda < \lambda_o$.

Therefore, the proof of B) is equivalent to proving that, uniformly in $\tau \geq 0$ over compacts, $W_\lambda \to 0$ in the norm of $\mathcal{L}(\mathcal{Q}, \mathcal{Q})$.

We use the Dyson expansion to write W_λ as a series of terms, which we interpret graphically.

We discuss first the estimates when V is assumed to be a Gaussian field (with mean zero) and give at the end a short indication about the general case.

The expressions which appear in $W_{(n)}^\lambda$ have the form

$$(-1)^n \lambda^{2n} \int_0^{\tau\lambda^{-2}} dt_1 \cdots \int_0^{t_{2n-1}} dt_{2n}\, E'[V(t_1), \cdots [V(t_{2n}), A] \cdots] \qquad (4.8)$$

The rules of Gaussian integration allow to associate to (4.8) 2^{2n} (planar) graphs (one for each choice in the commutators) each of which is composed of n links joining pairs of the points $t_1 \cdots t_{2n} \in R^+$.

The order in which the t's appear, from left to right, is determined
by the choice in the commutators and <u>not</u> by the chronological order.
A connected component of a graph Γ is a subgraph Γ_i such that $s(\Gamma_i)$,
its "shadow" (projection onto R^+), is connected as a subset of R^+
(here R^+ is seen as a half-line in the plane).
The apex in E' of (4.8) indicates that one should retain only those
graphs Γ in which t_1 belongs to the same connected component as t_{2n-1}
or t_{2n} (or both).
We call such graphs "tight". Notice that the factor $(1 - U_\lambda W_\lambda)^{-1}$
in (4.2) expresses the fact that every graph can be obtained putting
tight graphs in series.
The contribution from a generic tight graph of order n (i.e. with n
links) can be written, when operating on $A \in \mathcal{Q}$, as

$$C_\Gamma(p) = \pm \lambda^{2n} \int_0^{2\lambda^{-2}} dt_1 \cdots \int_0^{t_{2n-1}} dt_{2n} \int d^3 k_1 \cdots \int d^3 k_{2n-1} \exp \left\{ i \sum_{m=1}^{2n-1} K_m^2 (t_{i_m} - t_{i_{m+1}}) \right\} \cdot \prod_{i=1}^{n} \widetilde{G}(l_i(K)) \prod_{j=1}^{n-1} \delta(L_j(K)) \, A(L(K)) \tag{4.9}$$

where $K_\lambda \in R^3$ are "momenta" associated to the segments on R^+ which lie
between two consecutive t_i's.
The $l_i(K) \in R^3$, $L_j(K) \in R^3$ are linear homogeneous combinations of the
K's. The $l_i(K)$ can be looked at as momentum of the ith link, while
$\delta(L_j(K))$ reflects integration over space-variables which do not appear in the specific term in $E(V(x_1) \quad V(x_{2n}))$ which corresponds to
Γ by the rules of Gaussian integration (e.g. the term $G(x_1 - x_2)G(x_3 - x_4)$
in $E(V(x_1) \cdots V(x_4)$ does not depend on $x_2 - x_3$).
The l_i and L_j are completely determined by the requirement that momentum (of links or intervals) be conserved at each vertex of the graph
(vertices are the points on R^+ which are extreme point of a link).
Clearly C_Γ has the form, with

$$I_{2\lambda^{-2}} = \{ t \mid 2\lambda^{-2} \geq t_1 \geq \cdots \geq t_n \geq 0 \}$$

$$C_\Gamma = \int_{I_{2\lambda^{-2}}} \exp(i(\underline{K} \cdot Q \cdot \underline{K})) \widetilde{G}(K_1) \quad \widetilde{G}(K_n) \, A(L(K)) \, d^3 k_1 \quad d^3 k_n \tag{4.10}$$

where $\underline{K} = \{K_1 \cdots K_n\}$, $K_i \in R^3$, and Q is an $n \times n$ matrix with entries which are linear combinations of the t_i's.

The matrix Q is given explicitly by the following prescription (see also [5]).

Let, for a given Γ, $s_K = t_{i_K} - t_{i_{K+1}}$, where the t's have been ordered from left to right on R^+ (this is $\underline{not}$ the chronological order in general, as remarked above).

Let $s(\gamma)$ be the shadow of the arc $\gamma \in \Gamma$. Order the arcs according to the order of their left ends.

Then

$$Q_{ij} = \sum_{K \in M_{ij}} s_K$$
$$M_{ij} = \{ K \mid t_{i_K} \longmapsto t_{i_{K+1}} \subset (s(\gamma_i) \cap s(\gamma_j)) \}$$

We also remark that, if $\Gamma^{(M)}$ is obtained from Γ by deleting M arcs, the matrix $Q^{(M)}$ associated to $\Gamma^{(M)}$ is obtained from Q by deleting the corresponding rows and columns.

A crucial tool in our estimates is provided by the following lemma, which is proved in [7]

$\underline{\text{Lemma 2}}$ Let $G \in L_1 (R^3)$, and set $\|\|G\|\| = \max (\|G\|_1, \|\widetilde{G}\|_1)$. Let Q be any symmetric $N \times N$ matrix, and, for any $0 \le M \le N$, let $Q^{(M)}$ be any of the $M \times M$ matrices obtained from Q deleting N-M rows and columns, with the convention $Q^{(N)} = Q$, $Q^{(0)} = 1$.

Let $F_i \in L^\infty (R^3)$, $i = 1 \cdots S$

Let $l_i(p)$ be arbitrary linear combinations of the p_i's. Then, for all N, $0 \le M \le N$, $0 \le \alpha \le 1$, $0 \le \beta \le 1$, $0 \le \alpha + \beta \le 1$,

$$\left| \int \exp \left(i (p Q p) \right) \widetilde{G}(p_1) \cdots \widetilde{G}(p_N) F_1 (l_1(p)) \cdots F_S (l_S(p)) d^3 p_1 \cdots d^3 p_N \right| \le$$

$$\le C^N |F_1|_\infty \cdots |F_S|_\infty \|\|G\|\|^n |\det Q|^{-\frac{3\alpha}{2}} |\det Q^{(M)}|^{-\frac{3\beta}{2}} \tag{4.11}$$

for some $C > 0$.

$\underline{\text{Remark}}$. If the number of space-dimension were d instead of 3, one would have d/2 instead of 3/2 in (4.11). As we shall presently see, it is only

when $d \geq 3$ that estimate (4.11) can be used to give a proof of B).
We shall not reproduce here the proof of Lemma 2,but remark that,in
view of (4.10) ,it is obviously a natural tool .
Indeed,in view of Lemma 2 ,a strategy of proof of B) suggests itself.
a) Prove that the contribution of each graph converges to zero
b) Since det Q ,det Q^M are homogeneous polinomials in the t_κ's,use scaling to bound the contribution from a generic graph Γ_N of order N by
$C \lambda^{\beta(\Gamma_N)} \tau_0^{\delta(\Gamma_N)}$, where $\tau_0 = \max (\tau ,1)$,and in fact by $\frac{1}{N!} C_1^N \lambda^{\beta(\Gamma_N)} \tau_0^{\delta(\Gamma_N)}$,
using the fact that I_1 is a triangular domain.
c) Estimate the number of graphs of order N which contribute to W (this
should give a factor (2N-1)!! and especially prove that

$$\beta(\Gamma_N) \geq \beta_0 N \quad , \quad \delta(\Gamma_N) \leq \delta_0 N \qquad (4.12)$$

where β_0, δ_0 are independent of the graph.
This strategy is unfortunately a losing one,as it stands,mainly because
the second part of c) is <u>not</u> true . Also,the use of triangularity of I_1
is not totally trivial,since we expect a factor (N!) instead of (2N!) .
As we shall see presently,these two difficulties are strictly connected.

The proof of a) is very elementary,using Lemma 2 and the scaling behaviour of det Q , det $Q^{(M)}$.
We shall therefore concentrate on c),and explain where do the difficulties come from and how to cure them.
The problem comes from "rescattering graphs" ,i.e. graphs in which there
are links ("rescattering links") which have both these properties :
1) connect two vertices t_{i_m}, $t_{i_{m+1}}$ which are sequential in time (i.e. in
the chronological order).
2) be disjoint from the rest of Γ,i.e. not to intersect or shadow any
other link in Γ.
Indeed,for these graphs,one cannot in general choose α, β in Lemma 2 so
as to make the r.h. side integrable in $I_{\tau\lambda^{-2}}$, and at the same time satisfy (4.12) with some fixed β_0, δ_0. This is due to the fact that in general

one cannot choose Γ^M locally (in $I_{\tau_0 \lambda^{-2}}$) so that $|\det Q|^{-\gamma} |\det Q^{(M)}|^{-1+\gamma}$ is locally integrable.

The cure of the problem goes as follows : one considers first the contribution of all tight rescattering-free graphs ,i.e. graphs which do not contain rescattering links.

For these graphs ,α and β can be chosen in Lemma 2 so that the r.h. side is locally integrable in I_1 and (4.12) holds for some β_0 , δ_0 (e.g. $\beta_0 = 1/8$, $\delta_0 = 7/8$) .

To take now into account the graphs which are not rescattering-free , one uses the time - honored method of resummation (which is legitimate if $\lambda > 0$, since all series are absolutely convergent if $\lambda > 0$). This leads to the introduction of an "effective" propagator instead of G in (4.9). Let Γ be a rescattering-free graph (a "skeleton") and call ornament of Γ every graph (including itself) which differs from Γ at most by the addition of rescattering links.

One proves that,if Γ is a skeleton and C_Γ as given in (4.9) is its contribution to W_λ , the contribution of all ornaments of Γ introduces in the integrand in (4.9) an extra factor $B (K_1 \cdots K_n, t_1, \cdots t_{2n}, \lambda)$. One proves that there is a constant $C (\tau_0) > 0$ such that

$$| B (\cdot , t_1 \cdots t_{2n}, \lambda)|_\infty < C^n (\tau_0)$$

uniformly in $I_{\tau_0 \lambda^{-2}}$ and $0 \leq \lambda \leq 1$.

Here $\qquad | f (\cdot)|_\infty = \sup_{K_i \in R^3} |f(K_1 \cdots K_n)|$

Therefore,from Lemma 2,estimate (4.11) still holds for the sum of the contributions of all ornaments of a given skeleton Γ_n,possibly with a different constant C .

It is also evident that each tight graph is ornament of precisely one skeleton,and every tight graph can be obtained as ornament. The same estimate on C will hold in all those other sectors of $\{ \underline{t} \mid 0 \leq t_i \leq \tau \lambda^{-2} \}$ in which the graph Γ still appears as a skeleton (recall that Γ,even if it is a skeleton,will in general contains links which are disconnected from the rest of Γ;however,if t_{M_1}, t_{M_2} are end points,neither of

them can be chronological successor of the other ; and this conditions
may be violated if one looks at a different sector) .
It is easy to see that there are at least n! sectors in which Γ appears
as a skeleton . In each of them, the r.h. side of (4.11) is integrable,
with e.g. $\alpha = 1/2$ and $\beta = 5/12$, and in fact the integral is dominated
by $\frac{1}{n!} C_2^n$, for a suitably chosen C_2 (by sector we mean here a subset of
$\{ \underline{t} \mid 0 \leq t_i \leq \tau \lambda^{-2} \}$ defined by an ordering $\tau \lambda^{-2} \geq t_{i_1} \geq t_{i_2} \geq \cdots \geq t_{i_n} > 0$).
Therefore the contribution $\widetilde{C}_\Gamma$ of $\underline{all}$ graphs which are ornament of a
given skeleton Γ of order n can be estimated by

$$\widetilde{C}_\Gamma \leq \frac{1}{n!} C_2^n \, \lambda^{1/8 \, n} \, \tau_o^{7/8 \, n} \qquad , \text{ where } \tau_o = \max (\tau, 1) \text{ and } \lambda \leq 1 .$$

Since there are at most $(2n-1)!!$ graphs of order n , this is certainly
a bound on the number of skeletons of order n.
We conclude that

$$\left| (W_\lambda \cdot A)(p) \right| \leq \| A \| \sum_{n \geq 1} C_2^n \, 2^n \, \lambda^{\frac{n}{8}} \, \tau_o^{\frac{7n}{8}} \tag{4.12}$$

uniformly in $p \in R^3$.
Therefore $\| W_\lambda \| \to 0$ in the topology of $\mathcal{L}(\mathcal{a})$, uniformly in τ over com-
pacts of R^+ .
This concludes the outline of the proof of B) , and therefore of theorem
1 .
$\qquad$ $\square$

We conclude this section by giving some indications on the (minor) modifications in the proofs which are necessary if one considers more general random potentials V, which are not Gaussian .
Indeed, the assumption that V is Gaussian was made only to keep the proof
of part B) of theorem 1 to a minimum level of complexity.
One can repeat the same analisis as above if one assumes that, with
$W^n(x_1 \cdots x_n) = E(V(x_1) \cdots V(x_n))$, there exists functions W_c^n such
that

$$W(x_1 \cdots x_n) = \sum W_c^{n_1}(x_{k_1} \cdots x_{k_{n_1}}) \cdots W_c^{n_s}(x_{h_1} \cdots x_{h_{n_s}}) \tag{4.13}$$

where $\sum_{p=1}^{\hat{}} n_p = n$ and the sum is taken over all possible partitions of the points $x_1 \cdots x_n$ in subsets.

Moreover one requires

$$W_c^p (x_1 \cdots x_p) = U^{p-1} (x_1 - x_2 , \cdots , x_{p-1} - x_p) \qquad (4.13)'$$
$$U \in L_1 (R^{3(p-1)}) , \quad \tilde{U} \in L_1 (R^{3(p-1)}) \qquad (4.13)''$$

Remark that $W_c^1 (x) = 0$, $W_c^2 (x,y) = W (x,y)$.

Formulae (4.13),(4.13)',(4.13)" define what is known as Ursell expansion in Statistical Mechanics ,or expansion into irreducible components in Field Theory. The U's are known as Ursell functions. Such expansion is guaranteed under very minor mixing conditions on the random potential field. A Gaussian random potential is characterized by $U^p = 0$ for $p \geq 2$.

For a Gaussian field, (4.13) coincides with the rule of Gaussian integration (expansion in generalized Hermite polinomials) .

One can repeat the entire analisis of graphs given above,starting with (4.9) ,using (4.13) instead of Gaussian integration.

In general,one will have now for a generic term fewer δ-functions in (4.9 and correspondingly Q will be a matrix of higher rank than before. Thus the scaling properties,which were crucial to obtain (4.12) ,are no worse (in fact,better) than before. Also integration over allowed sectors is controlled as before. There is no change in the definition of tight and rescattering - free ;of course now a given point t_K can be end point of more than one link . The rescattering links are precisely as before,and therefore there is no change in the estimate of the contribution of all ornaments of a given skeleton. Also the estimate on the number of skeletons of a given order n can be done as before.

One finds that Theorem 1 still holds if $\exists\, a > 0$ such that

$$\| U^{(p)} \| \leq a^p \| U^{(2)} \|^{p/2}$$

where $\quad U^2 (x - y) = E (V(x) V(y))$

and $\quad \| F \| = \max (\| F \|_1 , \| \tilde{F} \|_1)$.

5 Convergence of finite-dimensional correlations, and classical limit

We want to outline here the proof of convergence of all finite-dimensional distributions to the corresponding distribution of the Markov process T_z .

The proof will rely, as in Theorem 1 , on an estimate of a suitably restricted Dyson series, using a description in terms of diagrams. We shall therefore be even more sketchy than in § 4 .

The only new ingredient is an identity which we state now.

Let $\{t_1 \dots t_N\} \in (R^+)^N$, $A^{(i)} \in \mathcal{Q}$, $i = 1 \dots N$, $U_\lambda(t)$ be as in § 3, for $t = t_1, \dots t_N, t_1 - t_2, \dots t_{N-1} - t_N$. From § 3 we know that there exist a set $\underline{\Omega} \subset \Omega$, of measure 1, such that for $\omega \in \underline{\Omega}$, $U_\lambda(t)$ can be defined.

Let $A_\lambda^{(j)}(t_j) = U_\lambda(t_j) A^{(j)} U_\lambda^{-1}(t_j)$

Then

$$\exp(-iH_0 t_1)\, A_\lambda^{(1)}(t_1) \cdots\cdots A_\lambda^{(N)}(t_N)\, \exp(iH_0 t_1) \;=\;$$

$$= \exp(-iH_0 t_N)\, U_\lambda^{t_1 - t_N}(t_N)\, \exp(-iH_0(t_1 - t_N))\, U_\lambda^{t_1 - t_{N-1}}(t_{N-1} - t_N) \cdots \exp(-iH_0(t_{N-1} - t_N))\, U_\lambda(t_1 - t_2) \cdot$$

$$\cdot A^{(1)} U_\lambda(t_1 - t_2)\, \exp(iH_0(t_{N-1} - t_N))\, A^{(2)} \cdots A^{(N)} U_\lambda^{t_1 - t_N}(-t_N)\, \exp(iH_0 t_N)$$

$$(5.1)$$

where $U_\lambda^\sigma(t) = \exp(-iH_0\sigma)\, U_\lambda(t)\, \exp(iH_0\sigma)$

This identity is an immediate consequence of the strong convergence results of   3 and of an identity of the same form as (5.1), but with $U_\lambda(t)$ replaced by $\exp(i(H_0 + \lambda V_{\varepsilon_n})t)$, which can be verified by direct computation.

One can integrate (5.1) over Ω with the measure μ , and one gets in this way an identity which holds for all $t_i \geq 0$, $\lambda \geq 0$

Recall now that, as we have seen in § 4, uniformly in $\mathcal{G}, \nu$ and uniformly in τ over compacts of R^+,

$$\lim_{\lambda \to 0} E\left(\exp(-iH_0 \nu \lambda^{-2}) \, U_\lambda^\rho(\tau \lambda^{-2}) \, A \, U_\lambda^\rho(-\tau \lambda^{-2}) \, \exp(iH_0 \nu \lambda^{-2}) \right) = T_\tau \cdot A \qquad (5.2)$$

where in fact the left-hand side is independent of ν since the random field is stationary in space.

From (5.2) and the unitarity of $U_\lambda^\rho(\tau)$ it follows, setting $\tau_\kappa = \lambda^2 t_\kappa$

$$\lim_{\lambda \to 0} E\left\{ \exp(-iH_0 t_N) \, U_\lambda^{t_1 - t_N}(t_N) \, E\left\{ \cdots \, E\left\{ \exp(-iH_0(t_{N-1} - t_N)) \, U_\lambda(t_1 - t_2) \cdot \right. \right. \right.$$

$$\cdot \, A^{(1)} \, U_\lambda(t_2 - t_1) \, \exp(iH_0(t_{N-1} - t_N))\} \, A^{(2)} \cdots \} A^{(N)} \, U_\lambda^{t_1 - t_N}(-t_N) \, \exp(iH_0 t_N)\} =$$

$$= T_{\tau_N} \cdot A^{(N)} \cdots T_{\tau_1 - \tau_2} A^{(1)} \qquad (5.3)$$

in the topology of $\mathcal{Q}$, uniformly in τ_κ over compacts of R^+.
We want to prove [7]

<u>Theorem 2</u>   If $\tau_1 \geq \tau_2 \geq \cdots \geq \tau_N$, one has, uniformly in τ_κ over compacts in R^+, in the topology of $\mathcal{Q}$

$$\lim_{\lambda \to 0} E\left(A_\lambda^{(1)}(\tau_1 \lambda^{-2}) \cdots A_\lambda^{(N)}(\tau_N \lambda^{-2}) \right) = \lim_{\lambda \to 0} E\left(A_\lambda^N(\tau_N \lambda^{-2}) \cdots A_\lambda^{(1)}(\tau_1 \lambda^{-2}) \right\} =$$

$$= T_{\tau_N} \cdot A^{(N)} \, T_{\tau_{N-1} - \tau_N} \cdot A^{(N-1)} \cdots T_{\tau_1 - \tau_2} A^{(1)} \qquad (5.4)$$

Sketch of proof: In view of (5.3), one has only to prove that, in the topology of $\mathcal{Q}$ and uniformly over compacts in τ_κ, the difference between the left-hand side of (5.3) and the expectation of the right-hand side of (5.1) converges to zero when $\lambda \to 0$.
It is not difficult to verify that such quantity can be written as

$$\sum_{j=1}^N \sum_{n_j=0}^\infty (-1)^{\sum_{i=1}^N n_j} \lambda^{2 \sum_{j=1}^N n_j} \int_{I_{\{\tau_1 \cdots \tau_N\}}} \prod_j \prod_{\kappa_j} dt_{\kappa_j}^{(j)} \, E\left\{ [V(t_1^{(1)}), [\cdots [V_{n_1}^{(t_1)}, [V_1^{(t_2)}, \cdots \right.$$

$$\cdots [V_N^{(t_N)}, A^{(N)}] \cdots] A^{(N-1)} \cdots] A^{(2)}] \cdots] A^{(1)}] \cdots] \right\} \qquad (5.5)$$

where $I_{\tau_1 \cdots \tau_N} \equiv \{ t^{(j)}_{K_j} \mid \tau_j \lambda^{-2} \geq t^{(j)}_1 \geq \cdots \geq t^{(j)}_{M_j} \geq \tau_{j+1} \lambda^{-2} \}$ $j = 1 \cdots N$
with $\tau_{N+1} = 0$

Using the same graphical representation as before, (5.5) can be estmated, in the topology of $\mathcal{L}(a^N, a)$, in terms of tight graphs which are rescattering-free.

In this context it is preferible (not necessary) not to consider as rescattering links those which connect two chronologically consecutive t_K's , if these times appear on opposite sides w.r. to one of the $\tau_i \lambda^{-2}$. With this precaution, (5.5) differs from the structure analized in §4 only for the presence of a large number of $A^{(i)}$; this does not alter the estimate in any essential way, in view of Lemma 2.
Estimates and conclusions are then as in § 4, and in this way Theorem 2 is proved (notice that, since the last expression in (5.4) is hermitian, and since the second is the adjoint of the first, equality of first and third implies equality of second and third.

Remark 1 We have denoted by a the algebra generated by the identity and by continuous functions of p vanishing when $|p| \to \infty$. Therefore, since $I \in a$, Theorem 1 is in fact a particular case of Theorem 2, obtained when $A^{(2)} = \cdots = A^{(N)} = I$

Remark 2 If, on the left-hand side of (5.4), one had taken the t_K's in an order which is neither chronological nor anti-chronological, the limit would have been different, and in general not simply connected with the Markov semi-group T_τ . Technically, this is so because there would be in this case an overlap of the ranges of the $t^{(j)}$'s in (5.5) for different j's , allowing for a more complicated structure of rescattering graphs.
Of course, more detailed assumptions about the $A^{(i)}$'s can still lead to a result corresponding to Theorem 2 in particular cases. But one should also notice that on the r.h. side of (5.3) the semi-group T_τ is defined on all a only when $\tau \geq 0$.

We now discuss briefly the convergence to the classical limit.

From the explicit expression (2.20) it is easy to verify that $\mathcal{L}_\hbar$ converges, when $\hbar \to 0$, to the diffusion generator $\mathcal{L}$ of the classical case, where $\mathcal{L}$ is defined in (2.10) , (2.11) .

Since for all ℓ , $\mathcal{L}_\ell$ is reduced on each sphere $S_c \equiv \{ p \mid |p| = c \}$,
and so is $\mathcal{L}$, it suffices to consider the restrictions of these
operators on S_c for all $c > 0$.

Consider the closed form φ_ℓ associated to the operator $\mathcal{L}_\ell$. One has,
with $f \in C(R^3)$

$$\varphi_\ell (f, f) = \frac{-1}{\ell^2} \int \Big(f (p + \ell k) - f (p) \Big) \delta \Big((p + \ell k)^2 - p^2) \frac{1}{\ell} \Big) \cdot$$

$$\cdot \Big(f (p + \ell k) - f (p) \Big) \cdot \widetilde{G} (k) \, d^3 k \, d^3 p \tag{5.6}$$

If $f \in C^1$ and has compact support not including the origin, one has,
from (5.6)

$$\lim_{\ell \to 0} \varphi_\ell (f, f) = -\frac{1}{2} \int k \cdot \nabla f (p) \; k \cdot \nabla f (p) \, \delta (k \cdot p) \, \widetilde{G} (k) \, d^3 k \, d^3 p \equiv \varphi (f, f) \tag{5.7}$$

which is the closed quadratic form associated to $\mathcal{L}$.

Let φ_ℓ^c , φ^c be the (positive, closed) quadratic forms associated to
$-\mathcal{L}_\ell^c$, $-\mathcal{L}^c$.

From (5.6) it follows that φ_ℓ^c is bounded for all $c, \ell > 0$, and, for
$c > 0$, $\varphi_\ell^c \to \varphi^c$ on a core for φ^c. Therefore $\mathcal{L}_\ell^c$ converge to $\mathcal{L}^c$
in strong resolvent sense, when $\ell \to 0$, as operators on $L^2(S_1)$, uniform-
ly for c in compact sets of $(0, \infty)$.

We have already remarked that $\mathcal{L}_\ell$ generates on each S_c a contraction
semi-group on the continuous functions, which has a unique invariant
distribution under mild assumptions on G (e.g. $\widetilde{G}(k) > 0$ $\forall k \in R^3$).
Under the same assumptions, the restriction of $\mathcal{L}$ to S_c is strictly
elliptic for all $c > 0$, and $\mathcal{L}_c$ generates a contraction semi-group on
all $L^p(S_c)$, $2 \leq p < \infty$, and also on continuous functions in the sup
norm.

One can also verify that $\exp t \mathcal{L}_\ell^c$ is a contraction on continuous
functions on S_1 , in the sup norm, uniformly in $0 < \ell < 1$, and in c
over compacts of $(0, \infty)$. Since strong semi-group convergence has already been proved in $L^2(S_1)$ (and in fact in $L^2(R^3)$, since continuous

functions with compact support not containing the origin are dense in
$L^2(R^3)$), it follows that, if f is a continuous function with support
in $R^3 \setminus \{0\}$, $\exp(t\mathcal{L}_{\hbar})\cdot f \to \exp(t\mathcal{L})\cdot f$ in the sup norm.
We close this section with the following remark.

In the proof of theorem 1 (and of theorem 2), convergence proofs in
A) relied on estimates which are in fact uniform in $\hbar$ for $\hbar \geq 0$. Therefore A) can also be proved for the classical case.

As for the estimates in B), as they stand they are certainly not uniform in $\hbar \geq 0$. One could recover uniformity if one were to make the
estimates after having summed the contribution of several skeletons,
corresponding to all possible choices in the multiple commutator
(4.8). The terms to be thus estimated would contain, in the limit
$\hbar \to 0$, a number of derivatives of $A(p) \in \mathcal{A}$ increasing with the order
of the graph considered. This is of course connected to the form taken
by the evolution of classical observables when written in the
"Liouville" form

$$f \mapsto f(t) = \exp((H_o + \lambda V)t)\cdot f \qquad (5.7)$$

where $H_o \cdot f = \{H_o, f\}$, $V \cdot f = \{V, f\}$ and $\{\,,\,\}$ denotes Poisson bracket.
The expression (5.7) is not adequate to perform estimates on the be-
haviour of $f(t)$, and the use of Hamilton's equation on phase space is
a much better candidate. We expect that this be a preliminary stage
in the proof of part B of Theorem 1 also in the classical case (and
in fact, uniformly for $\hbar \geq 0$), but we have so far no conclusive
results on this point.

R e f e r e n c e s

1 D. Hall, P. Sturrok   Phys. Fluids $\underline{10}$ 2620 (67)

   N. Shapiro   JETP Letters $\underline{2}$ 291 (65)

   N. Van Kampen   Phys. Rep. $\underline{24}$ 171 (76)

2 L. Van Hove   Physica $\underline{21}$ 517 (55)
                                    $\underline{23}$ 441 (57)

3 G. Emch, P. Martin   Helv. Phys. Acta $\underline{48}$ 59 (75)

4 H. Spohn   J. Stat. Physics $\underline{17}$ 6 (77)

5 G. Papanicolau, S. Varadhan Comm. Pure Applied Math. $\underline{26}$ 497 (73)

   E. Davies   Comm. Math. Phys. $\underline{39}$ 91 (74)

6 H. Kesten, G. Papanicolau   Comm. Math. Phys. $\underline{78}$ 19 (80)

7 G. F. Dell'Antonio   Submitted to Comm. Math. Phys.

8 Ph. Combe et al   Poisson processes on groups and the Feynmann
                             Integral Prepint Marseille 1979

9 M. Donsker, S. Varadhan   Comm. Pure Applied Math $\underline{28}$ 1 (75)

<u>LARGE DEVIATION ASYMPTOTICS AND THE POLARON</u> *

M.D. Donsker   and   S.R.S. Varadhan
Courant Institute of Mathematical Sciences
New York University

New York, NY   10012/USA

1.   Asymptotic Formulae and the Contraction Principle

In this note we describe in outline form the latest extension of our large deviation theory and how in particular the resulting asymptotics provide a solution to the polaron problem. The detailed proofs can be found in [1], [2], and [3].

Let Ω be a space of functions $\omega(\cdot)$ on $-\infty < t < \infty$ with values on a Polish space X. We assume Ω consists of functions with discontinuities only of the first kind, normalized to be right continuous, and with convergence induced by the Skorohod topology on bounded intervals. In this case Ω is itself a Polish space. Denote by Ω_t^+ the corresponding space of functions on $[t,\infty)$ with values in X. We denote by F_t^s the σ-field in Ω generated by $\omega(\sigma)$ for $s \leq \sigma \leq t$. We denote the translation map on Ω by θ_t , i.e., $(\theta_t\omega)(s) = \omega(s + t)$.

Let $P_{0,x}$ be a Markov process on Ω_0^+ starting from $x \in X$ satisfying the hypothesis that the mapping $x \to P_{0,x}$ is weakly continuous (which implies the Feller property for the process $P_{0,x}$).

Let $\omega \in \Omega$ and for each $t > 0$ define ω_t by

$$\omega_t(s) \quad = \quad \omega(s) , \qquad 0 \leq s < t$$

$$\omega_t(s+t) = \omega_t(s) , \qquad \text{all } s \in (-\infty,\infty) .$$

Now, $(\theta_s\omega_t)(\tau) = \omega_t(s+\tau), \; 0 \leq s \leq t$, and for any set $A \subset \Omega$ we define

$$(1.1) \qquad\qquad R_{t,\omega}(A) = \frac{1}{t} \int_0^t \chi_A(\theta_s\omega_t) \; ds .$$

If we let $M_s(\Omega)$ denote the space of stationary processes on Ω, we see that for each $\omega \in \Omega$ and each $t > 0$, $R_{t,\omega}(\cdot) \in M_s(\Omega)$. This is because, for any $\sigma > 0$,

---

* This work was supported in part by the National Science Foundation, under Grant No. NSF-MCS-80-02568.

$$R_{t,\omega}(\theta_\sigma A) = \frac{1}{t} \int_0^t \chi_{\theta_\sigma A}(\theta_s \omega_t) \, ds = \frac{1}{t} \int_0^t \chi_A(\theta_{s-\sigma}\omega_t) \, ds$$

$$= \frac{1}{t} \int_{-\sigma}^{t-\sigma} \chi_A(\theta_s \omega_t) \, ds = \frac{1}{t} \int_0^t \chi_A(\theta_s \omega_t) \, ds = R_{t,\omega}(A) \ .$$

Thus, for fixed $t > 0$, $R_{t,\omega}$ is a mapping of $\Omega \to M_S(\Omega)$ and as such is F_t^0 measurable. For each $t > 0$ and each $x \in X$ we use this mapping to induce a probability measure $\Gamma_{t,x}$ on $M_S(\Omega)$ by defining $\Gamma_{t,x} = P_{0,x} R_{t,\omega}^{-1}$, i.e., if $B \subset M_S(\Omega)$, then

$$(1.2) \qquad\qquad \Gamma_{t,x}(B) = P_{0,x}\{\omega \in \Omega: R_{t,\omega} \in B\} \ .$$

If $P_{0,x}$ is ergodic with invariant measure $\nu(dx)$ on X and if $\bar{Q} \in M_S(\Omega)$ is the stationary Markov process with ν as its marginal distribution, then, by the ergodic theorem, as $t \to \infty$,

$$(1.3) \qquad\qquad \Gamma_{t,x} \Rightarrow \delta_{\bar{Q}} \ , \qquad \text{for all } x \ .$$

Here, $\delta_{\bar{Q}}$ is the Dirac measure on $M_S(\Omega)$ concentrated at $\bar{Q}$. What we are concerned with are the probabilities of large deviations for the measure $\Gamma_{t,x}$. To see what we mean by this, suppose $G \subset M_S(\Omega)$ is open and contains $\bar{Q}$, then, as just noted, $\lim_{t\to\infty} \Gamma_{t,x}(G) = 1$. On the other hand, if A is a set whose closure does not contain $\bar{Q}$, then $\Gamma_{t,x}(A) \to 0$ as $t \to \infty$ and the question we are concerned with is the <u>rate</u> at which it goes to 0. As we will see shortly, knowing this rate allows one to determine the asymptotic behavior of a large class of function space integrals with respect to the Markov process $P_{0,x}$.

We define in Section 3 below, for any $Q \in M_S(\Omega)$, the entropy H(Q) of the stationary process Q with respect to the Markov process $P_{0,x}$. This entropy function governs the rate at which $\Gamma_{t,x}(A)$ of the preceding paragraph goes to zero. To be specific, we show in [2] that if the Markov process $P_{0,x}$ satisfies hypotheses (1)-(5) of [2], then for closed sets $C \subset M_S(\Omega)$

$$(1.4) \qquad\qquad \overline{\lim_{t\to\infty}} \frac{1}{t} \log \Gamma_{t,x}(C) \leq - \inf_{Q \in C} H(Q) \ .$$

Also we prove in [2] that, under hypotheses I-II of [2] on the Markov process $P_{0,x}$, for open sets $G \subset M_S(\Omega)$,

$$\text{(1.5)} \qquad \lim_{t \to \infty} \frac{1}{t} \log \Gamma_{t,x}(G) \geq - \inf_{Q \in G} H(Q) \ .$$

It turns out that $H(Q) \geq 0$ for all Q and $H(\bar{Q}) = 0$ so that (1.4) and (1.5) are consistent with (1.3).

The importance of (1.4) and (1.5) is that they imply respectively or any bounded continuous functional $F(Q): M_S(\Omega) \to R'$,

$$\text{(1.6)} \qquad \overline{\lim_{t \to \infty}} \frac{1}{t} \log E^{\Gamma_{t,x}}\!\left\{ e^{tF(Q)} \right\} \leq \sup_{Q \in M_S(\Omega)} [F(Q) - H(Q)]$$

and

$$\text{(1.7)} \qquad \lim_{t \to \infty} \frac{1}{t} \log E^{\Gamma_{t,x}}\!\left\{ e^{tF(Q)} \right\} \geq \sup_{Q \in M_S(\Omega)} [F(Q) - H(Q)] \ .$$

An equivalent formulation of (1.6) and (1.7), because of the definition of $\Gamma_{t,x}$ measure, is that (1.4) and (1.5) imply respectively

$$\text{(1.8)} \qquad \overline{\lim_{t \to \infty}} \frac{1}{t} \log E^{P_{0,x}}\!\left\{ e^{tF(R_{t,\omega})} \right\} \leq \sup_{Q \in M_S(\Omega)} [F(Q) - H(Q)] \ ,$$

and

$$\text{(1.9)} \qquad \lim_{t \to \infty} \frac{1}{t} \log E^{P_{0,x}}\!\left\{ e^{tF(R_{t,\omega})} \right\} \geq \sup_{Q \in M_S(\Omega)} [F(Q) - H(Q)] \ .$$

If, of course, the Markov process $P_{0,x}$ satisfies the hypotheses for (1.4) (hypotheses (1)-(5) of [2]) and also satisfies the hypotheses for (1.5) (hypotheses I-II of [2]) then we have all of (1.6), (1.7), (1.8) and (1.9) so that

$$\text{(1.10)} \qquad \lim_{t \to \infty} \frac{1}{t} \log E^{\Gamma_{t,x}}\!\left\{ e^{tF(Q)} \right\} = \sup_{Q \in M_S(\Omega)} [F(Q) - H(Q)] \ ,$$

or, equivalently,

$$\text{(1.11)} \qquad \lim_{t \to \infty} \frac{1}{t} \log E^{P_{0,x}}\!\left\{ e^{tF(R_{t,\omega})} \right\} = \sup_{Q \in M_S(\Omega)} [F(Q) - H(Q)] \ .$$

Before giving examples of these asymptotic formulae we want to make two remarks which will relate these formulae to our earlier results. First of all, our earlier results involved functions of the occupation distribution: for $\omega \in \Omega$, $t > 0$, and $A \subset X$, let

$$L_{t,\omega}(A) = \frac{1}{t} \int_0^t \chi_A(\omega(s)) \, ds \ ,$$

i.e., $L_{t,\omega}(A)$ is the proportion of time up to t that the particular path $\omega(\cdot)$ occupies the set $A \subset X$. For fixed $t > 0$ and $\omega \in \Omega$, $L_{t,\omega}(\cdot)$ is then a probability measure on X. We denote the space of probability measures on X by $M(X)$. Now, the relation between $L_{t,\omega} \in M(X)$ and $R_{t,\omega} \in M_S(\Omega)$ is made clear by observing that the marginal distribution of the stationary measure $R_{t,\omega}$ is precisely $L_{t,\omega}$.

Furthermore, in our earlier papers the asymptotic rate involving functionals of $L_{t,\omega}$ was governed by the I-function associated with the Markov process $P_{0,x}$ (see [1], especially III) and which was defined on $M(X)$. The relation between the I-function of these papers and the entropy function $H(Q)$ of these more recent results is that (proved in Section 6 of [2])

$$(1.12) \qquad \inf_{\substack{Q \in M_S(\Omega) \\ q(Q) = \mu}} H(Q) = I(\mu) ,$$

where in (1.12) the notation $q(Q) = \mu$ means the marginal of the stationary measure Q is μ. We refer to relation (1.12) as the contraction principle.

In Section 2 we will see how these asymptotic formuale and the contraction principle apply to the polaron problem, but first we end this section with an easier and more transparent example.

Let $V(\cdot)$ be a continuous real valued function on X and define $F(Q) = E^Q\{V(\omega(0))\}$. Then,

$$F(R_{t,\omega}) = \int_\Omega V(\omega'(0)) \, R_{t,\omega}(d\omega') = \frac{1}{t} \int_0^t V(\omega(s)) \, ds$$

and, assuming the Markov process $P_{0,x}$ satisfies the requisite hypotheses, (1.11) becomes

$$(1.13) \quad \lim_{t \to \infty} \frac{1}{t} \log E^{P_{0,x}}\left\{ e^{\int_0^t V(\omega(s)) \, ds} \right\} = \sup_{Q \in M_S(\Omega)} [E^Q\{V(\omega(0))\} - H(Q)] ,$$

where, in (1.13), $H(Q)$ is the entropy of the stationary process Q with respect to the Markov process $P_{0,x}$. Using (1.12), we see that (1.13) is the same as

$$(1.14) \quad \lim_{t \to \infty} \frac{1}{t} \log E^{P_{0,x}}\left\{ e^{\int_0^t V(\omega(s)) \, ds} \right\} = \sup_{\mu \in M(X)} [\int V(x) \, \mu(dx) - I(\mu)] ,$$

where in (1.14) $I(\mu)$ is the I-function associated with the Markov process $P_{0,x}$. This last formula agrees with earlier results obtained by the authors.

2. The Polaron Problem

Let P_t and $E^{P_t}\{\ \}$ denote respectively the probability measure and expectation with respect to three dimensional Brownian motion $\omega(\cdot)$ tied down at both ends, i.e., $\omega(0) = \omega(t) = 0$. For $\alpha > 0$, let

$$(1.15) \qquad G(\alpha,t) = E^{P_t}\left\{\exp\{\alpha \int_0^t \int_0^t \frac{e^{-|\sigma-s|}}{|\omega(\sigma)-\sigma(s)|}\, d\sigma\ ds\}\right\}.$$

A long standing problem in statistical mechanics [4], the "polaron problem", has been to show

$$(1.16) \qquad \lim_{t\to\infty} \frac{1}{t} \log G(\alpha,t) = g(\ )\quad \text{exists}$$

and moreover according to a conjecture of Pekar [5],

$$(1.17) \qquad \lim_{\alpha\pm\infty} \frac{g(\alpha)}{\alpha^2} = g_0\quad \text{exists}$$

with

$$(1.18)\ g_0 = \sup_{\substack{\phi \in L_2(R^3) \\ \|\phi\| = 1}} \left[2 \int \int \frac{\phi^2(x)\phi^2(y)}{|x-y|}\, dx\ dy - \frac{1}{2} \int |\nabla\phi|^2\, dx\right]$$

Using the results described above in Section 1 the authors showed (1.16) obtaining a variational formula for $g(\ )$ which was explicit enough to allow us to also prove (1.17) with g_0 being given indeed by (1.18). We now describe briefly how these results were obtained. The detailed proofs are in [2] and [3].

First of all we show that the asymptotic behavior of $G(\alpha,t)$ as $t \to \infty$ is the same if we replace tied down Brownian motion P_t by standard Brownian motion P. This is justified because for large t the terminal condition makes no essential difference. The proof of this assertion is in Section 3 of [3].

Next, we approximate the integrand in $G(\alpha,t)$ as follows:

$$\alpha \int_0^t \int_0^t \frac{e^{-|\sigma-s|}}{|x(\sigma)-x(s)|}\, d\sigma\ ds = 2\alpha \int_0^t \int_0^t \frac{e^{-(\sigma-s)}}{|x(\sigma)-x(s)|}\, d\sigma\ ds$$

$$(1.19) \quad \tilde{=} 2\alpha \int_0^t ds \int_s^\infty \frac{e^{-(\sigma-s)}}{|x(\sigma)-x(s)|}\, d\sigma = 2\alpha \int_0^t ds \int_0^\infty \frac{e^{-\tau}}{|x(s+\tau)-x(s)|}\, d\tau$$

$$= \int_0^t F^*(\theta_s\omega)\, ds$$

where $F^*(\omega) = 2\alpha \int_0^\infty \dfrac{e^{-\tau}}{|x(\tau)-x(0)|} \, d\tau$, $\omega = x(\cdot)$, and $\theta_s \omega = \omega(\cdot + s)$ is the translation operator introduced earlier. In [3] we show that the approximation introduced in (1.19) does not affect the determination of $g(\alpha)$ in (1.16) and indeed that problem now reduces to finding

$$(1.20) \qquad \lim_{t\to\infty} \frac{1}{t} \log E^P\left\{\exp\{ \int_0^t F^*(\theta_s \omega) \, ds\}\right\} .$$

The functional $F^*(\omega)$ is not bounded and so one introduces

$$F_\delta^*(\omega) = 2\alpha \int_0^\infty \frac{e^{-\tau}}{(\delta^2 + |x(\tau)-x(0)|^2)^{1/2}} \, d\tau$$

instead, and in [3] we show that we can recover (1.20) from the same limit with $F_\delta^*(\omega)$ replacing $F^*(\omega)$. Since this is just technical and not revealing of the results of Section 1 we neglect this feature in the rest of this note and refer the interested reader to [3].

We want to use (1.8) in order to get an upper estimate in (1.20). To do so requires that the Markov process $P_{0,x}$ satisfies the hypotheses that yield (1.4). For (1.20) the Markov process is standard Brownian motion in R^3 which does not satisfy hypotheses (1)-(5) of [2]. However, if for $\varepsilon > 0$ we let P_ε be the Ornstein-Uhlenbeck process in R^3, i.e., the process with infinitesimal generator $\frac{1}{2}\Delta - \varepsilon x \cdot \nabla$, then the hypotheses (1)-(5) of [2] are satisfied (proved in [2]) and one can apply (1.8) for the O-U process and estimate (1.20) in terms of it. For the lower estimate in (1.20) we want to use (1.9) and this we can do directly since standard Brownian motion satisfies hypotheses I-II of [2]. Thus, we succeed in proving (1.16) with

$$(1.21) \qquad g(\alpha) = \sup_{Q \in M_S(\Omega)} [E^Q\{F^*(\omega)\} - H(Q)] .$$

This is because (using the notation of (1.8))

$$F(Q) = E^Q\{F^*(\omega)\} ,$$

and so

$$t\, F(R_{t,\omega}) = t \int F^*(\omega') \, R_{t,\omega}(d\omega')$$

$$= \int_0^t F^*(\theta_s \omega) \, ds .$$

Thus, again using the notation of (1.8) on (1.9),

$$\sup_{Q \in M_S(\Omega)} [F(Q) - H(Q)] = \sup_{Q \in M_S(\Omega)} [E^Q\{F^*(\omega)\} - H(Q)]$$

which appears in the right of (1.21). Here, of course, $H(Q)$ is the entropy of the stationary process Q with respect to standard Brownian motion.

Now to prove Pekar's conjecture (1.17) and (1.18) we first observe that by using the scaling properties of Brownian motion,

$$(1.22) \qquad \lim_{\alpha \to \infty} \frac{g(\alpha)}{\alpha^2} = \lim_{\alpha \to \infty} \sup_{Q \in M_S(\Omega)} [E^Q\{2 \int_0^\infty \frac{e^{-\tau} \, d\tau}{|x(\alpha^2\tau) - x(0)|}\} - H(Q)] \; .$$

Letting $\lambda = 1/\alpha^2$ we see that proving (1.17) and (1.18) is the same as showing

$$(1.23) \qquad \lim_{\lambda \to 0} \sup_{Q \in M_S(\Omega)} [E^Q\{2 \int_0^\infty \frac{e^{-\lambda\sigma} \, d\sigma}{|x(\sigma) - x(0)|}\} - H(Q)]$$

$$= g_0 = \sup_{\substack{\phi \in L^2(R^3) \\ \|\phi\| = 1}} \left[2 \int \int \frac{\phi^2(x)\phi^2(y)}{|x-y|} \, dx \, dy - \frac{1}{2} \int |\nabla\phi|^2 \, dx \right] \; .$$

The details of the proof of (1.23) are in Section 4 of [3] but we can give a rough heuristic idea of why it is true. Going back to (1.22) we start with the fact (proved in [2]) that the entropy $H(Q)$ is actually linear in Q and therefore the supremum involved in (1.22) is over linear functionals of Q. Thus the supremum is attained at an extremal which is an ergodic stationary process. Using the ergodic theorem and letting μ_Q denote the marginal distribution of Q we show (for more precise formalism see [3])

$$(1.24) \qquad \lim_{\alpha \pm \infty} E^Q\left\{ 2 \int_0^\infty \frac{e^{-\tau} \, d\tau}{|x(\alpha^2\tau) - x(0)|} \right\} = 2 \int \int \frac{\mu_Q(dx) \, \mu_Q(dy)}{|x - y|} \; .$$

This means,

$$(1.25) \qquad g_0 = \lim_{\alpha \to \infty} \frac{g(\alpha)}{\alpha^2} = \sup_{Q \in M_S(\Omega)} \left[2 \int \int \frac{\mu_Q(dx) \, \mu_Q(dy)}{|x - y|} - H(Q) \right] \; .$$

Using the contraction principle (1.12) we find that

$$(1.26) \qquad g_0 = \sup_{\mu} \left[2 \int \int \frac{\mu(dx) \, \mu(dy)}{|x - y|} - I(\mu) \right]$$

where $I(\mu)$ is the I-function for Brownian motion. We showed earlier in [1] that this $I(\mu) < \infty$ if and only if $\mu(dx) = \phi^2(x) \, dx$ for some $\phi \in L^2(R^3)$ with $\|\phi\| = 1$ and $\int |\nabla\phi|^2 \, dx < \infty$. In such a case,

$$I(\mu) = \frac{1}{2} \int |\nabla\phi|^2 \, dx.$$ Using this in (1.26) gives finally

$$g_0 = \sup_{\substack{\phi \in L^2(R^3) \\ \|\phi\| = 1}} \left[2 \int \int \frac{\phi^2(x)\phi^2(y)}{|x-y|} \, dx \, dy - \frac{1}{2} \int |\nabla\phi|^2 \, dx \right]$$

3. Entropy

Using the notation of Section 1 we describe now how the entropy $H(Q)$ of the stationary process Q with respect to the Markov process $P_{0,x}$ is defined. All details are in [2].

As in Section 1, let $F_0^{-\infty}$ be the σ-field generated by $\omega(t)$ for $t \leq 0$. Let F_T^0 be the σ-field generated by $\omega(t)$ for $0 \leq t < T$. We can think of $F_0^{-\infty}$ and F_T^0 as sub σ-fields of F, the entire σ-field on which Q is a translation invariant measure.

Let Q_ω^T be the regular conditional probability of Q on F_T^0 given $F_0^{-\infty}$. Denote by $P_{\omega(0)}^T$ the measure on F_T^0 corresponding to some Markov process P starting at time 0 from $\omega(0)$. Let

$$h_T(\omega) = \int \log R_\omega^T(\omega') \, Q_\omega^T(d\omega')$$

be the entropy where $R_\omega^T(\cdot)$ is the Radon–Nikodym derivative of Q_ω^T with respect to $P_{\omega(0)}^T$ on F_T^0. We then define

$$h(T) = \int h_T(\omega) \, Q(d\omega) \ .$$

It turns out that either $h(T) = \infty$ for all $T > 0$ or $h(T) = H \cdot T$ for some constant H which depends on Q. We therefore write $h(T) = T \cdot H(Q)$ for some $0 \leq H(Q) \leq \infty$ and call $H(Q)$ the entropy of the stationary process Q with respect to the Markov process P.

References

1. Donsker, M. D. and Varadhan, S. R. S., _Asymptotic Evaluation of Certain Markov Process Expectations for Large Time I, II, III_, Comm. Pure Appl. Math. 28 (1975) 1–47; 28 (1975) 279–301; 30 (1976) 389–461.

2. Donsker, M. D. and Varadhan, S. R. S., _Asymptotic Evaluation of Certain Markov Process Expectations for Large Time IV_, to appear in Comm. Pure Appl. Math.

3. Donsker, M. D. and Varadhan, S. R. S., _Asymptotics for the Polaron_, to appear.

4. Feynman, R. P., _Statistical Mechanics_, W. A. Benjamin, 1972.

5. Pekar, S. I., _Theory of Polarons_, Zh. Eksperim. i Teor. Fiz. v. 19, 1949.

ALL THAT BROWNIAN MOTION

D.Dürr
Institut für Mathematik
Ruhr-Universität Bochum
463 Bochum,West-Germany

I shall present here some recent results in the mathematical rigorous study of nonequilibrium statistical mechanics. These results
have been obtained in collaboration with S.Goldstein and J.L.Lebowitz.

0. Introduction.

The prototype of a system modeling a typical nonequilibrium situation
on which many problems relevant to the foundations of nonequilibrium
statistical mechanics may be studied, may be described as follows:
Consider a classical system consisting of a special particle (molecule)
interacting with an infinite bath of identical particles (atoms). The
molecule may be distinguishable from the atoms by having a different
mass or different shape; it may also be a tagged particle, dynamically
indistinguishable from the atoms. We would like to include in the
notion of "bath particle" the case, where the bath consists of fixed
"soft" or "hard" scatterers; we may thus think of the atoms as having
infinite mass. The initial values of the positions and velocities of
the atoms are randomly distributed. The statistical character of the
system enters through this initial distribution and the initial distribution of the dynamical variables which describe the motion of the
molecule. For a given configuration of atoms and for given initial
data of the molecule the evolution of the system is deterministic and
governed by Newtonian laws of mechanics. We then study the motion of
the molecule, i.e. we observe its dynamical variables (symbolically
denoted by $A(t)$) describing the motion as functions of time. $A(t)$
will be a stochastic process on the probability space of the random
initial values and will be in general quite complicated so that in
many cases it is impossible to substract relevant information about
the (macroscopic) motion of the molecule. In certain "extreme" situations however $A(t)$ may be well approximated by Markovian stochastic
processes. These situations are mathematically treated by taking appropriate limits in which the evolution of $A(t)$ can then be described by
a diffusion process, thus yielding a relatively simple description
of the motion of the molecule.

To be more specific I shall give several examples some of which will
be discussed in more detail later on.

1) Let the bath be represented by an ideal gas of point particles of
mass m in R^ν. The molecule is a sphere of mass M interacting via elastic collisions with the atoms. The initial distribution of the system
is the conditional Gibbs measure where the molecule is at time t=0
at the origin ($\underline{Q}(0)=0$) say. The <u>hydrodynamical limit</u> corresponds to
observing the motion of the molecule on a very large (macroscopic)
time scale, simultaneously rescaling its position, i.e. we consider
$\underline{Q}_a(t)=a\underline{Q}(t/a^2)$ as a goes to zero. It is expected that

$$\text{w-}\lim_{a\to 0} \underline{Q}_a(t) = D^{\frac{1}{2}}\underline{W}_t \tag{0.1}$$

where $\underline{W}_t$ is the ν-dimensional Wiener process and the diffusion coefficient

$$D = 2 \int_0^\infty E(\underline{V}(0)\cdot\underline{V}(t))dt \quad . \tag{0.2}$$

Here $E(\cdot)$ denotes the expectation with respect to the initial measure
and $\underline{V}(t)$ is the velocity of the molecule. w-lim refers to the limit
in the sense of weak convergence of the corresponding process measures
on path space. As a side remark let us note that the more natural way
of defining the diffusion constant by

$$D = \lim_{t\to\infty} E(\underline{Q}(t)^2)/t$$

is indeed in many cases equivalent to (0.2). (Besides stationarity
one needs fast enough decay of the velocity autocorrelation function).

In one dimension (where the molecule can also be a point particle)
the result (0.1) has been proved for M=m by Spitzer $|1|$. (In this case
the molecule is a "labeled" atom).

Note that for a>0 neither $\underline{Q}_a(t), \underline{V}_a = a^{-1} \underline{V}(t/a^2)$ nor the pair
$(\underline{Q}_a(t),\underline{V}_a(t))$ constitute a Markov process since the molecule can recollide with atoms with which it has previously collided. Note further
that the recollisions do not become ineffective as a goes to zero;
they will contribute to the diffusion coefficient (0.2) which is determined by the unscaled process. However the number of collisions with
"fresh" particles on the time scale a^{-2} should be so large to produce
enough independence in the velocity process that (0.1) as a "central

limit theorem" for dependent variables is expected to hold in the general case.

2) We obtain a model of Brownian motion, the erratic motion of a small macroscopic particle suspended in a fluid, if we consider the previous system in the case where the mass M of the molecule is much larger than the mass m of an atom. We may assume that the molecule has initially a given velocity $\underline{V}^O$ and position $\underline{Q}^O$. It is enough to observe the velocity $\underline{V}(t)$, its position $\underline{Q}(t)$ may be obtained by integrating $\underline{V}(t)$ with respect to time. We want to study the Brownian limit or weak coupling limit which corresponds to observing

$$\underline{V}_a(t) = a^{-1} \underline{V}(t/a^2) \quad , \quad a^2 = m/M \ , \ \underline{V}_a(0) = \underline{V}^O \qquad (0.3)$$

in the limit as a goes to zero. The scaling is such that the effect of a single collision which is according to the collision laws proportional to $a^2 = m/M$ goes to zero, but on the macroscopic time scale a^{-2} enough collisions happen to produce a change in the scaled velocity. It is proved $|2|$ that

$$\underset{a\to 0}{\text{w-lim}} \ \underline{V}_a(t) = \underline{V}(t) \qquad (0.4)$$

where $\underline{V}(t)$ is an Ornstein-Uhlenbeck process with drift and diffusion coefficients determined by the parameters of the system. Note that contrary to the previous limit, the velocity exists and the position $\underline{Q}(t)$ is a stochastic process with differentiable paths. This is due to the fact that in the weak coupling limit also the physical parameters (the mass ratio) are scaled.

A perhaps more natural but equivalent way of defining $\underline{V}_a(t)$ is by scaling the bath parameters, leaving the molecule alone: The initial distribution of the atoms is Poisson in space with uniform density λ and each atom is given a velocity according to a velocity distribution $f(\underline{v})d\underline{v}$ (not necessarily Maxwellian) independently of its position and the other atoms. We consider the limit as m goes to zero, while the density of the atoms is scaled like $\lambda/m^{\frac{1}{2}}$ and the velocity distribution of the atoms is scaled like

$$f_m(\underline{v}) = m^{\nu/2} f(m^{\frac{1}{2}} \underline{v}) \quad . \qquad (0.5)$$

We thus obtain a family of stochastic processes $\underline{V}_m(t)$ for the velocity of the molecule and our result is that $\underline{V}_m(t)$ converges to an Ornstein-Uhlenbeck process $\underline{V}(t)$. From (0.5) follows that the average speed $<v>_m \sim m^{-\frac{1}{2}}$, while $\underline{V}_m$ is of order unity, i.e. in this scaling the atoms are very fast compared to the molecule. Note that for m>0 $\underline{V}_m(t)$ is not a Markov process because recollisions can occur but typically the molecule will only recollide with "slow" atoms which will not have much effect on the motion of the molecule. We will give more detailed information later; here we wish to emphasize that in the weak coupling limit of this model recollisions become ineffective and they will not contribute to the drift and diffusion coefficients of the Ornstein-Uhlenbeck process. One might therefore expect that there exists a"good" Markov approximation of $\underline{V}_m(t)$ if m is small enough.

3) The next example is the Landau model where the bath consists of uniformly randomly distributed soft scatterers (Poisson distribution with density λ). The scatterers are identical and one scatterer is represented by an isotropic smooth potential U(x) of finite range. The molecule is a point particle of mass M=1. We observe the velocity of the molecule $\underline{V}(t)$ (V(0)=1) and for each configuration $\omega = (\underline{r}_i)_{i\epsilon N}$ ($\underline{r}_i$ = center of a scatterer) of scatterers $\underline{V}(t)$ changes according to

$$\frac{d\underline{V}}{dt} = - \sum_\omega \nabla U(\underline{Q}(t) - \underline{r}_i) \, , \quad \frac{d\underline{Q}}{dt} = \underline{V}(t) \quad . \qquad (0.6)$$

We consider the weak coupling limit by scaling (in dimensions $\nu \geq 2$)

$$U_a(x) = aU(x/a^2)$$

$$\lambda_a = \lambda \, a^{-2\nu} \qquad . \tag{0.7}$$

This defines the family of processes $\underline{V}_a(t)$, ($V_a(0) = 1$), and we are interested in the limit a→0.

This model has been treated by the same method we used to show convergence to Brownian motion in the previous example; I shall describe this method in the next section. It is proved $|3|$ that $\underline{V}_a(t)$ converges to a Wiener process on the sphere of radius 1 with the diffusion coefficient

$$D \sim \int d^{\nu-1}\underline{k} \, |\hat{U}(k)|^2 \, k^2 \qquad ,$$

$\hat{U}(k)$ is the Fourier transform of U. The Fokker Planck equation corresponding to this process may be looked at as a special case of the Lan-
dau equation |4|. We adopted this name for the model.

The result for dimensions $\nu \geq 3$ was already obtained in |5| for
general force fields.

The convergence of the position process $\underline{Q}_a(t) = \underline{Q}^o + \int_o^t \underline{V}_a(t')dt'$
follows from the convergence of the velocity process by general theo-
rems. For a>0 neither $\underline{V}_a(t)$ nor the pair $(\underline{Q}_a(t),\underline{V}_a(t))$ is Markovian
because the molecule can "recollide" with scatterers (in two dimensions
it almost certainly will do so), and because the velocity changes
smoothly (cf.eq.(0.6)) in a scatterer. However the scaling is such
that the effect of a single scatterer on the velocity of the molecule
is of order a (use (0.7) in (0.6) and that the time of sejour in a
scatterer is $\sim a^2$) and the range of the scaled potential is $\sim a^2$, i.e.
in the limit a→0 the scatterer shrinks to a point and its effect goes
to zero. The density is now scaled in such a way that the number of
scattering events increases (λ_a · cross section of a scatterer $\sim a^{-2}$)
so that one obtains a non trivial limit. Note that the volume occupied
by scatterers is independent of a.

In dimensions $\nu \geq 3$ the paths of the limit position process do not
selfintersect, i.e. in this case recollisions become impossible. (For
this reason the result in |5| was valid only for $\nu \geq 3$). In two dimen-
sions this is not anymore true. However the regions of selfintersections
(regions where the molecule comes back to scatterers it has previously
encountered) become small as a→0. (The selfintersections will be "transversal"). Thus, although the possibility of selfintersections per-
sist in the limit, their number and hence the number of recollisions
is so small that they become ineffective.

4) This last example is to show that in the weak coupling limit recollisions do not always become irrelevant. Let us change the system
of example 2 into a semi-infinite 2-dimensional system by taking the
y-axis as a wall which elastically reflects the molecule and the atoms.
One might expect that in the weak coupling limit the velocity process
converges to the Ornstein Uhlenbeck process with reflection at the
y-axis. Near the boundary however the molecule has a shielding effect
on the atoms which might render the drift and diffusion coefficients
to become position dependent. The shielding effect is counteracted by

fast atoms bouncing back and forth between the molecule and the wall, i.e. recollisions play a substantial role and will influence the limit process. That the limit process will be Markovian is roughly speaking due to the fact that the time between a collision and recollision shrinks to zero since the scaled speed of the atoms goes to infinity so that the collision and possible recollisions between a fast atom and the molecule occur simultaneously. The situation is of course similar when there are more than one molecule (in the infinite system) present or when the surface of the molecule has concave parts.

In the next section I shall discuss example 2 more carefully and give the basic ideas for proving the convergence result. It is not clear whether the technique I present can be used successfully for proving convergence in the examples 1 and 4. I tried to indicate the differences of the examples; so far our method has been applied to situations where recollisions, i.e. non Markovian effects become negligible in the limit.

I. The Brownian Motion of a spherical molecule.

We describe the initial distribution of the ideal gas of point particles of mass m as a Poisson distribution in the one particle phase space R^6. Let

$$dL_m = \lambda \, m^{-\frac{1}{2}} d\underline{q} \, f_m(v) d\underline{v} \quad , \quad v = |\underline{v}| \tag{1.1}$$

where

$$f_m(v) = m^{3/2} \, f(m^{\frac{1}{2}} v) \tag{1.2}$$

is a probability density with at least four moments. For A,B Borel-sets of R^6 let N_A, N_B denote the number of atoms in A,B. $N_{(.)}$ are random variables on the probability space (Ω, σ, P_m) where $\omega \in \Omega$ represents a configuration of atoms of the ideal gas, i.e. $\omega = (\underline{q}_i, \underline{v}_i)$, $i \in \mathbb{N}$, $\underline{q}_i, \underline{v}_i$ denoting position and velocity of an atom. We have

$$P_m(N_A = k) = e^{-L_m(A)} \, (L_m(A))^k / \, k! \tag{1.3}$$

and N_A, N_B are independent if A,B are disjoint. (In (1.2) we made for simplicity the assumption that f(v) depends only on the magnitude of $\underline{v}$. In general we need the symmetry condition $\int v \, \underline{v} \, f(\underline{v}) d\underline{v} = 0$).

This Poisson field describes an equilibrium distribution of the ideal gas of spatial density λ_m where each atom is given independently a velocity $\underline{v}$ according to the distribution $f_m(v) d\underline{v}$.

We now place at time $t = 0$ into this ideal gas system the molecule, a sphere of radius r and mass M at a position $\underline{Q}^o$ (center of the sphere) removing of course all the atoms from the region to be occupied by the molecule. The initial velocity of the molecule is $\underline{V}^o$. The molecule changes then its velocity through elastic collisions with the atoms. With $\underline{v}_n(\underline{V}_n)$ denoting the projection of $\underline{v}(\underline{V})$ on the line through the center of the molecule and the collision point on the surface of the molecule we write

$$\underline{v} = \underline{v}_n + \underline{v}_t$$

$$\underline{V} = \underline{V}_n + \underline{V}_t$$

and we have for the post collision velocities

$$\underline{V}'_t = \underline{V}_t \quad , \quad \underline{v}'_t = \underline{v}_t \tag{1.4}$$

$$\underline{v}'_n = - \frac{M - m}{M + m} \underline{v}_n + \frac{2M}{M + m} \underline{V}_n \tag{1.5}$$

$$\underline{V}'_n = \frac{M - m}{M + m} \underline{V}_n + \frac{2m}{M + m} \underline{v}_n \quad . \tag{1.6}$$

Note, that the laws of collision do not resolve multiple collisions, i.e. collisions of the molecule with two or more atoms at the same time; also infinitely many collisions in a finite amount of time are problematical. One can show that on a set $\hat{\Omega} \subset \Omega$ with $P_m(\hat{\Omega}) = 1$ for all $m > 0$ the motion of the molecule is well defined, and for $\omega \in \hat{\Omega}$, $\underline{V}_m(t,\omega)$, the velocity of the molecule in the configuration ω, may be defined as a right continuous function of time. Thus we obtain the mechanical process $\underline{V}_{m,t}$, $t \in [0,T]$ on the probability space (Ω,σ,P_m). $\underline{V}_{m,t}$ can be represented also on $(D([0,T]),\sigma(D),\nu_m)$ - the space of right continuous functions equipped with the Skorohod topology and ν_m the measure induced by $\underline{V}_{m,t}$.

The process $\underline{V}_{m,t}$ is not a Markov process: Collisions between the atoms and the molecule disturb the bath distribution and change it locally into a non equilibrium distribution. Given the past trajectory $\underline{V}_m(s), s \leq t$, one has information about the position of atoms in phase space and the molecule can have recollisions both real (with atoms which have collided earlier) and virtual (i.e. collisions which cannot occur). It is very helpful for the understanding of the occurrence of recollisions to draw for a one dimensional system a space-time diagram in which the atoms move on $\pm 45°$ lines (velocity $v = \pm 1$) before collisions with the molecule. If the molecule starts out with velocity $\underline{V}^o = 0$ and $M > m$ only real recollisions can occur, if $M < m$ also virtual recollisions are possible (eqs.(1.5),(1.6)). One then understands easily the possibility of real or virtual recollisions if $M > m$ and the velocities of the atoms are absolutely continuously distributed. I wish to emphasize that the heart of the derivation of Brownian Motion from "first principles" is to take recollisions into account. They alone are responsible for the fact that $\underline{V}_{m,t}$ is a non Markovian process, and that is a "real" problem.

Instead of formulating the theorem now - the result we wish to obtain - I reverse the order and first give half of the proof, then the theorem, then the other half, hoping that the reader will then appreciate on the grounds of the previous discussion the necessity of the second half of the proof. First comes the physicist: Once the recollision business is understood one might try to approximate the mechanical process by a Markov process $U_{m,t}$, which is obtained from the mechanical one by "neglecting" recollisions. There are many ways to define such a Markov process, the most natural one is by observing the motion of a molecule in an ideal gas which is <u>always</u> in equilibrium. Let us turn this into a definition by considering the rates with which the molecule has collisions with Poisson distributed atoms. Suppose that at time t the molecule has velocity U and that it is surrounded by a bath of atoms having the Poisson distribution described in (1.3). The"probability" $p_m(dt,d\underline{v},d\underline{S},\underline{U})$ for the collision of an atom with velocity $\underline{v} \in d\underline{v}$ with the molecule in the surface element $d\underline{S} = r^2 \sin\theta\ d\theta d\phi\ \underline{e}_n$ ($\underline{e}_n$ is the surface normal vector pointing towards the center of the molecule) in $[t,t + dt]$ is given by

$$p_m(dt,d\underline{v},d\underline{S},\underline{U}) = \lambda\ m^{-\frac{1}{2}}(d\underline{S}\cdot(\underline{v} - \underline{U}))_+\ f_m(v)d\underline{v}dt \quad , \qquad (1.7)$$

where $(d\underline{S}\cdot(\underline{v} - \underline{U}))_+ = \max(d\underline{S}\cdot(\underline{v} - \underline{U}),0)$.
This follows easily from (1.3) for infinitesimal A.

We call

$$R_m(\underline{v},d\underline{S},\underline{U}) = \lambda\ m^{-\frac{1}{2}}(d\underline{S}\cdot(\underline{v} - \underline{U}))_+\ f_m(v) \qquad (1.8)$$

the collision rate (expected number of collisions per unit time) with which the molecule has collisions in its surface element $d\underline{S}$ with atoms of velocity $\underline{v}$.

$$R_m(\underline{U}) = \int\int R_m(\underline{v},d\underline{S},\underline{U})d\underline{v} \qquad (1.9)$$

is the total rate of collisions and $R_m(\underline{U})^{-1}$ is the mean waiting time between collisions.

All we need is (1.8) together with (1.4) and (1.6) to define in an informal way the "abstract" Markov process $U_{m,t}$:

$\underline{U}_{m,t}$ is the jump Markov process with exponential waiting time $R_m(\underline{U})^{-1}$ and jump probability $R_m(\underline{U})^{-1}R_m(\underline{v},d\underline{S},\underline{U})d\underline{v}$, where the size of the jump is determined by (1.4) and (1.6). Furthermore we set $\underline{U}_m(0) = \underline{v}^o$.

We may represent $\underline{U}_{m,t}$ on $(D,\sigma(D),\mu_m)$ where μ_m is the measure induced by $\underline{U}_{m,t}$.

Before discussing the quality of the approximation of the true mechanical process $\underline{V}_{m,t}$ by $\underline{U}_{m,t}$ let us consider the limit of $\underline{U}_{m,t}$ as $m \to 0$, or better the limit of the family of measures μ_m in the sense of weak convergence. (The weak convergence of measures can be rephrased as convergence in distribution " $\delta\to$ " of the corresponding processes). The weak convergence of a family of Markov processes is quite thoroughly studied and in the case where the limit process is a diffusion process the situation is very simple $|6|$: The weak convergence of the induced measures follows from the strong convergence of the generators of the Markov processes to the generator of the diffusion process on a sufficiently large class of functions (a core for the generator of the limit process which for simple diffusion processes are C^∞- functions of compact support.

The generator of a Markov process (here $\underline{U}_{m,t}$) is defined as

$$(A_m h)(\underline{U}^o) = \lim_{t\to 0} t^{-1} \{ E^U_{\underline{U}^o}(h(\underline{U}_{m,t})) - h(\underline{U}^o) \} \qquad (1.10)$$

for all functions h for which the limit exists in the sup-norm. $E^U_{\underline{U}^o}$ denotes the expectation corresponding to $\underline{U}_{m,t}$ starting at $\underline{U}^o$.

One "easily" computes (only "one collision" contributes)

$$(A_m h)(\underline{U}^o) = \iint h(\underline{U}^o + \frac{2m}{M + m} \underline{e}_n \cdot (\underline{v} - \underline{U}^o)) R_m(\underline{v},d\underline{S},\underline{U}^o)d\underline{v} -$$

$$- R_m(\underline{U}^o)h(\underline{U}^o) \quad . \qquad (1.11)$$

Here "easily" stands in quotation marks, because the limit in (1.10) has to be uniform in $\underline{U}^o$ which makes the computation in fact a bit technical. (One gets uniformity really easy if one uses a Markov process which is slightly modified for large velocities $|2|$).

A straight forward computation which mainly consists of doing a Taylor expansion of h around $\underline{U}^o$ up to third order and integration

yields (again one has to pay attention to uniformity, since we need
strong convergence of the generators)

$$\lim_{m \to 0} A_m h(\underline{U}) = Ah(\underline{U}) = (-a\underline{U} \cdot \frac{\partial}{\partial \underline{U}} + \tfrac{1}{2} D \frac{\partial^2}{\partial \underline{U}^2})h(\underline{U}) \qquad (1.12)$$

where

$$a = \frac{8}{3} \pi \lambda \frac{r^2}{M} (\smallint |\underline{v}_n| \, f(v)d\underline{v}) \qquad (1.13)$$

and

$$D = \frac{8}{3} \pi \lambda \frac{r^2}{M^2} (\smallint |\underline{v}_n|^3 \, f(v)d\underline{v}) \qquad . \qquad (1.14)$$

A is the generator of the Ornstein Uhlenbeck process $\underline{V}_t$ which equivalently may be defined by the stochastic differential equation

$$d\underline{V}_t = -a \, \underline{V}_t + D^{\frac{1}{2}} \, d\underline{W}_t \quad , \quad \underline{W}_t = \text{Wiener process,}$$

$$\underline{V}(0) = \underline{v}^o \quad . \qquad (1.15)$$

Thus we obtain for the Markov process $\underline{U}_{m,t}$ with $\underline{U}_m(0) = \underline{v}^o$, $t \, \epsilon \, [0,T]$

$$\underline{U}_{m,t} \xrightarrow{\delta} \underline{V}_t \quad , \quad \text{as } m \to 0 \qquad (1.16)$$

and $\underline{V}_t$ is defined by (1.15).

Let us remark that in case of a general velocity distribution we
need that the "net force" $\smallint v \, \underline{v} \, f(\underline{v})d\underline{v} = 0$, otherwise the limit would
not exist. The drift and diffusion coefficient in this case are given
by matrices.

All in all let us say that (1.16) is a standard result and the
easy step in proving the result we wish to obtain for the mechanical
process:

Theorem: For any T let on $[0,T]$ be given the Ornstein Uhlenbeck process $\underline{V}_t$ defined in (1.15) . Then

$$\underline{V}_{m,t} \xrightarrow{\delta} \underline{V}_t \quad \text{as } m \to 0 \qquad . \qquad (1.17)$$

We remark that the convergence of the position process

$$\underline{Q}_{m,t} = \underline{Q}^{o} + \int_{o}^{t} \underline{V}_{m,s} \, ds \quad \text{to} \quad \underline{Q}_{t} = \underline{Q}^{o} + \int_{o}^{t} \underline{V}_{s} \, ds \quad \text{follows then by continuity}$$

of the mapping $\underline{V} \rightarrow \underline{Q}$.

It is now the time to ask how good the Markov approximation $\underline{U}_{m,t}$ to the mechanical process $\underline{V}_{m,t}$ is: Since we already have that $\underline{U}_{m,t} \xrightarrow{\delta} \underline{V}_{t}$ we only need to show that $\underline{U}_{m,t}$ is in probability close to $\underline{V}_{m,t}$ as $m \rightarrow 0$ and the theorem follows [7]. This is the most difficult task in proving the result. Note however that $\underline{U}_{m,t}$ has been defined in an abstract way. What we attempt to do is to jointly represent $\underline{U}_{m,t}$ and $\underline{V}_{m,t}$ on the same probability space (Ω',σ',P'_{m}) (we call such a joint representation a <u>coupling</u> of $\underline{U}_{m,t}$, $\underline{V}_{m,t}$) such that, with $\underline{U}'_{m,t}$ and $\underline{V}'_{m,t}$ denoting the realisations of $\underline{U}_{m,t}$ and $\underline{V}_{m,t}$, for any $\varepsilon > 0$

$$\lim_{m \rightarrow 0} P'_{m}(\{ \ \omega' \ | \ \sup_{0 \leq t \leq T} \ | \ \underline{U}'_{m,t}(\omega')-\underline{V}'_{m,t}(\omega')| \ > \ \varepsilon \ \}) = 0 \ . \quad (1.18)$$

We call $(\underline{U}'_{m,t} \ , \ \underline{V}'_{m,t})$ then a <u>good coupling.</u>

The simplest coupling which of course is not good is the independent coupling, in which $\underline{U}'_{m,t}$ and $\underline{V}'_{m,t}$ evolve independently, i.e. $\Omega' = D \times D$ and $P'_{m} = \mu_{m} \times \nu_{m}$.

A coupling which is extremely good but unfortunately very special is the diagonal coupling in which $\underline{U}'_{m,t}$ and $\underline{V}'_{m,t}$ evolve in exactly the same way, i.e. $\underline{U}'_{m,t}$ and $\underline{V}'_{m,t}$ have to be copies of the same abstract process, which they are not in our case ($\underline{V}_{m,t}$ is not Markovian!) . However we can try to construct a coupling which is very close to a diagonal one: If we were to couple $\underline{V}_{m,t}$ to itself we could describe the diagonal coupling as follows:
We consider the motion of the molecule (ME) for some $\omega \ \varepsilon \ \Omega$ in the process $\underline{V}_{m,t}$ and define the coupled process $\underline{V}'_{m,t}$ mechanically by considering an identical molecule (ME') which moves exactly as ME , i.e. whenever ME suffers a collision from an atom with velocity $\underline{v}$ in the surface element $d\underline{S}$, ME' has exactly the same collision. (Note that in this case where ME is a sphere we only would have to specify $\underline{v}_{n}$ because it determines the point of the surface of ME (and ME') where the collision takes place.)

Following this prescription the motion of ME' defines of course a realisation $\underline{V}'_{m,t}$ of $\underline{V}_{m,t}$ and the coupling is clearly diagonal.

We now proceed similarily to construct a coupling of $\underline{V}_{m,t}$ and $\underline{U}_{m,t}$.
We introduce MA as a "Markov" molecule (identical to ME) whose motion
shall define $\underline{U}'_{m,t}$. ME and MA have the same initial conditions at
time t = 0. Let $\underline{V}$, $\underline{U}'$ be the generic variables for ME and MA. Given
now a configuration ω we specify the motion of MA in two steps:
We first construct a "diagonal part" of the coupling:
(i) Whenever ME suffers a collision from an atom with velocity $\underline{v}$ in
the surface element $d\underline{S}$ (i.e. normal component $\underline{v}_n$) the velocity of MA
is changed as if it too had suffered an identical collision, i.e.
according to (1.4), (1.6) with the <u>same</u> $\underline{v}_n$.

MA has thus at time t collisions with the rates $R_m(\underline{v},d\underline{S},\underline{V}(s \leqslant t))$
with which collisions in the mechanical process occur. Note that these
rates depend on the past trajectory $\underline{V}(s \leqslant t)$ of ME (otherwise the process would be Markovian). But we know that for $\underline{U}'_{m,t}$ to be a realisati-
on of $\underline{U}_{m,t}$ the collision rate for MA has to be $R_m(\underline{v},d\underline{S},\underline{U}')$ according
to the definition of $\underline{U}_{m,t}$. Therefore we have to add to the diagonal
part of the coupling described in (i) the "non diagonal" correcting
part:
(ii) To obtain the correct rates we either ignore some collisions of
(i) (so that they produce no effect on MA) or add some "extra-colli-
sions" depending on whether $R_m(\underline{v},d\underline{S},\underline{U}')$ is less or greater than
$R_m(\underline{v},d\underline{S},\underline{V}(s \leqslant t))$.

a)   The probability that a collision of (i) is <u>not</u> ignored is

$$P_m(\underline{v},d\underline{S},\underline{V}(s \leqslant t),\underline{U}') = \min \left\{ \frac{R_m(\underline{v},d\underline{S},\underline{U}')}{R_m(\underline{v},d\underline{S},\underline{V}(s \leqslant t)} \ , 1 \right\} \ . \quad (1.19)$$

b)   The rate for the occurence of extra-collisions is

$$R_m(\underline{v},d\underline{S},\underline{V}(s \leqslant t),\underline{U}') = \max\{ R_m(\underline{v},d\underline{S},\underline{U}') - R_m(\underline{v},d\underline{S},\underline{V}(s \leqslant t)),0\}. \quad (1.20)$$

(i) <u>together</u> with (ii) define the coupled process $\underline{U}'_{m,t}$ which indeed is equivalent to $\underline{U}_{m,t}$ since MA has the correct collision rate
$R_m(\underline{v},d\underline{S},\underline{U}')$.

The proof of the theorem is completed by showing that this coupling is good. Rather than going into the technical details I would like to discuss on a more heuristic basis why this coupling is good. I shall do this in a series of remarks.

<u>Remark 1:</u> Since $\underline{U}_{m,t} \xrightarrow{\delta} \underline{V}_t$ and $\underline{V}_t$ is a nice diffusion process we have that

$$\lim_{A \to \infty} \lim_{m \to 0} \mu_m (\ \underline{x}(t) \ \epsilon \ D \ ; \ \sup_{0 \le t \le T} |\ \underline{x}(t)\ | \ \ge A\ ) = 0 \quad .$$

Hence it will be enough to show the closeness of the paths of $\underline{U}'_{m,t}$ and $\underline{V}'_{m,t}$ on the set

$$G_A^m = \{\ \omega' |\ \sup_{0 \le t \le T} |\ \underline{U}'_{m,t}(\omega')\ | \ \le A\ \} \quad .$$

(Take $\Omega' = \Omega \times H$, where H represents the purely stochastic effects, i.e. collisions which are ignored or extra-collisions. Clearly $\underline{V}'_{m,t}(\omega') = \underline{V}'_{m,t}(\omega,h) = \underline{V}_{m,t}(\omega)$ and $P'_m(G_A^m) = \mu_m (\ \sup_{0 \le t \le T} |\ \underline{x}(t)| \ \le A)\ .)$

<u>Remark 2:</u> Let t_m^* be the stopping time

$$t_m^* = \inf_{t=0} \ \{\ t : |\underline{U}'_{m,t} - \underline{V}_{m,t}| \ \ge \epsilon\ \} \quad .$$

Then on G_A^m for $t \le t_m^*$ $|\ \underline{V}_{m,t}| \ \le 2A$ for ϵ small enough.

<u>Remark 3:</u> Let us call an atom "fast" if in a collision $|\underline{v}_n| \ge 8A$ and "slow" if $|\underline{v}_n| < 8A$. From (1.5) one easily finds that for $V \le 2A$ and $|\underline{v}_n| \ge 8A$ $|\underline{v}'_n| > 2A$ for m small enough, i.e. a fast atom colliding with ME bounces off with the normal speed $> 2A$.

<u>Remark 4:</u> We have thus the following facts about recollisions on G_A^m for $t \le t_m^*$. By remark 3 a fast atom cannot recollide with ME and a fast atom cannot have collided earlier with ME (follow the trajectories of ME and the atom from the collision point on backwards in time). Thus in particular a fast atom cannot become a slow atom and vice versa and the only atoms ME can recollide with are slow atoms.

<u>Remark 5:</u> On G_A^m for $t \le t_m^*$ by remark 4 the collision rate for ME with fast atoms is $R_m(\underline{v}, d\underline{S}, \underline{V})$ ("Markov rates") and the rates for "fast" extra-collisions are proportional to $\epsilon m^{-\frac{1}{2}}$ (use (1.19),(1.20) and (1.8)). Note that the "normal" collision rate is proportional to m^{-1} . The

effect of a fast collision is $\Delta V \sim m^{\frac{1}{2}}$, according to the scaling $|\underline{v}_n| \sim m^{-\frac{1}{2}}$ and (1.6) . Hence the total effect of extra-collisions is $O(\varepsilon)$.

<u>Remark 6:</u> The number of distinct slow atoms which collide with ME in G_A^m for $t \leq t_m^*$ is bounded by $const.m^{-\frac{1}{2}}$, the effect of one collision is proportional to m but the effect of one slow atom is because of possible recollisions not so simple to estimate. If one believes that the number of recollisions are finite (say 5) then the effect of slow atoms on the velocity of ME (and MA) is of order $m^{\frac{1}{2}}$ and goes to zero as $m \to 0$. The situation is unfortunately not so simple. Some argument is needed to estimate the effect of the slow atoms, i.e. the effect of recollisions.

<u>Remark 7:</u> The details of how one puts the things together to show that

$$\lim_{m \to 0} P_m'(\, t_m^* \leq T \,) = 0$$

may be found in $|2|$.

<u>Remark 8:</u> I wish to emphasize the importance of the diagonal part of the coupling.This remark is historical : Instead of defining an abstract Markov process via "collision rates" one can also try to define a Markov approximation directly on Ω for example by giving the prescription that atoms in ω which are slow in a collision don't have an effect on the molecule MA. The Markov process would then be automatically defined on Ω and one would try to compare the motion of ME and MA on Ω . But as soon as ME and MA are in different positions with respect to a configuration ω (which because of slow atoms will certainly happen) the fast atoms will collide with ME and MA at different times and at different points on the surfaces, so ME and MA suffer collisions with different normal velocities of one fast atom. The error due to this is very difficult to handle in higher dimensions. Holley proceeded in this way in his work on the one dimensional model $|8|$. Through the diagonal part of the coupling "many" fast atoms collide exactly at the same time on the same surface points of MA and ME. By proving that the coupling is good we can replace"many" by "nearly all".

The final section is only a short communication that the"obvious" generalisation of the above model has been done.

II. The Brownian Motion of a convex body.

Instead of being a ball we can also consider ME to be a massive particle having a convex surface. The motion of the molecule is then one of translation and rotation described respectively by the velocity $\underline{V}_m$ of its center of mass and the angular velocity $\underline{Y}_m$ with respect to its center of mass. The limit process is a coupled rotational and translational Brownian motion. The stochastic differential equation reads $|9|$

$$d\underline{Z}_t = - (\ \frac{\underline{Y}_t \times \underline{V}_t}{\underline{\underline{I}}^* \cdot (\underline{Y}_t \times \underline{\underline{I}} \cdot \underline{Y}_t}\ )\ dt - 2\lambda\ (\int f(v)|\underline{v}_n|d\underline{v})\ \underline{\underline{M}}^{-1}\ \underline{\underline{D}} \cdot \underline{Z}_t\ dt\ +$$

$$+\ (2\lambda\ \int f(v)|\underline{v}_n|^3 d\underline{v}\ )^{\frac{1}{2}}\ \underline{\underline{M}}^{-1}\ \underline{\underline{D}}^{\frac{1}{2}} \cdot d\underline{W}_t$$

where $\underline{Z}_t = (\underline{V}_t, \underline{Y}_t)$, $\underline{\underline{I}}^*$ denotes the inverse of the moment of inertia tensor $\underline{\underline{I}}$,

$$\underline{\underline{M}}^{-1} = \begin{pmatrix} M^{-1}\ \underline{\underline{1}} & 0 \\ 0 & \underline{\underline{I}}^* \end{pmatrix} \qquad ,\quad \underline{\underline{1}} = 3\times 3 \text{ unit matrix,}$$

$$\underline{\underline{D}} = \int dS\ \begin{pmatrix} \underline{e}_n \\ \underline{r} \times \underline{e}_n \end{pmatrix} \begin{pmatrix} \underline{e}_n \\ \underline{r} \times \underline{e}_n \end{pmatrix}$$

and $\underline{r}$ is the vector from the center of mass to the surface element $d\underline{S}$ with normal vector $\underline{e}_n$. Here everything is described within a body fixed frame of reference (principle axis system). The drift contains besides the linear friction term the non linear systematic term due to the"free" motion. The proof of course consists of the two steps described in the previous section. First define the "natural" abstract Markov process which converges to the limit process then construct a good coupling of the Markov process and the mechanical process. The diagonal part of the coupling however can only be described in a body fixed frame of reference. To estimate the effect of slow atoms (remark 6) is essentially different from the case of the spherical molecule. For this we "needed" that the surface of the molecule is smooth (no edges). Let us finally remark that the convexity of the surface is essential for the result; a molecule with concave parts could "catch" fast atoms so that recollisions with fast atoms are possible. It is not clear how to construct a good coupling in this case.

References

|1| F. Spitzer, J.Math.Mech. $\underline{18}$, 973 (1969).

|2| D. Dürr, S. Goldstein, and J.L. Lebowitz, Commun.Math.Phys. $\underline{78}$, 507 (1981).

|3| D. Dürr, S. Goldstein, and J.L. Lebowitz, The Landau Model in dimensions greater or equal than two. In preparation.

|4| De Lenner and Resibois, Kinetic Theory of Fluids, John Wiley, 1978.

|5| H. Kesten and G. Papanicolaou, Commun.Math.Phys. $\underline{78}$, 19 (1981).

|6| Th. Kurtz, Ann.Probab. $\underline{4}$, 618 (1975).

|7| P. Billingsley, Convergence of probability measures, New York: John Wiley and Sons 1968.

|8| R. Holley, Z. Wahr. Verw. Geb. $\underline{17}$, 181 (1971).

|9| D. Dürr, S. Goldstein, and J.L. Lebowitz, A mechanical model for the Brownian motion of a convex body. In preparation.

THE DIFFUSION EQUATION AND CLASSICAL MECHANICS: AN ELEMENTARY FORMULA

K.D. Elworthy   and   A. Truman
Mathematics Institute   Department of Mathematics
University of Warwick   Heriot-Watt University
COVENTRY CV4 7AL   EDINBURGH EH14 4AS

§1   INTRODUCTION

Let M be an n-dimensional Riemannian manifold and Δ its Laplace-Beltrami operator. We shall examine the limiting behaviour as $\mu \to 0$ of the diffusion equation

$$\frac{\partial g_t^\mu}{\partial t} = \frac{1}{2} \mu^2 \Delta g_t^\mu + \frac{1}{\mu^2} V g_t^\mu \qquad 0 \le t < \infty$$

$$g_o^\mu(x) = T_o(x) \exp \{-S_o(x)/\mu^2\} \tag{1}$$

for $g_t^\mu : M \to \mathbb{R}$, given $T_o : M \to \mathbb{R}$ and $S_o : M \to \mathbb{R}$. Using the Girsanov formula we shall obtain an exact path integral expression for the minimal solution g_t^μ to (1) which shows clearly how the solution behaves in relation to its W.K.B. approximation. It is valid up to the time when caustics appear, if they do, and requires some smoothness conditions on V, S_o and T_o. We also give the first 'post W.K.B' term when $T_o \equiv 1$. Having done this we briefly mention some further developments, in particular indicating how to obtain an exact expression for the propagator for a restricted class of manifolds.

A study of the limiting behaviour of (1) by path integral techniques when $M = \mathbb{R}^n$ was begun by Donsker in connection with a problem in hydrodynamics discussed in (Hopf 1950), followed up with a Laplace asymptotic expansion by Schilder (1965), and generalized by Varadhan (1966) with further extensions by Donsker & Varadhan. There is a straightforward treatment using the Cameron-Martin formula by Truman (1977) which lead to the rather complicated discussion by stochastic differential geometry of the Riemannian case in (Elworthy & Truman 1981). In the Riemannian case the asymptotic behaviour of the fundamental solution, or propagator, corresponding to (1), was a celebrated problem especially for $V \equiv 0$. For analytical discussion see (Berger et al 1971) or (Pinsky 1978). It was examined in detail by Molchanov (1975) by stochastic methods, using the deep results of Varadhan. A thorough treatment and further references can be found in (Azencott 1981). These discussions have also led to attempts to obtain analogous results for the corresponding Schrödinger equation using Feynman path integrals: (DeWitt-Morette et al 1979; Elworthy & Truman 1981), with a rigorous principle of stationary phase for path integrals in the $\mathbb{R}^n$ case in (Albeverio & Hoegh-Krohn 1977).

The Girsanov-Cameron-Martin theorem has played a major role in most of this work and it is very surprising that the simple formula which we give does not seem to be in the literature. If it is we apologize for not giving the reference, but have no apologies for advertising it!

§ 2   PATH INTEGRAL SOLUTIONS VIA GIRSANOV'S THEOREM

A. To set up the path integrals we shall need some stochastic differential geometry. The relevant parts are discussed in more detail in (Elworthy 1978) and (Elworthy & Truman 1981), with a full treatment in (Malliavin 1978), (Ikeda & Watanabe 1981) and (Elworthy 1982).

Let $\pi : O(M) \to M$ denote the orthonormal frame bundle of M. The Levi-Civita connection of M determines a map

$$X : O(M) \times \mathbb{R}^n \to TO(M)$$

into the tangent space to $O(M)$, which trivializes the horizontal tangent bundle to $O(M)$. Suppose $\{\underset{\sim}{A}_t : 0 \le t \le T\}$ is a C^1 time dependent vector field on M. Let $\tilde{A}_t : O(M) \to TO(M)$ be its horizontal lift: so

$$T\pi \, (\tilde{A}_t(u) \, = \, \underset{\sim}{A}_t(\pi(u)) \qquad\qquad 0 \le t \le T, \qquad u \in O(M)$$

where $T\pi : TO(M) \to TM$ is the derivative map of π.

For $x_o \in M$ take $u_o \in \pi^{-1}(x_o)$ and consider the Stratonovich stochastic equation for $u : [0, \xi_A) \times \Omega \to O(M)$ with $u(o,\omega) = u_o$,

$$du_t^\mu = \mu X(u_t^\mu) dB_t + \tilde{A}_t(u_t^\mu) dt \tag{2}$$

where $\{B_t : t \ge 0\}$ is an n-dimensional Brownian motion defined on the probability space (Ω, F, P), and ξ_A is the explosion time of the solution: $0 < \xi_A(\omega) \le \infty$ for all ω in Ω. We say the system is *complete* if $\xi_A \equiv \infty$, and that M is *stochastically complete* if $\xi_A \equiv \infty$ for $A \equiv 0$.

Set $x_t^\mu = \pi(u_t)$. Then x has the Markov property and the Markov process has the semigroup which is the minimal sub-Markovian semigroup associated to the equation

$$\frac{\partial f_t}{\partial t} \, = \, \tfrac{1}{2} \, \mu^2 \Delta f_t \, + \, A_t(f_t)$$

for $f_t : M \to \mathbb{R}$, $t > 0$. When $A \equiv 0$ and $\mu = 1$ the process x is just a Brownian motion on M. In fact for most of this article the precise construction of x^u is not essential. The important point is that it is Markov and has generator $\tfrac{1}{2}\mu^2\Delta + A_t$. For simplicity we shall *assume throughout that M is stochastically complete,* so that its Brownian motion is defined for all time. This holds when the Ricci curvature of M is bounded below (Yau 1978) and in particular when M is compact.

A map $Y:[O,T] \times M \to \mathbb{R}$ will be said to be of class $C^{1,2}$ if it is C^1 and has continuous second partial derivatives in the manifold variable. We will write Y_t for $Y(t,-):M \to \mathbb{R}$.

Throughout $\| \ \|$ and $<,>$ will be the norms and inner products of the tangent spaces to M, determined by the Riemannian metric.

<u>Proposition 2</u> Suppose V is continuous and bounded above and g_o^μ is bounded and measurable. Then if $Y:[O,T] \times M \to \mathbb{R}$ is any $C^{1,2}$ map which is itself bounded and has uniformly bounded gradient $\{\nabla Y_t : O \le t \le T\}$, the solution g_t^μ to (1) is given by

$$g_t^\mu(x_o) = \mathbb{E}\{g_o^\mu(x_t^\mu) M_t \exp \frac{1}{\mu^2} \int_o^t V(x_s^\mu)\,ds\} \qquad O \le t \le T \qquad (3)$$

where x^μ is as above for $A_t = \nabla Y_t$ and

$$M_t = \exp[\frac{1}{\mu^2}(Y_o(x_o) - Y_t(x_t^\mu)) + \frac{1}{\mu^2}\int_o^t (\frac{\partial Y_s}{\partial s}(x_s^\mu) + \tfrac{1}{2}\|A_s(x_s^\mu)\|^2)\,ds$$

$$+ \tfrac{1}{2}\int_o^t \Delta Y_s(x_s^\mu)\,ds].$$

<u>Proof</u> The Girsanov and Feynman-Kac formulae combine to give the truth of (3) for

$$M_t = \exp[-\frac{1}{\mu}\int_o^t <A_s(x_s^\mu), u_s^\mu\ dB_s> - \frac{1}{2\mu^2}\int_o^t \|A_s(x_s^\mu)\|^2\,ds].$$

On the other hand the Itô formula yields

$$Y_t(x_t^\mu) = Y(x_o) + \mu\int_o^t <A_s(x_s^\mu), u_s^\mu\ dB_s> + \int_o^t \frac{\partial Y}{\partial s}(x_s^\mu)\,ds$$

$$+ \int_o^t <A_s(x_s^\mu), A_s(x_s^\mu)>ds + \tfrac{1}{2}\mu^2\int_o^t \Delta Y_s(x_s^\mu)\,ds$$

which gives the required answer after solving for $\int_o^t <A_s(x_s^\mu), u_s^\mu dB_s>$ and substituting in the expression for M_t. //

Proposition 2 suggests that we choose Y so that the expression (3) becomes particularly simple or illuminating. To do this we need to recall some classical mechanics; detailed proofs of the special cases we mention can be found in (Elworthy & Truman 1981).

§ 3 <u>CLASSICAL MECHANICS</u>

A. Let V and S_o be C^1 and consider the classical mechanical system for $\Phi_s(a) \in M$:

$$\ddot{\phi}_s(a) = -\nabla V(\phi_s(a)) \qquad\qquad s \geq 0$$
$$\left.\right\} \qquad (4)$$
$$\dot{\phi}_o(a) = \nabla S_o(a), \phi_o(a) = a$$

for each a in M.

A priori $\phi_s(a)$ may only exist for s > O in some $\frac{1}{2}$-open interval $[O, \xi_a)$, but it will exist for all time when ∇V is bounded and M is complete. Moreover in (Elworthy & Truman 1981) a criterion is given for $a \to \phi_s(a)$ to be a diffeomorphism $\phi_s : M \to M$ for all $s \in [O, \tau]$ with explicit lower bounds on the possible τ i.e. a lower bound on the time at which 'caustics' first appear. In particular the following proposition is proved there:

<u>Proposition 3A</u> *If M is compact, or if M is complete and ∇V, $\nabla^2 V$, ∇S_o and the curvature tensor R are all bounded on M then there exists $\tau > O$ so that $\phi_s : M \to M$ is a diffeomorphism for $O \leq s \leq \tau$.* //

For the moment we shall only assume:

no caustics condition: there exists $\tau > O$ such that $\phi_s : M \to M$ exists and is a diffeomorphism for $O \leq s \leq \tau$.

B. With this assumption define

$$S : [O, \tau] \times M \to \mathbb{R}$$

by

$$S(t,a) = S_o(\phi_t^{-1}(a)) + \frac{1}{2}\int_O^t |\dot{\phi}_s \circ \phi_t^{-1}(a)|^2 ds - \int_O^t V(\phi_s \circ \phi_t^{-1}(a)) ds. \quad (5)$$

Then S satisfies the Hamilton-Jacobi equation

$$\frac{1}{2} \| \nabla S_t(a) \|^2 + V(a) + \frac{\partial S_t}{\partial t}(a) = O \qquad O \leq t \leq \tau \qquad (6)$$

with the given initial function S_o.

Moreover

$$\dot{\phi}_t(a) = \nabla S_t(\phi_t(a)) \qquad O \leq t \leq \tau \qquad (7)$$

We shall need to run the flow backwards. For this fix $t \in (O, \tau]$ and set

$$\theta_s(a) = \phi_{t-s}(\phi_t^{-1}(a)) \qquad O \leq s \leq t, \quad a \in M.$$

Then

$$\frac{\partial}{\partial s} \theta_s(a) = -\nabla S_{t-s}(\theta_s(a)) \qquad (8)$$

C. Also define $\phi : [0,\tau] \times M \to (0,\infty)$

by

$$\phi(s,a) = |\det T_a \, \phi_s^{-1}| \tag{9}$$

i.e. the modulus of the Jacobian determinant of the inverse of Φ_s, computed using orthonormal bases with respect to the Riemannian inner product on the tangent spaces to M at a and $\Phi_s^{-1}(a)$. There is the continuity equation which follows from (7):

$$\frac{\partial \phi_s}{\partial s} + \operatorname{div}(\phi_s \, \nabla S_s) = 0 \qquad\qquad 0 \le s \le \tau \tag{10}$$

From this, since $\Delta = \operatorname{div} \nabla$, we see that for V and S_o of class C^2

$$\Delta S_t = - \frac{\partial}{\partial t} \log \phi_t - \langle \nabla \log \phi_t, \nabla S_t \rangle . \qquad 0 \le t \le \tau \tag{11}$$

In particular

$$\Delta S_{t-s}(\Theta_s(a)) = \frac{\partial}{\partial s} \log \phi_{t-s}(\Theta_s(a)) \qquad 0 \le s \le t, \; a \in M \tag{12}$$

4. THE FIRST FORMULA

In Proposition 2 take

$$Y(s,a) = -S(t-s,a) \qquad\qquad 0 \le s \le t \tag{13}$$

with t = T fixed, as above. Using the precise form of the initial function g_o^μ together with the Hamilton-Jacobi equation (6) and using (7) to bound ∇S_t we obtain:

<u>Formula A</u>   *Suppose that V and S_o are C^2 with V bounded above and S_o bounded below, and that T_o is bounded and measurable. Assume that the classical flow Φ_t given by (4) satisfies the no caustics conditions for $0 \le t \le \tau$ and has $\dot{\Phi}_t(a)$ uniformly bounded on $0 \le t \le \tau$ and $a \in M$. Then for $\mu \ne 0$ and $0 \le t \le \tau$ the solution g_t^μ to (1) is given by*

$$g_t^\mu(x_o) = \exp\left[- \frac{1}{\mu^2} S_t(x_o) \right] \mathbb{E} \left\{ \exp\left[- \frac{1}{2} \int_0^t \Delta S_{t-s}(x_s^\mu) \, ds \right] T_o(x_t^\mu) \right\}$$

$$0 \le t \le \tau$$

where x^μ is as in §2 with $A_s = -\nabla S_{t-s}$, $0 \le s \le t$ and S is given by (5).

From Proposition 3A we see that all the conditions on Φ are verified for some $\tau > 0$ when V and S_o are C^2, V is bounded above, S_o is bounded below and ∇V, $\nabla^2 V$, ∇S_o and the curvature tensor R are all bounded on M, and M is complete.

Now as $\mu \to 0$ so x_s^μ converges to the solution x_s^o to

$$\frac{dx_s^o}{ds} = A_s(x_s^o) \equiv -\nabla S_{t-s}(x_s^o) \quad \Big\}$$

$$x_o^o = x_o \tag{13}$$

in probability, uniformly in $s \in [0,t]$: for example see (Elworthy 1982). This solution is just $\Theta_s(x_o)$ by (8). Therefore we can apply the dominated convergence theorem and (12) to obtain the well known 'W.K.B.' result:

<u>W.K.B. Limit</u>   Under the conditions of Formula A if also T_o is continuous and ΔS_t is bounded below on M uniformly in $0 \le t \le \pi$, then for all $x_o \in M$ and $0 \le t \le \tau$

$$\lim_{\mu \to 0} \exp[-\frac{1}{\mu^2} S_t(x_o)]g_t^\mu(x_o) = \sqrt{|\det T_{x_o} \phi_t^{-1}|}\, T_o(\phi_t^{-1}(x_o)).$$

Note that we had no need to appeal to the Donsker-Varadhan theory, or the Ventsel-Freidlin theory.

§5   AN IMPROVED FORMULA

With slightly stronger conditions on our coefficients we can go a step further. This time we will take

$$Y(s,a) = -S(t-s,a) + \frac{1}{2} \mu^2 \log \phi(t-s,a) \tag{14}$$

in Proposition 2. A brief computation using (6) and (7) yields

<u>Formula B</u>   *Assume that V and S_o are C^3 with all the conditions of Formula A and also that $\nabla \log \phi_t$ is bounded on M uniformly for $0 \le t \le \tau$. Then for $\mu \neq 0$ and $0 \le t \le \tau$*

$$g_t^\mu(x_o) = \exp[-\frac{1}{\mu^2} S_t(x_o)]\sqrt{\phi_t(x_o)}\,E\{\exp[\frac{1}{2}\mu^2 \int_o^t \phi_{t-s}(x_s^\mu)^{-1}\Delta\phi_{t-s}^{\frac{1}{2}}(x_s^\mu)\,ds]T_o(x_t^\mu)\}$$

where x^μ is as in §2 with $A_s = -\nabla S_{t-s} + \frac{1}{2}\mu^2\nabla \log \phi_{t-s}$. //

After encouragement by Cecile DeWitt-Morette we can give the next term after the 'W.K.B' term, at least when $T_o \equiv 1$:

<u>First post-W.K.B. term</u>   *Assume all the conditions of Formula B and furthermore that $\phi_t(a)^{-\frac{1}{2}}\Delta\phi_t^{\frac{1}{2}}(a)$ is bounded for $a \in M$, uniformly in $t \in [0,\tau]$. Then if $T_o \equiv 1$, as $\mu \to 0$ so for $x_o \in M$ and $0 \le t \le \tau$*

$$\frac{1}{\mu^2} \{ \exp[\frac{1}{\mu^2} S_t(x_o)] g_t^\mu(x_o) - \sqrt{\phi}_t(x_o)$$

$$- \frac{1}{2} \mu^2 \int_o^t \phi_{t-s}(\Theta_s(x_o))^{-\frac{1}{2}} \Delta\phi_{t-s}^{\frac{1}{2}}(\Theta_s(x_o)) ds \} \rightarrow 0$$

__Proof__   Set $\alpha_t^\mu = \frac{1}{2} \int_o^t \phi_{t-s}(x_s^\mu)^{-\frac{1}{2}} \Delta\phi_{t-s}^{\frac{1}{2}}(x_s^\mu) ds$.   By hypothesis α_t^μ is

bounded uniformly in μ and $t \in [0,\tau]$.   Now, for any real α

$$e^{\mu^2 \alpha} - 1 - \mu^2 \alpha = \mu^4 \alpha^2 \int_0^1 (1-\theta) e^{\mu^2 \alpha\theta} d\theta,$$

and so Formula B gives

$$\exp[\frac{1}{\mu^2} S_t(x_o)] g_t^\mu(x_o) = \sqrt{\phi}_t(x_o) \mathbb{E} \{(1+\mu^2\alpha_t^\mu+\mu^4(\alpha_t^\mu)^2\int_0^1 (1-\theta) e^{\mu^2\alpha_t^\mu\theta} d\theta) T_o(x_t^\mu)\}$$

and the result follows by the dominated convergence theorem together with
the uniform convergence in probability of x_s to $\Theta_s(x_o)$.   //

For general T_o it seems that we should have to expand x_t^μ in powers of
μ rather as in (Elworthy & Truman 1981) and there would be extra terms
in the derivatives of T_o.

§6   FURTHER DEVELOPMENTS

A.   The technique we described of combining the Hamilton-Jacobi theory
with the Girsanov theorem can be varied to take into account the different limits which are to be investigated.   For example suppose we want
to examine the fundamental solution $p_t(x,y)$, or 'propagator' of the
equation

$$\frac{\partial g_t^\mu}{\partial t} = \frac{1}{2} \mu^2 \Delta g_t^\mu + \mu^2 V g_t^\mu \tag{15}$$

Then

$$p_t(x_o,y_o) = \lim_{\lambda \downarrow 0} (2\pi\mu^2\lambda)^{-n/2} f_t^\lambda(x_o) \tag{16}$$

when $n = \dim M$ and f^λ is the solution to

$$\frac{\partial f_t}{\partial t} = \frac{1}{2} \mu^2 \Delta f_t^\lambda + \mu^2 V f_t^\lambda$$

with $\qquad\qquad\qquad\qquad\qquad\qquad\qquad\qquad\qquad\qquad \Big\} \tag{17}$

$$f_o^\lambda(x) = T_o(x) \exp\left[- \frac{d(x,y_o)^2}{2\lambda\mu^2}\right]$$

where d is the Riemannian distance on M and T_o is any bounded continuous function with $T_o(y_o) = 1$. At least this is so when we make the assumption:

Assumption on y_o and M: from now on y_o is assumed to be a pole for M, i.e. there is a unique geodesic joining each point of M to y_o.

This is so if M is simply connected and with non-positive sectional curvature, and complete. For non-simply connected complete manifolds with non-positive curvature we can work on the universal cover.

To get a useful path integral expression for f_t^λ defined by (17) we take normal coordinates at y_o. Then y_o becomes the origin of $\mathbb{R}^n$ and the geodesics from y_o are just the straight lines emanating from the origin. This time we take the classical mechanical system $\ddot\Phi_t^\lambda(a) = 0$ with $\dot\Phi_o^\lambda(a) = \nabla S_o^\lambda(a)$ for $S_o^\lambda(a) = d(x,y_o)^2/(2\lambda)$. In our normal coordinates

$$\Phi_s^\lambda(a) = \frac{\lambda+s}{\lambda}\, a$$

and the corresponding solution S^λ to the Hamilton-Jacobi equation is given by

$$S_t^\lambda(a) = \tfrac{1}{2}\, \frac{d(x,y_o)^2}{\lambda+s}\ .$$

The Jacobian determinant ϕ^λ corresponding to (9) is best expressed in terms of Ruse's invariant $\theta_{y_o} : T_{y_o} M \to (0,\infty)$ which is the Jacobian determinant of $\exp_{y_o} : T_{y_o} M \to M$: see (Besse 1978) Chapter 6 or (Elworthy & Truman 1981) section 4 where it is related to the Van-Vleck determinant. In fact in our coordinates

$$\phi_s^\lambda(a) = (\tfrac{\lambda}{\lambda+s})^n\, \theta_{y_o}(a)^{-1}\, \theta_{y_o}(\tfrac{\lambda}{\lambda+s}\, a).$$

We will not give the details here, but a useful expression for f_t^λ is obtained via Proposition 2 by taking

$$Y(s,a) = -\frac{1}{2}\frac{\|a\|^2}{\lambda+t-s} - \frac{1}{2}\mu^2 \log \theta_{y_o}(a) + \frac{1}{2}\mu^2 \log \theta_{y_o}(\tfrac{\lambda a}{\lambda+t-s}),$$

given suitable bounds. Given additional bounds on V and $\theta_{y_o}^{\frac{1}{2}}\,\Delta\theta_{y_o}^{-\frac{1}{2}}$ we obtain

<u>Formula C</u> $p_t(x_o,y_o) = (2\pi\mu^2 t)^{-n/2}\, \theta_{y_o}(x_o)^{-\frac{1}{2}}\, \exp\left[-\frac{d(x_o,y_o)^2}{2t\mu^2}\right]$

$$\times\ \mathbb{E}\{\exp \tfrac{1}{2}\mu^2 \int_o^t (\theta_{y_o}^{\frac{1}{2}}(x_s)\Delta\theta_{y_o}^{-\frac{1}{2}}(x_s) + 2V(x_s))ds\}$$

where x_s is as in §2 for $0 \le s < t$ with

$$A_s(a) = -\frac{a}{t-s} - \frac{1}{2}\mu^2 \nabla \log \theta_{y_o}(a) \qquad\qquad 0 \le s < t$$

and

$$x_t = y_o. \quad //$$

The corresponding formula for the special case $M = \mathbb{R}^n$ can be found in (Simon 1979) Chapter II, §6. The 'bridge' process x from x_o to y_o is not as exotic as it looks: if $r_s = d(x_s, y_o)$ and $x_o \ne y_o$, $n \ge 2$ an application of Itô's formula shows that $\{r_s : 0 \le s \le t\}$ has the same law as the radial component of the Brownian bridge from x_o to 0 in $\mathbb{R}^n$.

Full details about Formula C and some of its consequences will be discussed in a future article. However before concluding, we would like to mention briefly a possible physical application to the thermodynamic limit of the free Bose gas. In recent papers (Van den Berg and Lewis 1981; Pulè 1982) the barometric formula for the free Bose gas in an external potential has been calculated in the thermodynamic limit by using the technology of Wiener integrals. If this thermodynamic limit is obtained by dilating about the origin, as is usually the case, the limit can actually be obtained by letting $\hbar \to 0$ in the functional integral expression for the quantum mechanical partition function, giving the barometric formula in a simple way. The formulae established in the present paper should be applicable to calculating this thermodynamic limit for the free Bose gas in a weak external field in a curved space background, enabling one to determine explicitly the effects of curvature on the barometric formula. These barometric formulae could be of physical interest for instance for thin films of liquid helium at low temperature.

References

Albeverio, S. & Hoegh-Krohn, R. (1977). Oscillatory integrals and the method of stationary phase in infinitely many dimensions with applications to the classical limit of quantum mechanics I. Invent. Mathem., 40, 59-106.

Azencott, R. et al. (1981). Géodésiques et diffusions en temps petit. Séminaire de probabilités, Université de Paris VII. Astérique 84-85 Société mathématique de france.

Van den Berg, M. & Lewis, J.T. (1981). On the free Bose gas in a weak external potential. Comm. Math. Phys., 81, 475-494.

Berger, M., Gauduchon, P. & Mazet, E. (1971). Le Spectre d'une Variété Riemannienne. Lecture Notes in Math. 194. Berlin, Heidelberg, New York: Springer-Verlag.

Besse, A.-L. (1978). Manifolds all of whose Geodesics are closed.
  Ergebnisse der Mathematik 93. Berlin, Heidelberg, New York:
  Springer-Verlag.

DeWitt-Morette, C., Maheshwari, A., & Nelson, B. (1979). Path integration in non-relativisitic quantum mechanics. Physics Reports, 50,
  no. 5, 255-372. Amsterdam: North-Holland.

Elworthy, K.D. (1978). Stochastic dynamical systems and their flows.
  In Stochastic Analysis, ed. A. Friedman & M. Pinsky, 79-95. London,
  New York: Academic Press.

Elworthy, K.D. (1982). Stochastic Differential Equations on Manifolds.
  London Math. Soc. Lecture Notes in Mathematics. Cambridge University
  Press (to appear: Autumn 1982)

Elworthy, K.D. & Truman, A. (1981). Classical mechanics, the diffusion
  (heat) equation and the Schrödinger equation on a Riemannian manifold.
  J. Math. Phys. 22, no. 10, 2144.

Hopf, E. (1950). The partial differential equation $u_t + uu_x = \mu u_{xx}$.
  Comm. Pur. Appl. Math., 3, 201-230.

Ikeda, N. & Watanabe, S. (1981). Stochastic Differential Equations
  and Diffusion Processes. Tokyo: Kodansha. Amsterdam, New York,
  Oxford: North-Holland.

Malliavin, P. (1978). Géométrie Differentielle Stochastique. Séminaire
  de Mathématiques Supérieures. Université de Montreal.

Molchanov, S.A., (1975). Diffusion processes and Riemannian geometry.
  Usp. Math. Nauk, 30, 3-59. English translation: Russian Math. Sur-
  veys, 30, 1-63.

Pinsky, M. (1978). Stochastic Riemannian geometry. In Probabilistic
  Analysis and Related Topics, 1, ed. A.T. Bharucha Reid, London, New
  York: Academic Press.

Pulè, J. (1982). Free Bose gas in a weak external potential. Journ.
  Math. Phys. to appear.

Schilder, M. (1965). Some asymptotic formulas for Wiener integrals.
  Trans. Amer. Math. Soc. 125, 63-85.

Simon, B. (1979). Functional Integration and Quantum Physics. London, New York: Academic Press.

Truman, A. (1977). Classical mechanics, the diffusion (heat) equation,
  and the Schrödinger equation. J. Math. Phys., 18, 2308-2315.

Varadhan, S.R.S. (1966). Asymptotic probabilities and differential
  equations. Comm. Pure Appl. Math., 19, no. 3, 261-286.

Yau, S.-T. (1978). On the heat kernel of a complete Riemannian manifold. J. Math. pures et appl., 57, 191-201.

Additional Remarks

We would like to thank Cécile DeWitt-Morette for bringing to our
attention the two references below where related asymptotics are described for the Green's function and Feynman propagator of a scalar
field in a curved space time in the presence of a Yang-Mills field.
In particular she drew our attention to equations (17.59) of the first
article, and (7.18) of the second, where the recurrence relation for
the Hadamard coefficients is given; pointing out the relationship between the resulting equation for the second coefficient a_1 and our
formula for the "first post-W.K.B. term".

We would also like to thank P.J. Baxendale for correcting some errors
in the first version of this article.

1. DeWitt, B.S. (1965) Dynamical Theory of Groups and Fields.
      Documents in Modern Physics. New York, London, Paris: Gordon
      and Breach
2. DeWitt, B.S. (1977) Quantum theory of gravity III. Applications
      of the covariant theory. Phys. Rev., <u>162</u>, 1239-1256.

<u>STOCHASTICITY IN NON-EQUILIBRIUM STATISTICAL MECHANICS</u>

Gérard G. Emch

Depts.of Mathematics and of Physics

The University of Rochester, Rochester, NY 14627, USA

To say that some form of stochasticity is intrinsic to any presentation of the ideas
of non-equilibrium statistical mechanics verges on the pleonasm. To ask where, precisely, stochasticity does in fact (or even should) emerge from the initially deterministic mechanical models turns out however to be a much harder question which runs
in filigree through the bulk of the literature. Rather than attempting to give an
overview of a century of active research, this lecture will more modestly restrict itself to the outline of a <u>proposal</u> which evolved from the study of a few specific,
exactly solvable models. To focus the discussion, we recall that the original aim of
non-equilibrium statistical mechanics is to give a description of macroscopic, dissipative, transport phenomena, based on microscopic, mechanistic models. We are thus
to analyze the following diagramme:

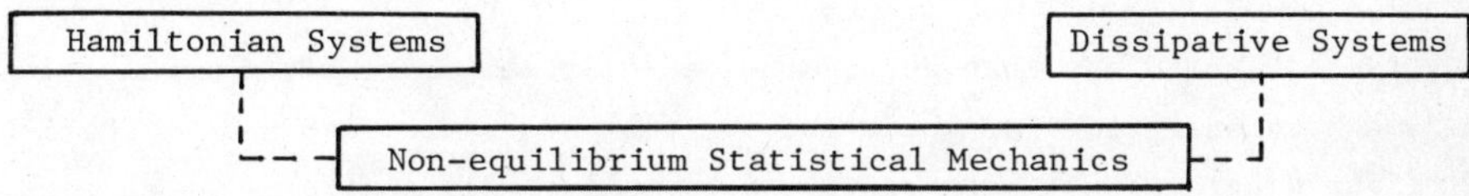

The dotted line in this diagramme involves several manipulations and intermediate
structures which one has to specify. Towards this aim, the following diagramme shows
a tentative route, or programme, which this lecture will outline.

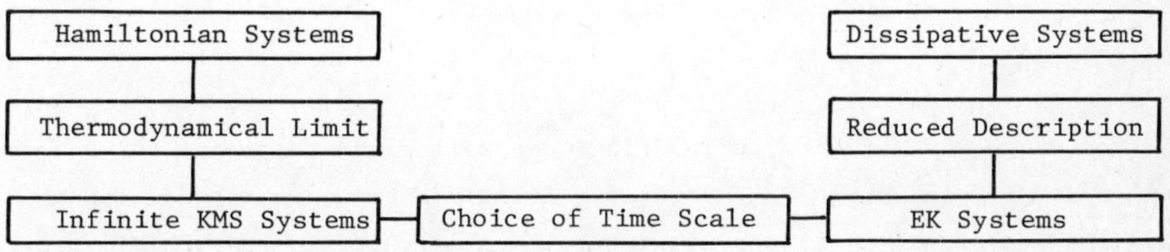

By the upper left-hand box, labelled "<u>Hamiltonian Systems</u>", we mean that structure
which is embodied in the usual axioms of Quantum Mechanics for systems involving only
finitely many degrees of freedom. In particular, we have a Hamiltonian H_g^Λ depending
on the volume Λ in which the system is confined, and which is typically of the form
$H_g^\Lambda = H_o^\Lambda + g\, V^\Lambda$.
In the "<u>Thermodynamical Limit</u>" the number of degrees of freedom, recognized to be very
large, is let go to infinity, typically leaving fixed some intensive variables such
as the density. The advantages of taking this limit are two-fold. Firstly, it sepa-
rates bulk properties from surface effects; secondly, it simplifies the discussion of

asymptotic dynamics.

The lower left-hand box, labelled "Infinite KMS Systems", represents situations where the quasi-local observables form a C*-algebra $A = \bigcup_\Lambda A(\Lambda)$ defined as the inductive limit of the local C*-algebras $A(\Lambda)$. On A is defined a priviledged state ϕ_g obtained, for instance, as the limit of Gibbs states with respect to H_g^Λ. Technically, it is advisable at this point to work in a definite Hilbert space representation adapted to the physical situation at hand; the Gelfand-Naimark-Segal construction based on ϕ_g provides such a representation, the weak-operator closure of which, denoted by N, is a von Neumann algebra on which ϕ_g is supposed (or better proved) to extend as a faithful normal state, again denoted by ϕ_g. Through the thermodynamical limit H_g^Λ determines on N an evolution $\alpha_g : t \in R \to \alpha_g(t) \in \mathrm{Aut}(N)$ with respect to which ϕ_g is supposed (or better proved) to satisfy the Kubo-Martin-Schwinger condition, namely: for every pair (A,B) of elements of N there exists a function F_{AB}, analytic inside, and continuous on the boundaries of the strip $0 < \mathrm{Im}(z) < \beta$, such that:

$$F_{AB}(t) = < \phi_g \; ; \; A \, \alpha_g(t)\{B\} > \quad \text{and} \quad F_{AB}(t+i\beta) = < \phi_g \; ; \; \alpha_g(t)\{B\} \, A > .$$

In other words $\alpha_g(R)$ is the modular group associated to ϕ_g by the Tomita-Takesaki theory [1].

Let it suffice here to say that during the last fifteen years a wide collection of models have been analyzed, for which the limiting procedure involved in the left-hand column of our diagramme has been proven to be fully under control (for an early example, see ref. [2]).

The upper box of the right-hand column of our diagramme, labelled "Dissipative Systems" contains thermodynamical systems showing an irreversible, exponential approach to equilibrium. These are mathematically characterized by differential equations such as the diffusion equation, or any similar transport equation. The first principal feature of these phenomenological equations, on which we want to focus our attention here, is that the evolution they describe is governed by a dissipative semi-group. Surely enough, there are more complicated equations one will ultimately want to consider, but the former are common enough in laboratory situations to warrant some special attention. A second feature of the equations we want to consider here is that they involve explicitly much fewer degrees of freedom than the systems described in the left-hand column (typically less than 10, as opposed to 10^{23}).

The fundamental question, the essence of which is summarized by the labelling "Reduced Description", is whether it is possible that these equations be a reflection, at a very "coarse-grained" level, of a finer description akin to those described in the left-hand column. First of all, one needs to give this problem a mathematical formulation, realizing however that even if that mathematical problem is solved the question will still remain as to whether one can find a physical link between the left- and right-hand column of our diagramme. This second question will be addressed later on in this lecture; at this point, we first propose a mathematical formulation [3] of the process involved in the right-hand column alone. For sake of definiteness, we propose to capture some of the essence of the structure involved in the upper right-hand box as follows. Let M be the von Neumann algebra of observables relevant to this level of the description; we assume that on M are defined a faithful normal state ψ and a *-weakly continuous semi-group $\{ \gamma(\tau) | \tau \in R^+ \}$ of completely positive maps of M into itself such that : (i) $\psi \cdot \gamma(\tau) = \psi$ $\forall$ $\tau \in R^+$; (ii) $\gamma(\tau)\{I\} = I$ $\forall \tau \in R^+$; (iii) there exists a semi-group $\{ \nu(\tau) | \tau \in R^+ \}$ of CP maps of M into itself such that for every pair (A,B) of elements of M and every $\tau \in R^+$ we have : $< \psi ; \gamma(\tau)\{A\} B > = < \psi ; A \nu(\tau)\{B\} >$.

To proceed down the right-hand column of the diagramme, we ask whether, given the structure just described, there exist : a von Neumann algebra N ; an injection $i : M \to N$; a faithful normal state ϕ on N ; a conditional expectation $E : N \to M$ such that : $E \cdot i = id$ and $\phi = \phi \cdot i \cdot E$; and finally a *-weakly continuous group $\{ \alpha(\tau) | \tau \in R \}$ of automorphisms of N such that $\gamma(\tau) = E \cdot \alpha(\tau) \cdot i$ $\forall$ $\tau \in R^+$ and $\phi \cdot \alpha(\tau) = \phi$ $\forall$ $\tau \in R$. If these objects exist, we say that $\{ M, \psi, \gamma(R^+) \}$ is a <u>reduced description</u> from $\{ N, \phi, \alpha(R) \}$. Mathematically our problem is thus to determine whether there exist a conservative dynamical system $\{ N, \phi, \alpha(R) \}$ and maps (E , i) such that a <u>given</u> dissipative system $\{ M, \psi, \gamma(R^+) \}$ appears as a reduced description. In case M (and N) is abelian, this is exactly the old question of whether every Markov semi-group comes from a Markov process; this question is known to admit a positive answer through the Kolmogorov-Daniell construction.

To convince a physicist that there is something for him in the above formulation of the problem, let us give here two examples.

The first example is the Bloch equations describing the relaxation of a spin-½ in a thermal bath. These phenomenological equations read :

$$\frac{d}{d\tau} \{ \sigma_j(\tau) - \varepsilon_j I \} = \Sigma_{k=1}^3 \lambda_{jk} \{ \sigma_k(\tau) - \varepsilon_k I \} \qquad j = 1,2,3.$$

The mathematical conditions we imposed in the above paragraph translate[4] in physical terms by the following restrictions on λ_{jk} and ε_j :

$$\frac{d}{d\tau} \sigma_1(\tau) = - \lambda \sigma_1(\tau) - \omega \sigma_2(\tau)$$

$$\frac{d}{d\tau} \sigma_2(\tau) = \omega \sigma_1(\tau) - \lambda \sigma_2(\tau)$$

$$\frac{d}{d\tau} \{ \sigma_3(\tau) - \varepsilon I \} = - \mu \{ \sigma_3(\tau) - \varepsilon I \}$$

with $-1 < \varepsilon < 1$ and $0 \leq \mu \leq 2 \lambda$, i.e. $T \leq 2 T$.

Another typical example is that of a quantum particle diffusing[5] in a harmonic well $V(x) = \omega^2 x^2 / 2$. Here M is the von Neumann algebra of all bounded operators on the hilbert space $L^2(R,dx)$, and is the closed linear span (in the weak operator topology) of the Weyl operators $W(z = \omega^{\frac{1}{2}} a + i \omega^{-\frac{1}{2}} b) = \exp\{-i(aP + bQ)\}$ where P and Q are the usual momentum and position operators. The canonical state ψ at natural temperature β and the semi-group $\gamma(R^+)$ of completely positive maps of M are then determined by their restriction to $\{ W(z) | z \in C \}$, namely :

$$< \psi ; W(z) > = \exp (- \Theta |z|^2 / 4)$$

$$\gamma(\tau)\{ W(z) \} = W(e^{-\lambda\tau}z) \exp\{-\Theta|z|^2(1-e^{-2\lambda\tau})/4\}$$

with $\Theta = \coth(\beta\omega/2)$ and $\lambda > 0$. The physical interpretation of ψ and $\gamma(\tau)$ is obtained through the following computations : (i) $< \psi ; W(z) > = \mathrm{Tr}\{\rho_\beta W(z)\}$ with $\rho_\beta = \exp(-\beta H)/\mathrm{Tr}\ \exp(-\beta H)$ and $H = (P^2 + \omega^2 Q^2)/2$; (ii) for every density matrix ρ and every direction (a,b) in the classical phase space R^2 , the probability density $\rho_{a,b}(\cdot,\tau)$ defined by :

$$\mathrm{Tr}\ \rho\ \gamma(\tau)\{ e^{-ik(aP+bQ)} \} = \int d\xi\ \rho_{a,b}(\xi\ \tau)\ e^{-ik\xi}$$

satisfies the classical diffusion equation

$$\{ \partial_\tau - D_{a,b}\big(\partial_\xi^2 + \beta\ V'_{a,b}(\xi)\ \partial_\xi + \beta\ V''_{a,b}(\xi)\big) \}\ \rho_{a,b}\ (\xi\ \tau)\ =\ 0$$

in the effective potential $V_{a,b}(\xi)\ =\ \Omega^2_{a,b}\ \xi^2/2$ where

$$\Omega^2_{a,b} = 2\ \{\ \beta(\omega a^2 + \omega^{-1}b^2)\}^{-1}\ \tanh(\beta\omega/2).$$

The diffusion constant appearing in this diffusion equation is linked to the parameters λ, ω and β , defining ψ and $\gamma(\tau)$, by the relation $D_{a,b} = \lambda / \Omega^2_{a,b}\beta$.
For both of these examples one knows how to construct, in a minimal and canonical manner, the conservative dynamical systems of which they are a reduced description.
For illustrative purpose, let us indicate how this look for the second of our examples.
Here N is the von Neumann algebra obtained as the closed linear span (in the weak-operator topology) of the Weyl operators $\{ W(f)|\ f \in T = L^2(R,dx)\}$ defined in the GNS representation corresponding to the faithful normal state determined by the functional

$$\phi : f \in T \rightarrow\ < \phi ; W(f) >\ =\ \exp(-\ \Theta\ \|f\|^2/4)\ .$$

The group $\alpha(R)$ of automorphisms of N is then defined by

$$\alpha(\tau)\{W(f)\} =\ W(u_\tau f)\ \text{with}\ (u_\tau f)(x) = \exp(-ix\tau)f(x)\ .$$

Note that this evolution is <u>different</u> from the modular group $\sigma(R)$ associated to ϕ and given by:

$$\sigma(\tau)\{W(f)\} =\ W(u^o_\tau f)\ \text{with}\ (u^o_\tau f)(x) = \exp(-i\omega\tau)f(x)\ .$$

To embed M in N , we single out the vector $f_\lambda \in T$ defined by $f_\lambda(x)=\{\lambda/\pi(\lambda^2+x^2)\}^{\frac{1}{2}}$ and we define the injection $i : M \rightarrow N$ through $i\{W(z)\} = W(zf_\lambda)$. Upon noticing that $N = i\{M\} \times N_R$ where $N_R = \{W(f)\ |\ f \in T , (f,f_\lambda) = 0\}''$, we define $E : N \rightarrow M$ through $: i\{M\} \times N_R \rightarrow\ < \phi ; N_R > M$ for all $M \in M$ and $N_R \in N_R$. It is then a simple exercise to check that $\{M,\psi,\gamma(R^+)\}$ is, through (E,i), a reduced description of $\{N,\phi,\alpha(R)\}$. We have thus proven the consistency of the definitions involved in the right-hand column of our diagramme.
It is interesting to open here a parenthesis, analyzing some further structures which manifest themselves in this example. Let indeed T_τ be the closed linear subspace of T spanned by the elements of the form $u_t f_\lambda$ with $t \leq \tau$, and A_τ be the von Neumann subalgebra of N obtained as $A_\tau = \{ W(f)\ |\ f \in T_\tau \}'' = \alpha(\tau)\{A\}$ with $A = A_o$.

One then verifies that A_τ satisfy the following conditions:

(i) $A_t \subseteq A_\tau$ whenever $t \leq \tau$

(ii) $V_\tau \ A_\tau = N$

(iii) $\bigcap_\tau \ A_\tau = \{ \lambda I \mid \lambda \varepsilon C \}$

(iv) $\sigma(t)\{A_\tau\} = A_\tau$ for all t and τ in R .

These properties generalize to the case of non-commutative dynamical systems the notion of a Kolmogorov flow; accordingly, any $\{N,\phi, \alpha(R),A\}$ satisfying these conditions is refered to as an extended Kolmogorov flow[6], or simply an __EK-system__. These systems enjoy very strong __ergodic properties__; in particular, the algebra of elements of N invariant under the time evolution $\alpha(R)$ is trivial. Moreover, one checks easily for the explicit example constructed above that with A (resp. C) defined as the C*-algebra generated by $\{A_\tau \mid \tau \varepsilon R \}$ (resp. $\{A_\tau^c \mid \tau \varepsilon R \}$ where $A^c = N \bigcap \alpha(\tau)\{A\}'$), one has : (i) A and C are dense in N with respect to the strong operator topology ; (ii) for all $A \ \varepsilon \ A$ and all $C \varepsilon C$:

$$\lim_{|\tau| \to \infty} \ || \left[A \ , \ \alpha(\tau)\{C\} \right] || = 0 .$$

A wide class of "weakly reversible" EK-systems enjoying this property has been identified. This __asymptotic abelianness__ of the time evolution $\alpha(R)$ responsible for the approach to equilibrium has in turn very strong spectral consequences.

Two questions now remain. The first is how to link the right- and left-hand columns of our diagramme, i.e. how to give a microscopic interpretation of the construction involved in the right-hand column. The second question, to which a tentative answer will be proposed in the last paragraph of this lecture, is to locate where and how stochasticity manifests itself in this picture.

The first question would be trivially answered if it were possible that the infinite KMS systems obtained by the limiting process described in the left-hand column could be simply made to coincide with the EK-systems obtained by the extension process described in the right-hand column of the diagramme. A warning that this is in general not possible without the introduction of further physical ingredients is provided by the fact that, in the left-hand column, the evolution $\alpha_g(R)$ is the modular group associated to the state ϕ_g , whereas in the right-hand column, the evolution $\alpha(R)$, governing the approach to equilibrium in EK-systems, is definitely distinct from the modular group $\sigma(R)$ associated to the state ϕ .

Some intuitive insight into this phenomenon can be gained from an heuristic consideration of the idealized model, known as the classical Lorentz gas, in which light particles bounce elastically against heavy, recoilless, randomly distributed spherical obstacles of radius σ . Since the collisions are recoilless, the speed $|v|$ of the light particles is constant along the trajectories. The mean free path λ of these particles is given by the relation $\pi \sigma^2 \lambda = \Lambda / N$ where Λ is the total volume available, and N is the number of the obstacles in Λ . As usual, we place ourselves in the situation where $N \to \infty$, with $d = N / \Lambda$ kept fixed. Suppose then that one wishes to consider a limit where $\sigma / \lambda \to 0$. Among the many possibilities to do that, two have retained the attention of physicists. The first consists in keeping λ fixed. To achieve $\sigma / \lambda \to 0$, we must therefore have that the radius σ of the obstacles tend to zero, whereas their density d tends to infinity, while keeping $\sigma^2 d$ fixed. This is the Grad limit, the value of which is well-known in the study of the Boltzmann equation. Another way to achieve the same result is to keep d fixed. The limit $\sigma / \lambda \to 0$ is then obtained by letting simultaneously σ tend to 0 and λ tend to ∞ , while keeping $\sigma^2 \lambda$ fixed. In quantum mechanical situations where $H_g = H_o + g V$, this procedure can be reinterpreted to indicate that one should consider a limiting process where $g \to 0$ and $t \to \infty$ while $g^2 t = \tau$ is kept fixed. This limit has actually been controlled[7] in a model where non-relativistic particles are moving according to the laws of Hamiltonian quantum mechanics on a three-dimensional lattice on which impurities are distributed in such a manner that they act on the particles as a classical, static, translation invariant, gaussian random field. A markovian quantum master equation then results, generating an evolution described by a contraction semi-group. This model of a quantum Lorentz gas thus vindicates the prescription originally proposed by van Hove[8], in which time is rescaled according to an inverse square power of the interaction strength. This <u>van Hove limit</u> thus extracts from the microscopic Hamiltonian evolution $\alpha_g (R)$ the long-time, cumulative effects of the interaction . Since the state ϕ_g is invariant under $\alpha_g (R)$, ϕ_g approaches in this limit a state ϕ_o whose modular group $\sigma(R)$ has henceforth no reason to coincide with the evolution $\alpha(R)$ governing the approach to equilibrium.

The final step of the proposal described by our diagramme is thus to suggest that one should look in general into such time rescaling, based on a physically motivated "<u>choice of a time-scale</u>", to link the left- and right-hand column of the diagramme. Such a procedure is actually involved[9] in the Ford-Kac-Mazur model[10] of an infinite chain of coupled harmonic oscillators. This model does in fact produce the diffusion equation for a quantum particle in an harmonic well which we used as our motivation for the construction of EK-systems; here the algebra of observables of interest is provided by the observables relative to a single oscillator of the chain, whereas the other oscillators form the thermal bath, the algebra of observables of which we denoted by N_R in the description of this example given above.

When the manipulations involved in our proposal can thus actually be implemented, we are in position to give a definite answer to the question of how and where stochasticity does manifest itself. Indeed for EK-systems one can define[6] a quantum generalization of the concept of dynamical entropy (first introduced by Kolmogorov in a classical context) which estimates the information gained by carying out a measurement when identical measurements were repeated at infinitely many regular intervals in the past. A large class of quantum EK-systems has been identified, where this entropy is strictly positive. Moreover, this entropy has been computed[6], and found to be infinite (as is the Kolmogorov entropy of the classical flow of Brownian motion) , for the EK-systems obtained from the quantum diffusion model described above. In that sense, now made precise, the non-equilibrium statistical mechanics of this type of models is very stochastic indeed.

References

1) M. Takesaki, Tomita's Theory of Modular Hilbert Algebra and its Applications, Lecture Notes in Mathematics No.128, Springer, New York, 1970.
2) H. Araki, Commun.math.Phys. $\underline{14}$ (1969) 120-157.
3) G.G. Emch, $\underline{in}$ C*-Algebras and Applications to Physics, H.Araki & R.V.Kadison,eds., Lecture Notes in Mathematics No.650, Springer, New York, 1978.
4) G.G. Emch & J.C. Varilly, Lett.Math.Phys. $\underline{3}$ (1979) 113-116.
5) G.G. Emch, Acta Phys. Austriaca, Suppl. $\underline{XV}$ (1976) 79-131.
6) G.G. Emch, Commun.math.Phys. $\underline{49}$ (1976) 191-215.
7) Ph. Martin & G.G. Emch, Helv.Phys.Acta $\underline{48}$ (1975) 59-78.
8) L. van Hove, Physica $\underline{21}$ (1955) 517-540 ; $\underline{23}$ (1957) 441-480.
9) E.B. Davies, Commun.math.Phys. $\underline{27}$ (1972) 309-325.
10) G.W. Ford, M. Kac & P. Mazur, Journ.Math.Phys. $\underline{6}$ (1965) 504-515.

A Stochastic Picture of Spin[+]

William G. Faris
Department of Mathematics
University of Arizona
Tucson, Arizona 85721 USA

Abstract: Dankel has shown how to incorporate spin into stochastic mechanics. The resulting non-local hidden variable theory gives an appealing picture of spin correlation experiments in which Bell's inequality is violated.

[+]This material is based upon work supported by the National Science Foundation under Grant No. MCS-8002945.

1. Introduction

The quantum theory does not attempt to assign values to observable
quantities apart from those values brought out by the act of measurement. This
may be regarded either as a virtue or as a defect of the theory, but there is no
doubt that it makes it difficult to form a picture of the world on the atomic
scale. The temptation to replace quantum mechanics by a more intuitive
stochastic theory is strong, but such theories are still in the stage of
proposals and speculations rather than established as serious competitors to
quantum mechanics.

Stochastic mechanics [1] should not be looked on as a genuine competitor to
quantum mechanics, but rather as a radical reinterpretation of quantum mechanics
that might help to suggest such a competitor. In stochastic mechanics a solution
of the Schrödinger equation determines a time-dependent Markov process that
governs particle motion. Each particle follows a continuous path in space. The
process is such that the joint probability distribution of the particle positions
at fixed time coincides with that in quantum mechanics. For this reason no
conventional scattering experiment can distinguish the two theories.

There are serious difficulties in constructing genuine competitors to
quantum mechanics. (See [2] for a recent survey.) The purpose of the present
lecture is to review some of these difficulties and to evaluate stochastic
mechanics in their light. We shall see that stochastics mechanics is a non-local
hidden variable theory of a particularly natural form. The advantages of the
stochastic mechanics point of view will be discussed in the conclusion.

The most dramatic illustration of the problems with stochastic theories is
through experiments involving spin correlations. Dankel [3] has derived the
stochastic mechanics of spin and so it is possible to examine spin correlation
experiments in this framework. A recent article [4] has carried this out in
detail. This lecture may be taken as an introduction to these ideas.

2. Bell's inequality

The relation between probability and quantum mechanics is most dramatically
illustrated by Bell's inequality. This inequality is a constraint that must be
satisfied by any purely probabilistic model of discrete spin. The interest of
the inequality is that it is violated in quantum mechanics. Quantum mechanics
has no underlying probability model and Bell's inequality brings this out in so
simple a context that the implications are difficult to evade.

We shall give two derivations of the inequality. The first is in the
language of random variables and expectations and the second is in the language
of events and probabilities. The first derivation is somewhat more sophisticated
but has the advantage that it makes explicit the point at which quantum mechanics

departs from probability.

We begin by recalling the derivation of the Schwarz inequality for the expectation of a product of random variables. Let X and Y be random variables normalized so that $E(X^2) = E(Y^2) = 1$. (Here E denotes expectation.) Then

$$0 \leqslant (X + Y)^2 = X^2 + Y^2 + 2XY. \qquad (2.1)$$

Take expectations, divide by two, and subtract 1 from each side. This gives

$$-1 \leqslant E(XY). \qquad (2.2)$$

The general case of the Schwarz inequality follows by scaling and change of sign. Notice that the inequality has an analog in quantum mechanics if the products are symmetrized.

The Schwarz inequality has an obvious generalization to three random variables. Let X, Y, Z be random variables with $E(X^2) = E(Y^2) = E(Z^2) = 1.$ Then

$$0 \leqslant (X + Y + Z)^2 = X^2 + Y^2 + Z^2 + 2XY + 2YZ + 2ZX. \qquad (2.3)$$

As before we obtain

$$-3/2 \leqslant E(XY) + E(YZ) + E(ZX). \qquad (2.4)$$

Again this has an analog in quantum mechanics.

Bell's inequality requires the further assumption that the random variables X, Y, Z assume only the values ±1, but the conclusion is sharper. The following proof of the inequality is due to Selleri [5]. Note that X + Y + Z is always an odd integer. Thus

$$1 \leqslant (X + Y + Z)^2 = 3 + 2XY + 2YZ + 2ZX. \qquad (2.5)$$

Even before we take expectations we have, for every experimental outcome,

$$-1 \leqslant XY + YZ + ZX. \qquad (2.6)$$

When we take expectations we have Bell's inequality:

$$-1 \leqslant E(XY) + E(YZ) + E(ZX). \qquad (2.7)$$

We shall see in a moment that Bell's inequality is false in quantum mechanics. It is worth pausing a moment to see exactly where the derivation breaks down. The problem comes when we want to use the fact that $X + Y + Z$ is an odd integer. In quantum mechanics, even though X, Y, and Z each can assume only the values ± 1, it is forbidden to talk about their assuming these values simultaneously. As a consequence the spectrum of the sum need not be contained in the sum of the spectra.

Bell's inequality may also be stated in the language of events and probabilities. If A is an event we let $\bar{A}$ be the event that A does not occur. The probability of an event is denoted $P(A)$, so that, for instance, $P(\bar{A}) = 1 - P(A)$. The probability formulation of Bell's inequality is:

$$P(A \ \& \ \bar{B}) + P(B \ \& \ \bar{C}) + P(C \ \& \ \bar{A}) \leqslant 1. \qquad (2.8)$$

The proof is to note that the three events are mutually exclusive. The following picture makes this fact memorable:

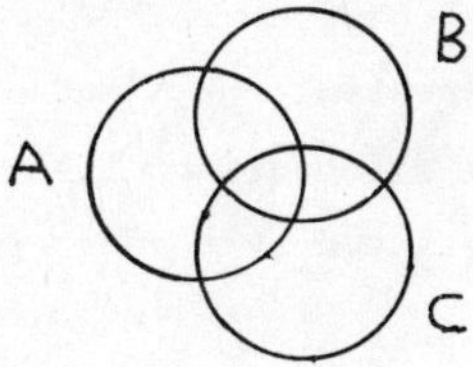

The left hand side is therefore just

$$P(\ (A \ \& \ \bar{B}) \ \text{ or } \ (B \ \& \ \bar{C}) \ \text{ or } \ (C \ \& \ \bar{A}) \ ) \leqslant 1. \qquad (2.9)$$

The two versions of the inequality are really the same. To see this, let A, B, and C be random variables that assume only the values 0 and 1. These may be identified with corresponding events. Let X = 2A-1, Y = 2B-1, and Z = 2C-I. Then Bell's inequality in the form (2.6) implies that

$$A(1 - B) + B(1 - C) + C(1 - A) \leqslant 1 . \qquad (2.10)$$

Take expectations to obtain (2.8).

3. The EPRB experiment

The Einstein–Rosen–Podolsky (EPR) experiment is a thought experiment designed to bring out the peculiar features of quantum mechanics. Bohm proposed a more practical version of the experiment that involves spin. We shall refer to this version as the EPRB experiment. The EPRB experiment has been performed a number of times, but this does not diminish its status as a thought experiment.

For simplicity I shall describe the EPRB experiment as if the spinning particle were a proton. The spin component of a proton in any particular direction may be measured by letting the particle pass through an apparatus involving an inhomogeneous magnetic field appropriate to this direction. The deflection of the particle is taken to be a measure of its spin. The remarkable fact is that the spin component can take on only two values (± 1 in units of h/2), no matter what direction is chosen.

The EPRB experiment is to place two protons in a state of total spin zero. (Their spins are opposite). The protons are then separated in space. Two directions, represented by unit vectors a and b, are chosen. One apparatus measures the spin component $X^{(1)}$ along a of the first proton, and another apparatus measures the spin component $Y^{(2)}$ along b of the second proton. The measurements are repeated many times and the averages of $X^{(1)} Y^{(2)}$ gives a good estimate of the expectation $E(X^{(1)} Y^{(2)})$.

The value for this expectation predicted by quantum mechanics is

$$E(X^{(1)} Y^{(2)}) = -\cos \Theta, \qquad (3.1)$$

where Θ is the angle between the vectors a and b. In particular,
when $\Theta = 0$, the expectation is -1. The spin components of the two particles
along direction b are always opposite. In our notation

$$Y^{(2)} = -Y^{(1)}. \tag{3.2}$$

In order to compare this result with Bell's inequality one needs three
directions. Let a, b, and c be the three directions
and $X^{(1)}$, $Y^{(1)}$, $Z^{(1)}$ and $X^{(2)}$, $Y^{(2)}$, $Z^{(2)}$ be the spin components of the two
protons along the three directions. One can repeat the whole process to
measure $E(X^{(1)} Y^{(2)})$, $E(Y^{(1)} Z^{(2)})$, and $E(Z^{(1)} X^{(2)})$.

If probability theory were applicable we could conclude from Bell's
inequality (2.7) and from (3.2) that

$$E(X^{(1)} Y^{(2)}) + E(Y^{(1)} Z^{(2)}) + E(Z^{(1)} X^{(2)}) \leqslant 1. \tag{3.3}$$

However this is false in quantum mechanics. Consider the configuration of
three direction vectors in which each angle $\Theta = 2\pi/3$. The picture is
symmetric:

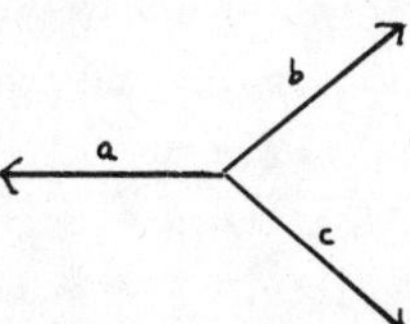

Since $\cos \Theta = -1/2$ the left hand side of (3.3) is $-3 \cos \Theta = 3/2$. Bell's
inequality is violated.

The same point can be made in probability language. Consider two
directions a and b. Let $A^{(1)}$ be the event that the a spin component of
the first proton is $+1$ and let $B^{(2)}$ be the event that the b spin
component of the second proton is $+1$. The frequency of the event
$A^{(1)}$ & $B^{(2)}$ in many repetitions of the experiment is an estimate of the
probability $P(A^{(1)}$ & $B^{(2)})$.

Quantum mechanics gives the value

$$P(A^{(1)} \text{ \& } B^{(2)}) = (1/2)\sin^2(\Theta/2). \qquad (3.4)$$

The probability that the two components are -1 is the same, so the probability that they have the same sign is $\sin^2(\Theta/2)$. The probability of opposite sign is $\cos^2(\Theta/2)$. (This gives the previous value for the correlation $\sin^2(\Theta/2) - \cos^2(\Theta/2) = -\cos\Theta$.) In particular, the spin components along the b axis are always opposite. Thus

$$B^{(2)} = \bar{B}^{(1)}. \qquad (3.5)$$

With three directions a, b, c, and corresponding events $A^{(1)}$, $B^{(1)}$, $C^{(1)}$ and $A^{(2)}$, $B^{(2)}$, $C^{(2)}$, Bell's inequality (2.8) together with (3.5) would give

$$P(A^{(1)} \text{ \& } B^{(2)}) + P(B^{(1)} \text{ \& } C^{(2)}) + P(C^{(1)} \text{ \& } A^{(2)}) < 1. \qquad (3.6)$$

But when $\Theta = 2\pi/3$, $(1/2)\sin^2(\Theta/2) = 3/8$ and the left hand side is $9/8$. The extra $1/8$ is hard to explain away.

The way quantum mechanics evades a contradiction is by arguing that while expressions like $X^{(1)} Y^{(2)}$ and $A^{(1)} \text{ \& } B^{(2)}$ make perfectly good sense, other combinations such as $X^{(1)} Y^{(1)}$ and $A^{(1)} \text{ \& } \bar{B}^{(1)}$ have no direct meaning. But it is just such combinations that enter in Bell's inequality. In the words of Peres [6], "unperformed experiments have no results." In some sense this is a denial of the existence of a real world.

4. Stochastic mechanics

There is another way to explain the EPRB experiment. If we are willing to allow instantaneous action at a distance, then the act of measuring particle 1 can influence the state of particle 2, or vice versa. The definition of the experiment would then depend on the decision of how the magnets used in the measurement are oriented. Changing the orientations would change the probabilities. This would mean that the probability measure in Bell's inequality is not well-defined. There are actually different probability measures in each term, and so the proof would no longer work. Such an explanation is characteristic of a non-local hidden variable theory.

Such theories are not hard to construct, but stochastic mechanics is a particularly natural example. The construction starts with the solution ψ of the time-dependent Schrödinger equation on a Riemannian manifold. The solution is decomposed into a density ρ and phase ϕ, so that

$$\psi = |\psi| \, \exp(i \, \phi) \qquad (4.1)$$

$$\rho = |\psi|^2 .$$

The gradient of the phase is used to define the current velocity vector field

$$v = \hbar \, \nabla \, \phi . \qquad (4.2)$$

Here $\hbar$ is Planck's constant. The logarithmic derivative of the density defines an osmotic velocity vector field

$$u = \hbar \, \nabla|\psi| / |\psi| = (\hbar/2) \, \nabla\rho/\rho . \qquad (4.3)$$

The sum $v + u$ of these two vector fields is the drift term that defines a time-dependent Markov diffusion process. There is also a random noise term proportional to $\sqrt{\hbar}$, so the particle follows a random path in space. Notice that the definition of these two vector fields may be combined in a convenient way using complex notation:

$$v - iu = -i\hbar \, \nabla\psi/\psi . \qquad (4.4)$$

It should be emphasized that only real quantities go into the definition of the process. The drift velocity of the particle at any time is $v + u$ plus a random term that averages to zero.

The process is the natural one to associate with a solution of the

Schrödinger equation. The current velocity term v in the drift ensures that the particle will have the position probability density ρ predicted by quantum mechanics. The osmotic velocity term u permits the particle to wander about and explore the possible positions in a way consistent with this density. It is also possible to start with the process and define v and u as certain conditional expectations, and thus recover the quantum mechanical ψ (up to a time-dependent multiple).

It may help to describe the distinction between the two types of drift velocity in terms of time reversal. The current velocity changes sign under time reversal, much like the velocity in fluid mechanics. The osmotic velocity does not change sign under time reversal. Its role is to damp out fluctuations and attempt a return to equilibrium, and this works the same with either direction of time. The sum v + u thus has mixed time reversal properties, which would be a fatal asymmetry were it not for the noise term. The noise at any moment is independent of the past of the process, but of course it influences and hence is dependent on the future. This introduces a compensating asymmetry that restores symmetry to the description of the process.

The correspondence between the Schrödinger equation and Markov processes is particularly appealing when we start with an eigenfunction ψ of the Schrödinger operator. Then v and u are independent of time and the Markov process is time-independent and stationary with invariant probability density ρ.

5. Spin

Stochastic mechanics works with any number of particles. Of course the drift for each particle depends on the positions of the other particles, so the particle motions can be highly correlated, even over long distances. The same should be true in the stochastic mechanics of spin.

Dankel [3] has shown in detail how to incorporate spin into stochastic mechanics. The idea is to describe each particle by position and orientation. The orientation is given by an element of the covering group of the rotation group. The Schrödinger equation makes sense on this bigger space and the construction goes exactly as before. There are solutions that correspond to spins of any positive integer multiple of $\hbar/2$. The corresponding velocity vector fields now have orientation components that are actually angular velocity vector fields.

The spins in Dankel's theory result when these angular velocities are expressed in terms of the Lie algebra of the rotation group. They are functions of the orientations of the particles. They are not discrete random

variables, but it is possible for one component to be discrete. In particular
the measured component becomes discrete after passing through the measuring
apparatus.

There is no reason to repeat here the full description of spin in
stochastic mechanics. However the physical picture in Dankel's theory is so
attractive that it deserves an illustration. The example we present is a
single particle of spin $\hbar/2$.

The rotation group describes the orientations of a rigid body. Its
covering group carries one extra piece of topological information, and it is
the covering group that is actually used in the theory. This covering group
may be identified with the group $SU(2)$ of special (determinant 1) unitary
2 by 2 matrices. A typical element of the group is

$$U = \begin{pmatrix} \alpha & \beta \\ -\beta^* & \alpha^* \end{pmatrix} , \qquad (5.1)$$

where α and β are complex numbers with

$$\det U = |\alpha|^2 + |\beta|^2 = 1. \qquad (5.2)$$

The elements of the group for which $|\alpha| = 1$ represent a body rotated about the
vertical axis by an angle equal to the phase of α^2. The
quantity $|\alpha|^2 - |\beta|^2$ is the cosine of the angle by which the body is tilted
from the vertical axis. Thus when $|\beta| = 1$ the body is upside down.

The Lie algebra of $SU(2)$ may be identified with the tangent space at
the identity. This space consists of skew-adjoint matrices with trace zero.
Let σ_1, σ_2, σ_3 be the self-adjoint Pauli matrices

$$\sigma_1 = \begin{pmatrix} 0 & 1 \\ 1 & 0 \end{pmatrix}, \quad \sigma_2 = \begin{pmatrix} 0 & -i \\ i & 0 \end{pmatrix}, \quad \sigma_3 = \begin{pmatrix} 1 & 0 \\ 0 & -1 \end{pmatrix} . \quad (5.3)$$

Then $(i/2)\, \sigma_1$, $(i/2)\, \sigma_2$, $(i/3)\, \sigma_3$ form a basis for the Lie algebra.

These elements of the Lie algebra define tangent vector fields on all
of $SU(2)$. The tangent space at U in $SU(2)$ is spanned
by Y_1, Y_2, Y_3, where

$$Y_j = (1/2) U \sigma_j .$$

(5.4)

Let

$$D = \begin{pmatrix} \partial/\partial\alpha & \partial/\partial\beta \\ -\partial/\partial\beta^* & \partial/\partial\alpha^* \end{pmatrix} .$$

(5.5)

The directional derivatives along the directions Y_j are the differential operators

$$L_j = tr (Y_j D) .$$

(5.6)

The gradient of a function f on $SU(2)$ may be calculated by using the basis Y_1, Y_2, Y_3 according to

$$\nabla f = (L_1 f)Y_1 + (L_2 f)Y_2 + (L_3 f)Y_3 .$$

(5.7)

In particular we may compute the current and osmotic angular velocities using (4.4). We shall refer to the (complex) coefficients

$$s_j = -i\hbar (L_j \psi)/\psi$$

(5.8)

as the spin components. Thus the angular velocities are given in terms of the spin by

$$v - iu = s_1 Y_1 + s_2 Y_2 + s_3 Y_3 .$$

(5.9)

We consider the orientation of a single free particle. Its motion will be such that its expected spin is in the vertical direction. The solution ψ of the Schrödinger equation in this case is a function on $SU(2)$

that is constant in time. For spin $\hbar/2$ in the vertical direction we take ψ to be the function

$$\psi = \sqrt{2}\,\alpha . \tag{5.10}$$

This is the proper normalization since

$$\rho = |\psi|^2 = 2|\alpha|^2 \tag{5.11}$$

has the same integral over $SU(2)$ as $|\alpha|^2 + |\beta|^2 = 1$, by symmetry. Thus

$$\int \rho\, d\nu = \int 2|\alpha|^2 d\nu = 1 . \tag{5.12}$$

The spin is calculated by (5.6), (5.8), and (5.10) to be

$$\begin{aligned}
s_1 &= (\hbar/2)\ \beta/\alpha \\
s_2 &= i(\hbar/2)\ \beta/\alpha \\
s_3 &= \hbar/2 \qquad\qquad .
\end{aligned} \tag{5.13}$$

The Y_3 component of the spin is a constant $\hbar/2$, so the Y_3 component of the current angular velocity is $\hbar/2$ and the Y_3 component of the osmotic angular velocity is zero. The total spin is

$$s^2 = |s_1|^2 + |s_2|^2 + |s_3|^2 = |v|^2 + |u|^2 = (\hbar^2/4)\,[1 + 2|\beta|^2/|\alpha|^2]. \tag{5.14}$$

From these expressions we see that when the body is oriented up ($\beta = 0$) the tendency is simply to rotate with constant angular momentum $\hbar/2$. However when the body wanders into a nearly upside down position ($\alpha = 0$), the restoring forces become intense and the body is soon righted. The resulting equilibrium favors a nearly upright position, as shown in (5.11).

The expected spins in equilibrium are

$$E(s_1) = \hbar \int \beta \, \alpha^* d\nu = 0$$
$$E(s_2) = i\hbar \int \beta \, \alpha^* d\nu = 0 \qquad (5.15)$$
$$E(s_3) = \hbar/2 \; .$$

The expected current angular velocity is vertical and the expected osmotic angular velocity is zero.

The expected value of $|s_j|^2$, $j = 1,\ 2$, is given by

$$E(|s_j|^2) = (\hbar^2/2) \int |\beta|^2 \, d\nu = \hbar^2/4 \; . \qquad (5.16)$$

Thus the expected total spin is

$$E(s^2) = 3\hbar^2/4, \qquad (5.17)$$

in agreement with quantum mechanics.

6. Spin measurement

The description of the EPRB experiment in stochastic mechanics involves two moving rigid bodies. Initially they are governed by a process in which the spins of the two particles are always opposite. The particles move far apart and pass through suitably arranged magnetic fields. Then each particle follows one of two possible paths, with probability 1/2 for either one. The joint probabilities are given by the same expression (3.4) as in quantum mechanics. (However, the conditional probability of a particle following a particular path given its orientation depends on the spin, and it is in this sense that the spin is a hidden variable.) After the particles have passed through this measuring apparatus the process looks like a mixture of single particle processes of the type described above. Furthermore, the expected spin is along the measurement direction, and the spin component in this direction is discrete.

Bell's inequality fails at all times. Before the measurement the spins are not even discrete random variables. After the measurement the spin distributions depend on the spin components one has decided to measure. The

discrete spin components have come into being as a result of the
measurement. Stochastic mechanics gives a perfectly consistent probabilistic
description of the EPRB experiment.

7. Conclusions

There are two ways to look at stochastic mechanics: as a way of
picturing quantum mechanics and as a proposal for an alternate description of
the world. Both points of view have something to be said for them.

There is no doubt that one can associate Markov processes with solutions
of the Schrödinger equation as a mathematical device. This allows a picture
of quantum mechanics in terms of particle motion. One can imagine making a
film in which a number of sample paths were shown. The particle would move
respecting the quantum mechanical position probabilities. The stochastic
features of the process would ensure that even a single sample path would give
an idea of the possible position configurations allowed in quantum mechanics.

The film project would also work with several particles. One could see
the correlations between particles and the effects of the exclusion
principle. The incorporation of spin would also be possible; one would only
have to turn the particles into video game creatures whose orientations can be
followed by the eye. It could be arranged to have some of these creatures
gobble up others to illustrate principles of quantum field theory.

It is irrelevant to object that this gives a false picture of quantum
mechanics. If quantum mechanics is true, then there is no conceivable picture
of the real world, so every picture must be misleading. This does not mean
that we must do without pictures. Pictures are necessary for public
understanding of modern physical science. If there is no real world, then
there is only more freedom to imagine it.

The other possibility is that quantum mechanics is false and there is an
objective description of the world. The EPRB experiment seems to require
instantaneous action at a distance, at least on the level of hidden
variables. But action at a distance is no stranger than the denial of an
objective world. A decade ago the violation of special relativity, even on
the level of hidden variables, would have been upsetting to many physicists.
However now we may see it as just another broken symmetry.

References

1. E. Nelson, _Dynamical Theories of Brownian Motion_, Princeton University Press, Princeton, N. J., 1967.

2. F. Selleri and G. Tarozzi, "Quantum mechanics, reality, and separability," _Revista del Nuovo Cimento_ 4, no. 2 (1981), 1-53.

3. T. G. Dankel, "Mechanics on manifolds and the incorporation of spin into Nelson's stochastic mechanics," _Arch. Rational Mech. Anal._ 37 (1970), 192-221.

4. W. G. Faris, "Spin correlation in stochastic mechanics," _Foundations of Physics_ 12 (1982), 1-26.

5. F. Selleri, "On the consequences of Einstein locality," _Foundations of Physics_ 8 (1978), 103-116.

6. A. Peres, "Unperformed experiments have no results," _Am. J. Phys._ 46 (1978), 745-747.

ON THE STATISTICAL MECHANICS OF SURFACES[§]

J. Fröhlich[1,*], C.E. Pfister[2] and T. Spencer[3,**]

[1] I.H.E.S., 91440 Bures-sur-Yvette, France

[2] Mathématiques, Ecole Polytechnique Fédérale,
CH-1007 Lausanne, Suisse

[3] C.I.M.S., N.Y.U., New York, N.Y. 10012, U.S.A.

0. Introduction.

The purpose of these notes is to report some recent results and speculations concerning the statistical mechanics of surfaces or interfaces and to try to convey an impression of the beauty of and interest in a mathematical theory of random surfaces.

Random surfaces and their statistical mechanics appear in many different physical contexts among which one might mention :

(i) Crystal growth and the statistical mechanics of crystal surfaces in a solution.

(ii) Interfaces between different phases of a physical system; (e.g. Bloch walls, or the liquid-vapor interface in water, etc.)

(iii) Gauge theories; (the high temperature expansion expresses a lattice gauge theory as a theory of random surfaces; the low temperature expansion expresses a four-dimensional lattice gauge theory with discrete gauge group as a theory of two-dimensional vortex sheets.)

(iv) Dual resonance models; (string theory in its Euclidean formulation can be formulated as a theory of random surfaces. It may be viewed as a generalization of Brownian motion, from random paths to random surfaces.)

Needless to say that random surfaces appear in other problems of condensed

§ Lecture given by J.F.
* Address after August 1982 : Theoretical Physics, ETH, CH-8049 Zürich, Switzerland.
** Work supported in part by N.S.F. Grant DMR 81 00 417.

matter physics, in geophysics (surfaces of mountains),... .

In the following, we briefly review some rigorous results concerning random surfaces and interfaces. We discuss :

1. The interface in the three-dimensional Ising- and rotator model [1].

2. The solid-on-solid model [2].

3. Self-avoiding random surfaces and string theories [3].

4. Lattice gauge theories [4].

We refer to the literature quoted here and in the following for information concerning the physical situations described by these models, detailed statements of results and proofs.

We have profitted from collaboration and/or discussions with M. Aizenman, J. Bricmont, J.L. Lebowitz and E. Seiler.

1. The interface in the Ising- and rotator model.

We start by recalling the definition of the Ising- and the rotator (classical XY-) model on a simple, (hyper) cubic lattice $\mathbb{Z}^d$, $d \geq 3$: With each site $j \in \mathbb{Z}^d$ we associate a spin S_j , and

1) $S_j = \pm 1$ in the Ising model;

2) $S_j \in S^1$, in the rotator model, i.e. S_j can be parametrized by an angle $\theta_j \in [0, 2\pi)$.

We use the convention

$$
S_j = \pm \quad \Leftrightarrow \quad
\begin{cases}
S_j = \pm 1 \text{ , in the Ising model} \\
\\
\theta_j = 0, \pi \text{ , in the rotator model.}
\end{cases}
$$

Let Λ be some finite sublattice of $\mathbb{Z}^d$, e.g.

$$
\Lambda = \Lambda_{L,T} = \{ j \in \mathbb{Z}^d : -T \leq j_1 \leq T, -L \leq j_\alpha \leq L , \alpha = 2, \ldots, d \} \quad .
$$

The energy of a configuration $S_\Lambda = \{S_j\}_{j \in \Lambda}$ of spins in Λ , given a fixed configuration $S_{\Lambda^c} = \{S_j\}_{j \in \Lambda^c}$ of spins in the complement, Λ^c , of Λ , is given by the Hamilton function

$$H_\Lambda = - \sum_{(ij)\subset\Lambda} S_i \cdot S_j + W(S_\Lambda | S_{\Lambda^c}) \; , \tag{1}$$

where W is a boundary term defined by

$$W(S_\Lambda | S_{\Lambda^c}) = - \sum_{\substack{(ij) \\ i\in\Lambda, j\in\Lambda^c}} S_i \cdot S_j \; , \tag{2}$$

and (ij) indicates that i and j are nearest neighbors. The equilibrium state for a spin system in Λ with Hamilton function H_Λ given by (1), (2) and some fixed b.c. S_{Λ^c}, at inverse temperature β, is defined to be

$$d\mu_\beta(S_\Lambda | S_{\Lambda^c}) = Z_{\beta,\Lambda}(S_{\Lambda^c})^{-1} e^{-\beta H_\Lambda(S_\Lambda | S_{\Lambda^c})} \prod_{j\in\Lambda} dS_j \; , \tag{3}$$

where dS is the counting measure on $\{-1,1\}$, in the Ising model, and the Lebesgue measure on S^1, in the rotator model. Furthermore

$$Z_{\beta,\Lambda}(S_{\Lambda^c}) = \int e^{-\beta H_\Lambda(S_\Lambda | S_{\Lambda^c})} \prod_{j\in\Lambda} dS_j$$

is the partition function. We shall impose the following kinds of boundary conditions:

(+b.c.) $S_j = +$, for all $j \in \Lambda^c$

($\pm$b.c.) $S_j = +$, for all $j \in \Lambda^c$ with $j_1 \geq 0$

            $S_j = -$, for all $j \in \Lambda^c$ with $j_1 < 0$

(step b.c.) $S_j = +$ if $j_1 > 0$, or $j_1 = 0$ and $j_2 \geq 0$

            $S_j = -$, otherwise.

Let $A(S)$ be some continuous function depending only on finitely many S_j. The equilibrium expection of A in the thermodynamic limit, for X b.c. $(X = +,\pm, \text{step})$, is given by

$$\langle A \rangle_{\beta,X} = \lim_{L,T\to\infty} \int A(S) d\mu_\beta(S_{\Lambda_{L,T}} | X) \; . \tag{4}$$

The limit is known to exist for $+$ b.c., but some limit can always be obtained by passing to subsequences. We define $\langle (\cdot) \rangle_{\beta,-}$ by

$$\langle A(S) \rangle_{\beta,-} = \langle A(-S) \rangle_{\beta,+} \ .$$

The spontaneous magnetization, $M(\beta)$, is given by

$$M(\beta) = \langle S_j \rangle_{\beta,+} \ . \tag{5}$$

It is known that for $d \geq 3$

$$M(\beta) \neq 0 \ , \text{ for large enough } \beta \ .$$

In two dimensions this remains true in the Ising model, but the two-dimensional rotator model does not exhibit spontaneous magnetization, except at $\beta = \infty$, (a well-known theorem due to Mermin.) However, this model shows a Kosterlitz-Thouless transition, from a high temperature phase with exponentially decaying correlations to a low temperature phase where correlations have only power law decay. This has been rigorously established in [5]. This transition appears to be closely related to the roughening transition in the three-dimensional Ising model, (see Sect. 2). It is essentially the same phenomenon as the roughening transition in the solid-on-solid model described in the next section.

Next, we define thermodynamic functions :

(a) The <u>free energy</u>

$$f(\beta) = \lim_{L,T\to\infty} (TL^{d-1})^{-1} \log Z_{\beta,\Lambda_{L,T}}(S_{\Lambda^c}) \tag{6}$$

which is independent of the b.c. that are imposed.

(b) The <u>surface tension</u> (or surface free energy)

$$\tau(\beta) \equiv f^{(1)}(\beta) \equiv \lim_{L\to\infty} \lim_{T\to\infty} L^{1-d} \log \frac{Z_{\beta,\Lambda_{L,T}}^{(+)}}{Z_{\beta,\Lambda_{L,T}}^{(\pm)}} \tag{7}$$

(c) The <u>step free energy</u>

$$\sigma(\beta) \equiv f^{(2)}(\beta) \equiv \lim_{L\to\infty} \lim_{T\to\infty} L^{2-d} \log \frac{Z_{\beta,\Lambda_{L,T}}^{(\pm)}}{Z_{\beta,\Lambda_{L,T}}^{(step)}} \tag{8}$$

Similarly, $f^{(k)}(\beta)$, $k = 3,\ldots,d-1$, can be defined.

<u>Theorem 1.</u>

1) [6] <u>In the</u> $d \geq 2$ <u>dimensional Ising model</u>

$$\tau(\beta) > 0 \Leftrightarrow M(\beta) > 0 \ , \ (\text{i.e.} \ \ \beta > \beta_c \ .)$$

<u>(See also</u> Sect. 4.)

2) [7] <u>In the rotator model</u>

$$\tau(\beta) = 0 \ , \ \underline{\text{for all}} \ \ \beta < \infty \ \ \underline{\text{and arbitrary}} \ \ d \ .$$

3) [7] <u>If</u> $\tau(\beta) = 0$ <u>then there is no interface,</u> <u>in the sense that</u>

$$\langle(\cdot)\rangle_{\beta,\pm} = 1/2 \ \langle(\cdot)\rangle_{\beta,+} + 1/2 \ \langle(\cdot)\rangle_{\beta,-} \ ,$$

(<u>provided</u> $\langle(\cdot)\rangle_{\beta,\pm}$ <u>is invariant under translations in directions perpendi-</u>
<u>cular to the 1-direction.</u>)

We define the roughening temperature $T_R = \beta_R^{-1}$ as the smallest temperature
for which $\langle(\cdot)\rangle_{\beta,\pm} = 1/2 \ \langle(\cdot)\rangle_{\beta,+} + 1/2 \ \langle(\cdot)\rangle_{\beta,-}$. It was first proven by Dobrushin
[8] (see also [9,10] for simplifications and extentions) that β_R is finite for
the Ising model in three or more dimensions, i.e. the Ising model in $d \geq 3$ dimensions has non-translation-invariant equilibrium states at sufficiently low temperatures. In two dimensions, all equilibrium states of the Ising model are convex com-
binations of $\langle(\cdot)\rangle_{\beta,+}$ and $\langle(\cdot)\rangle_{\beta,-}$, hence translation-invariant. This result is
due to Aizenman [11].

It is conjectured that

$$\beta_R > \beta_c \ , \ \text{in} \ \ d = 3 \ ;$$
$$\beta_R = \beta_c \ , \ \text{in} \ \ d \geq 4 \ . \tag{9}$$

A theoretical argument for the truth of this conjecture is described in the next
section.

Next, we introduce some order parameters for the roughening transition in the
Ising model. (Our discussion serves mainly as a preparation for the considerations
in Sect. 4.) A convenient order parameter to locate the interface is

$$D(\beta,n) = \langle S_{(n,\vec{0})} \cdot S_{(-n-1,\vec{0})}\rangle_{\beta,\pm} \ , \tag{10}$$

where $\vec{j} = (j_2,\ldots,j_d)$, and

$$D(\beta) = \lim_{n\to\infty} D(\beta,n) \ . \tag{11}$$

For $\beta < \beta_R$,

$$D(\beta) = M(\beta)^2 \geq 0$$

Moreover,

$$D(\beta) \to -1 \ , \quad \text{as} \quad \beta \to \infty \ ,$$

in dimension $d \geq 3$. We conjecture that for all $\beta > \beta_R$

$$D(\beta) < M(\beta)^2 \ ,$$

in fact, that $D(\beta,n)$ is negative, for n large enough. (We are not aware of any proof of this very plausible conjecture.) We define

$$\beta_R' = \inf\{\beta : D(\beta) < M(\beta)^2\} \geq \beta_R \tag{12}$$

Another convenient "order parameter" for the roughening transition might be the step free energy, $\sigma(\beta)$, defined in (c) above. For $\beta < \beta_c$,

$$\sigma(\beta) = 0 \ .$$

In dimension $d \geq 3$

$$\lim_{\beta\to\infty} \beta^{-1}\sigma(\beta) = -2 \ .$$

It is conjectured that

$$\sigma(\beta) = 0 \ , \quad \text{for } \beta \leq \beta_R$$

$$\sigma(\beta) > 0 \ , \quad \text{for } \beta > \beta_R \ . \tag{13}$$

We set

$$\beta_R'' = \inf\{\beta : \sigma(\beta) > 0\} \tag{14}$$

<u>Theorem 2.</u>

$$\beta_c(d=3) \leq \beta_R, \beta_R', \beta_R''(d=3) \leq \beta_c(d=2) \ .$$

For β_R and β_R' this result was proven by van Bejeren [12], for β_R'' it has recently been established in [13]. The expected result would be

$$\beta_R = \beta_R' = \beta_R'' \quad , \quad \text{for all} \quad d$$

$$\beta_c(d=3) < \beta_R(d=3) < \beta_c(d=2) \tag{15}$$

$$\beta_c = \beta_R \quad , \quad d \geq 4 \ .$$

Next, we sketch a suitable notion of an <u>interface</u> in an Ising model : We notice that if $\pm$ b.c. are imposed at the boundary of $\Lambda_{L,T}$ then there is a (Peierls) contour $\overset{\circ}{\Sigma}_L$ decomposing $\Lambda_{L,T}$ into two disjoint subsets, $\Lambda_{L,T}^+$ and $\Lambda_{L,T}^-$, such that $(T,\vec{0}) \in \Lambda_{L,T}^+$, $S_j = +$ if $j \in \Lambda_{L,T}^+$ borders $\overset{\circ}{\Sigma}_L$, $S_j = -$ if $j \in \Lambda_{L,T}^-$ borders Σ_L and

$$\partial\overset{\circ}{\Sigma}_L = \partial\Lambda_{L,T} \cap \{x : x_1 = -1/2\} \ .$$

We define the interface Σ_L to consist of the union of $\overset{\circ}{\Sigma}_L$ and all closed contours which are $*$ connected with $\overset{\circ}{\Sigma}_L$.

In two dimensions, the interface, Σ_L , has finite width, uniformly in L , and has long wave length fluctuations on a scale of $\sqrt{L}$, provided the temperature is small enough, $(\beta > \beta_c)$. This result is due to Gallavotti [14].

In $d \geq 3$ dimensions the conjectured behaviour of the interface is as follows:

For $d = 3$ and $\beta > \beta_R$, or for $d \geq 4$ and all $\beta > \beta_c$, the interface Σ_L is well localized near $\{x : x_1 = -1/2\}$ and very rigid and thin, uniformly in L ; $D(\beta,n)$ is negative, for n large enough - presumably for all $n \geq 1$. (For rigorous results valid at large β see [8,9,10,12].)

For $d = 3$ and $\beta < \beta_R$ (but β close to β_R) the interface Σ_L still has finite width but fluctuates on a <u>logarithmic scale</u>. (A theoretical argument supporting this claim is reviewed in the next section.) As β is decreased, some of the following phenomena may occur : Interlacing chains of $-$ spins will start to percolate into Λ^+ and, as a consequence, the interface grows many <u>handles</u>. In addition, short wave length fluctuations may cause a lot of wrinkles on the interface. Hence the interface fattens. When β approaches β_c , the interface might approach some self-similar surface, and below β_c it will become "space-filling". Unfortunately, there are no rigorous results, except for very large β . (See also [13] for some speculations.)

We now turn to the discussion of the <u>rotator model</u> : By Theorem 1, parts 2) and 3), <u>the rotator model never exhibits an interface</u>.

176

Does this mean that all equilibrium states are translation-invariant ? Before attempting to answer this question we quote a result that characterizes the translation-invariant equilibrium states of the Ising- and the rotator models. For the rotator model, define $\langle(\cdot)\rangle_{\beta,\theta}$ by

$$\langle A(S)\rangle_{\beta,\theta} = \langle A(R(\theta)S)\rangle_{\beta,+} , \tag{16}$$

where $R(\theta)$ rotates each spin S_j through an angle θ . If $M(\beta) = 0$ (i.e. $\beta < \beta_c$) the states $\langle(\cdot)\rangle_{\beta,\theta}$ coincide with $\langle(\cdot)\rangle_{\beta,+}$, for all $\theta \in [0,2\pi)$. Indeed, for $\beta < \beta_c$, $\langle(\cdot)\rangle_{\beta,+}$ is the unique translation-invariant equilibrium state. This result is valid in the Ising- and the rotator model [15].

<u>Theorem 3.</u>

$\beta.^{1)}$ <u>Let</u> $\beta > \beta_c$ <u>be such that the free energy is continuously differentiable at</u> <u>Then</u>

1) [16] <u>In the Ising model, every translation invariant equilibrium state is a convex combination of</u> $\langle(\cdot)\rangle_{\beta,+}$ <u>and</u> $\langle(\cdot)\rangle_{\beta,-}$.

2) [7] <u>In the rotator model, every translation-invariant equilibrium state has a representation</u>

$$\int d\rho(\theta) \langle(\cdot)\rangle_{\beta,\theta} , \tag{17}$$

<u>where</u> ρ <u>is some probability measure.</u>

We now return to the question as to whether all equilibrium states of the rotator model are translation-invariant. The physical reason why there are no interfaces in the rotator model, as remarked, is quite obvious : One might wish to measure the profile of an interface in the rotator model in terms of

$$D_{L,T}(\beta,n) = \int S_{(n,\vec{0})} \cdot S_{(-n-1,\vec{0})} d\mu_\beta(S_{\Lambda_{L,T}} | \pm b.c.) .$$

But

$$\lim_{L,T\to\infty} D_{L,T}(\beta,n) = \langle S_{(n,\vec{0})} \cdot S_{(-n-1,\vec{0})}\rangle_{\beta,+} ,$$

---

1) Since $f(\beta)$ is concave, this condition is satisfied for almost all values of β .

<u>for all</u> β <u>and</u> n ; (see [7] for a precise statement.) Thus, the interface becomes very wide (fat.) Since the model has a continuous symmetry, this is no surprise : In order to fulfil ± b.c., it suffices to turn the spins upside down extremely slowly as one moves from $j_+ = (T,\vec{0})$ down to $j_- = (-T,\vec{0})$. Interfaces (Bloch walls) are <u>not</u> among the "topologically stable" defects of this model[1]

The role of Bloch walls (Peierls contours) or interfaces in the Ising model is really played, in the rotator model, by another type of "topologically stable defects", the <u>vortices</u>. They are characterized by an integer winding number of the spin configuration, and, since the spin takes values in the unit circle, must have co-dimension 2. The easiest way of describing vortex configurations in the rotator model proceeds by applying a <u>duality transformation</u>, i.e. Fourier transformation in the angular variables (see [5,17] , and refs. given there.)

Let

$$r_\beta(\theta) := \exp[\beta\cos\theta] \quad (\text{or} := \sum_{n\in\mathbb{Z}} \exp[-\frac{\beta}{2}(\theta+2\pi n)^2] \quad ,$$

the <u>Villain approximation</u>.) Let $\hat{r}_\beta(n)$ denote the nth Fourier coefficient of r_β . The equilibrium state of the rotator is given by the measure

$$d\mu_\beta(\theta) = Z^{-1} \prod_{(ij)} r_\beta(\theta_i - \theta_j) \prod_j d\theta_j \tag{18}$$

The Fourier coefficients of μ_β are thus given by

$$\hat{\mu}_\beta(n) = Z^{-1} \prod_{(ij)} \hat{r}_\beta(n_{ij}) \prod_j \delta_{(\delta n)_j,0} \quad , \tag{19}$$

where ij is the <u>oriented</u> bond pointing from i to j ,

$$(\delta n)_j \equiv \sum_{b\ni j} n_b \quad , \quad \text{and} \quad n_b \equiv -n_{-b} \quad .$$

The factor $\prod_j \delta_{(\delta n)_j,0}$ arises by integrating the factors $\exp[i\theta_j(\delta n)_j]$ over θ_j , for all j . It imposes the constraint

$$\delta n = 0$$

which is solved (Poincaré's lemma) by

$$n = \delta m \quad ,$$

where $m : p \to m_p \in \mathbb{Z}$ is defined on oriented unit squares (plaquettes), $p \in \mathbb{Z}^d$. We may write

$$m_p \equiv \alpha_c \in \mathbb{Z} \quad ,$$

where c is the oriented $(d-2)$-cell in $(\mathbb{Z}^d)^*$ dual to p .

In the Villain approximation,

$$\hat{r}_\beta(n) = \exp[-\frac{1}{2\beta} n^2] \quad .$$

Applying now the Poisson summation formula, we conclude that the Villain approximation to the rotator model is isomorphic to a model whose equilibrium state is given by

$$d\mu_\beta(\alpha) = \hat{Z}^{-1} \prod_{c \in (\mathbb{Z}^d)^*} (\sum_{\varphi_c \in \mathbb{Z}} e^{i\varphi_c \alpha_c})d\mu_G(\alpha) \quad ,$$

where $d\mu_G$ is a Gaussian measure on the space of "orbits" $[\alpha]$, where $[\alpha]$ is the equivalence class $\{\alpha_c + \sum_{c' \in \partial c} \chi_{c'}\}$, and $\chi : c' \to \chi_{c'} \in \mathbb{R}$ is a function defined on $(d-3)$-cells, c' , of $(\mathbb{Z}^d)^*$. The inverse covariance of $d\mu_G$ is $\beta^{-1}\delta d$. It follows from the (gauge invariance) properties of $d\mu_G$ that all configurations $\varphi = \{\varphi_c \in \mathbb{Z} : c \subset (\mathbb{Z}^d)^*\}$ must satisfy the constraint

$$\delta\varphi = 0 \quad .$$

This shows that the connected components of each configuration φ can be interpreted as closed, $(d-2)$-dimensional vortices with integer winding numbers prescribed by $\{\varphi_c\}$. (Back in the rotator model φ corresponds to vortices in the spin field.) By choosing appropriate, non-translation-invariant boundary conditions one can force an open vortex into the system which extends to the boundary (where it is "closed off" by the b.c.) and plays the role of an interface, Σ , in the Ising model. This vortex might cause a breakdown of translation invariance in the thermodynamic limit. In the next section we sketch theoretical arguments supporting the following

<u>Conjecture</u>. [7]

1) In $d \leq 3$ dimensions, all equilibrium states of the rotator model are translation-invariant and are given by formula (17) of Theorem 3.

2) In $d \geq 5$ dimensions, the rotator model has non-translation-invariant equilibrium states for all $\beta > \beta_c$.

3) In $d = 4$ dimensions there exists an inverse temperature $\beta_R > \beta_c$ such that for $\beta > \beta_R$ there exist non-translation invariant equilibrium states while for $\beta < \beta_R$ all states are of the form (17).

The idea behind this conjecture is that in dimension $d \leq 3$ vortices have dimension 0 or 1 and are therefore unstable against long wave length fluctuations, no matter how large β is. (For $d = 3$, results analogous to the ones of Gallavotti [14] should hold.) For $d = 4$, vortices are two-dimensional. They are therefore likely to be rigid for very large β , but are expected to have logarithmic fluctuations above a roughening temperature; see Sect. 2. Finally, vortices of dimension ≥ 3 are expected to have finite fluctuations, as long as $\beta > \beta_c$; (Sect. 2.)

2. The solid-on-solid model.

In this section we review some recent rigorous results on an approximate, statistical theory of (lattice) surfaces, like the interface in the Ising model, the vortex sheets in the four-dimensional rotator model or the electric flux "world sheets" in a lattice gauge theory. We also show that the same approximation yields an uninteresting theory of one- or three- and higher dimensional objects : One dimensional objects (strings) fluctuate on a scale of $\sqrt{L}$, as expected on the basis of the central limit theorem, while three-dimensional objects ("bags") have uniformly bounded fluctuations. See Theorem 4, below.

The approximation considered in this section involves the following elements :

1) Only surfaces (or strings, or bags) which are graphs of functions are admitted as elements of the statistical ensemble, E .

2) The statistical weight of a surface is a local functional of the surface, e.g. its area.

Specifically, the models which we consider are defined as follows : As our parameter space we choose some finite, rectangular array of sites, Λ , in the lattice $\mathbb{Z}^d$, $d = 1,2,3,\ldots$; (the interesting case is $d = 2$.) Each (hyper-) surface in our statistical ensemble $E \equiv E_\Lambda$ is given by the graph of a function, $\vec{\phi}_\Lambda$, assigning to each site $j \in \Lambda$ an m-tuple of integers, $\vec{\phi}_j = (\phi_j^1,\ldots,\phi_j^m)$ interpreted as the coordinates ("heights") of the (hyper-) surface in the directions transverse to the parameter directions, in such a way that $(j^1,\ldots,j^d, \phi_j^1,\ldots,\phi_j^m)$ are the coordinates of the center of a d-cell in the surface described by ϕ_Λ . We assume, temporarily, that

180

$$\vec{\phi}_j = 0 \; , \; \text{for} \; j \notin \Lambda \; , \; (0 \; \text{b.c.}).$$

The statistical weight, $w_\beta(\vec{\phi}_\Lambda)$, of the surface described by $\vec{\phi}_\Lambda$ is defined by

$$w_\beta(\vec{\phi}_\Lambda) = Z_{\beta,\Lambda}^{-1} e^{-\beta A(\vec{\phi}_\Lambda)} \; , \tag{20}$$

where the "action" $A(\vec{\phi}_\Lambda)$ is given by the total d-dimensional volume of $\vec{\phi}_\Lambda$ (or an approximation thereof), in particular $A(\vec{\phi}_\Lambda)$ is the area of the surface when $d = 2$, and the partition function, $Z_{\beta,\Lambda}$, is chosen such that

$$\sum_{\vec{\phi}_\Lambda} w_\beta(\vec{\phi}_\Lambda) = 1 \; .$$

For $m = 1$,

$$A(\phi_\Lambda) = |\Lambda| + \sum_{(jj')} |\phi_j - \phi_{j'}| \; , \tag{21}$$

where the sum ranges over all nearest neighbor pairs. The factor $\exp(-\beta|\Lambda|)$ can be absorbed in a redefinition of $Z_{\beta,\Lambda}$. The model so obtained is called the <u>solid-on-solid</u> (s-o-s) <u>model</u> [2]. It describes the statistical mechanics of the interface of a limiting, d-<u>dimensional Ising model</u> with $\pm$ b.c. which is obtained by letting the nearest neighbor couplings in the 1-direction tend to ∞ while keeping them fixed in the other directions.

When $m > 1$ it is difficult to analyze the models with actions given by the volume of d-dimensional hypersurfaces in $\mathbb{Z}^{d+m}$; (see Sect. 3.) We shall consider, instead, e.g. the small fluctuation approximation to the volume, given by

$$A(\vec{\phi}_\Lambda) \approx |\Lambda| + \frac{1}{2} \sum_{(jj')} (\vec{\phi}_j - \vec{\phi}_{j'})^2 \; , \tag{22}$$

but approximate actions like

$$|\Lambda| + \sum_{(jj')} |\vec{\phi}_j - \vec{\phi}_{j'}| \tag{23}$$

can be analyzed, too.

We let $\langle(\cdot)\rangle_{\beta,\Lambda}$ denote the expectation defined by (20), with $A(\vec{\phi}_\Lambda)$ as in (21) or (22), (23). [We shall usually think of the s-o-s model corresponding to (21), but most results described in the following remain valid for the models with actions (22), (23), as follows from the analysis in [5].] Let $F(\vec{\phi})$ be an arbitrary continuous, polynomially bounded function of $\{\phi_j - \phi_{j'}\}$, where (jj') are nearest neighbor pairs belonging to some finite subset of $\mathbb{Z}^d$. On this class of functions a thermodynamic limit

$$\langle F\rangle_\beta = \lim_{j\to\infty} \langle F\rangle_{\beta,\Lambda_j} \quad ,$$

$\Lambda_j \nearrow \mathbb{Z}^d$, as $j \to \infty$, can be constructed by a compactness argument; (for the action in (22) it exists by correlation inequalities [18], and, in all cases, it exists for large enough values of β .) We have

<u>Theorem 4.</u>

<u>Consider the models defined in</u> (20) - (23). <u>Then</u>

1) <u>For</u> $d = 1$,

$$\langle(\vec\phi_o-\vec\phi_x)^2\rangle_\beta \sim c_1(\beta)|x| \ , \ \underline{as} \ |x| \to \infty \ .$$

2) <u>For</u> $d = 2$,

$$\langle(\vec\phi_o-\vec\phi_x)^2\rangle_\beta \le c_2(\beta) \quad ,$$

<u>uniformly in</u> x , <u>provided</u> β <u>is large enough. When</u> β <u>is small enough,</u>

$$c_3(\beta) \log|x| \le \langle(\vec\phi_o-\vec\phi_x)^2\rangle_\beta \le c_4(\beta) \log|x| \tag{24}$$

3) <u>For</u> $d \ge 3$,

$$\langle(\vec\phi_o-\vec\phi_x)^2\rangle_\beta \le c_5(\beta) \ ,$$

<u>for all</u> β .

$\square$

<u>Remarks.</u>

(1) Part 1) is a standard consequence of the central limit theorem : The random variables $\vec\phi_j-\vec\phi_{j'}$, where (jj') ranges over the bonds (nearest neighbor pairs) of $\mathbb{Z}$, are independently distributed!

The first half of part 2) follows from a standard low-temperature (Peierls contour) expansion, (as observed in [19].) The deepest result is the lower bound in (24) which was established in [5] by a rather difficult analysis. The upper bound in (24) and part 3) are standard consequences of infrared bounds [20] which are applicable, because the functions $\exp(-\beta|\phi|)$ and $\exp(-\frac{\beta}{2}\phi^2)$ are of positive type. Part 3) has recently been noticed in [13].

(2) The model with $d = 2$, $m = 1$ and $A(\phi_\Lambda)$ given by the r.s. of (22) is dual to the Villain approximation of the two-dimensional rotator model; see Sect. 1, (18), (19), etc. The behaviour described in part 2) of Theorem 4 is, in this case, related to the Kosterlitz-Thouless transition [5].

(3) The transition described in part 2) is a model of the <u>roughening transition</u>: For large β typical lattice surfaces are <u>rigid</u>, i.e. have uniformly bounded fluctuations. When β drops below some critical value, β_R , then typical surfaces are rough and exhibit logarithmic fluctuations. This is the universal behavior of continuum surfaces. A roughening transition occurs only in ensembles of lattice surfaces, because the lattice breaks the continuous group of translations transverse to the surface. At high temperatures, this symmetry is restored, i.e. "enhanced at large distances" in the models considered above, [5,21]; (see also [22].)

Next, we sketch a few ideas in the proofs of parts 2) and 3) of Theorem 4. For simplicity we consider the action (22) with $m = 1$, but the results hold in general [5]. We start with the lower bound in (24).

Let $d\mu_{\beta,\Lambda}(\phi)$ be the Gaussian measure with mean 0 and covariance $(-\beta\Delta_\Lambda)^{-1}$, where Δ_Λ is the finite difference approximation of the Laplacean with 0 Dirichlet data at the boundary of Λ . The equilibrium state of our model can be rewritten as follows :

$$\tilde{Z}_{\beta,\Lambda} dw_\beta(\phi_\Lambda) = \prod_{j\in\Lambda} (\sum_{n\in\mathbb{Z}} \delta(\phi_j - n)) d\mu_{\beta,\Lambda}(\phi)$$

$$= \prod_{j\in\Lambda} (1 + 2 \sum_{q_j=1} \cos(2\pi q_j \phi_j)) d\mu_{\beta,\Lambda}(\phi) . \qquad (25)$$

There are three basic steps in the proofs [5] of the lower bound in (24).

$1°$ The first step is a combinatorial identity : Let ρ denote an arbitrary function on $\mathbb{Z}^2$ of finite support with values in $2\pi\mathbb{Z}$; ρ is called a "charge density". We say that ρ is neutral iff $\sum_j \rho_j = 0$. Let $\phi(\rho) \equiv \sum_j \phi_j \rho_j$. It is proven in [5] by means of an inductive construction extending over all distance scales of 2^n , $n = 0,1,2,\dots$, that, for all $\Lambda \subset \mathbb{Z}^2$,

$$\prod_{j\in\Lambda} (1 + 2 \sum_{q_j=1}^{\infty} \cos(2\pi q_j \phi_j)) = \sum_{N\in F_\Lambda} c_N \prod_{\rho\in N} (1 + K(\rho)\cos\phi(\rho)) , \qquad (26)$$

where F_Λ is a finite family of collections, N , of <u>neutral</u> charge densities, ρ , with the property that two densities, ρ and $\rho' \neq \rho$, in each N have disjoint supports which are so far separated that $\cos \phi(\rho)$ and $\cos \phi(\rho')$ are "almost independent". Furthermore, $c_N > 0$ for all $N \in F_\Lambda$. The constant $K(\rho)$ is an entropy factor which can be bounded by $\exp(cA(\rho))$, where

$$A(\rho) = \sum_{n=o}^{\infty} (A_n(\rho)-1) \; ,$$

and $A_n(\rho)$ is the number of $2^n \times 2^n$ squares needed to cover the support of ρ .

$2°$ The second step consists of a "block spin integration" which allows us to extract "self-energies" of the densities, ρ , providing convergence factors which compensate the constants $K(\rho)$. In the simplest case (namely for the partition function) it results in the following identity : For all $N \in F_\Lambda$,

$$\int \prod_{\rho \in N} (1+K(\rho)\cos \phi(\rho))d\mu_{\beta,\Lambda}(\phi)$$

$$= \int \prod_{\rho \in N} (1+e^{-\beta\widetilde{E}(\rho)}K(\rho)\cos \phi(\overline{\rho}))d\mu_{\beta,\Lambda}(\phi) \; , \tag{27}$$

where $\widetilde{E}(\rho) \approx \text{const.} \sum_{i,j} \rho_i(-\Delta_\Lambda)^{-1}_{ij} \rho_j$ is related to the electrostatic energy of the charge density ρ , and the renormalized charge densities, $\overline{\rho}$, are still neutral but have "magnified" supports.

A key estimate consists in showing that

$$\widetilde{E}(\rho) \geq \varepsilon A(\rho) \; ,$$

for some $\varepsilon > 0$. Thus, for large β ,

$$z(\rho) \equiv e^{-\beta\widetilde{E}(\rho)}K(\rho) \leq e^{-\beta/2 \; \ell n d(\rho)} << 1 \; ,$$

where $d(\rho)$ is the diameter of the support of ρ . Thus, for large β ,

$$dw_{\text{ren.}}(\phi) \equiv \widetilde{Z}^{-1}_{\beta,\Lambda} \sum_{N \in F_\Lambda} c_N \prod_{\rho \in N} (1+z(\rho)\cos \phi(\overline{\rho}))d\mu_{\beta,\Lambda}(\phi) \; , \tag{28}$$

is a <u>positive measure</u> which is, formally, <u>invariant under the continuous symmetry</u>

$$\phi_j \to \phi_j+c \; , \tag{29}$$

where c is an arbitrary real constant; for

$$\cos \phi(\overline{\rho}) = \cos[(\phi+c)(\overline{\rho})] \; ,$$

as $\sum_j c \overline{\rho}_j = 0$, by the neutrality of $\overline{\rho}$; moreover $d\mu_{\beta,\Lambda}(\phi)$ is clearly formally invariant under the symmetry (29), except that the b.c. imposed on $d\mu_{\beta,\Lambda}(\phi)$ break (29).

3° Since, for large β , the measure $dw_{ren.}$ given by (28) is positive and formally invariant under the continuous group of symmetries (29) which, however, is always broken by the b.c. imposed at $\partial\Lambda$, we may apply a Mermin-type argument to conclude that

$$\lim_{\Lambda \nearrow \mathbb{Z}^2} \int dw_{ren}(\phi_\Lambda)\,|\hat{\phi}(k)|^2 \geq \frac{const.}{k^2} \qquad (30)$$

for all $k \neq 0$. Here $\hat{\phi}(k) \equiv (2\pi)^{-2} \sum_{j\in \mathbb{Z}^2} \phi_j e^{ik\cdot j}$. From this one can deduce the lower bound in (24), (by Fourier transformation.)

Next, we comment on the proof of the upper bound in (24) and part 3) of Theorem 4. Part 3) was previously proven in [13]. Here we sketch a slightly different argument which gives a stronger result. For technical convenience we interpret the state $\langle(\cdot)\rangle_\beta$ as a limit of finite volume states $\langle(\cdot)\rangle_{\beta,\Lambda}$ with <u>periodic b.c.</u> . [In order to define the periodic b.c. state, one replaces the counting measure on $\{\phi_j \in \mathbb{Z}\}$ by $\exp(-\epsilon\phi_j^2) \times$ the counting measure. One first takes $\Lambda \nearrow \mathbb{Z}^d$ and subsequently $\epsilon \searrow 0$.] As explained in [20], the upper bound in (24) and part 3) follow from estimates of the form

$$\langle\exp(\epsilon \sum_j h_j(\partial_\alpha\phi)_j)\rangle_{\beta,\Lambda} \leq \exp[c(\beta)\epsilon^2\|h\|_2^2] \quad , \qquad (31)$$

with ϵ small enough, $(\epsilon\|h\|_2^2 \leq \epsilon_o$, for some $\epsilon_o > 0$.) Here ∂_α is the αth component of the finite difference gradient, and h is an arbitrary real-valued function on Λ . Inequality (31) is proven by using a transfer matrix in the α-direction of the lattice. As explained in [20], the transfer matrix formalism reduces the problem to estimating the quadratic form with integral kernel

$$e^{\epsilon h(x-x')} e^{-F_\beta(x-x')} \quad ,$$

where

$$F_\beta(x) = \begin{cases} \beta|x| \ , \text{ or} \\ \\ (\beta/2)x^2 \end{cases}$$

from above in terms of the quadratic form with integral kernel $\exp[-F_\beta(x-x')]$. This is accomplished by using Fourier transformation : For $F_\beta(x) = \beta|x|$, the required bound follows by noticing the inequality

$$|[(k+i\epsilon h)^2+\beta^2]^{-1}| \leq e^{c(\beta)\epsilon^2 h^2} [k^2+\beta]^{-1} \quad ,$$

for $|\epsilon h| < \beta/2$. For $F_\beta(x) = (\beta/2)x^2$, one uses

$$\left|\exp[-(1/2\beta)(k+i\epsilon h)^2]\right| \le e^{\epsilon^2 h^2/2\beta}\,\exp[-(1/2\beta)k^2]\ ,$$

for arbitrary ϵ and h .

From these inequalities (31) follows. The upper bound in (24) and part 3) of Theorem 4 follow from (31) by expanding to second order in ϵ , dividing by ϵ^2 and taking ϵ to 0 .

In dimension $d \ge 3$, we expect that a result much stronger than part 3) of Theorem 4 holds. For __all__ $\beta > 0$,

$$|\langle\vec{\phi}_o\cdot\vec{\phi}_x\rangle_\beta| \le c(\beta)e^{-m(\beta)|x|}\ , \tag{32}$$

for some constants $c(\beta) < \infty$ and $m(\beta) > 0$. This would imply that all correlations between distant pieces of three- or higher dimensional random hypersurfaces decay exponentially. For $d = 3$, $(m = 1)$ and the action $A(\phi_\Lambda)$ given by (22), the bound (32) has recently been proven by Göpfert and Mack in [23].

Next, we review some results on __surface__ (or __step__) __free energies__ in the solid-on-solid models : Let $Z_{\beta,\Lambda}$ be the usual partition function of the model with 0 b.c. defined in (20). Let $Z_{\beta,\Lambda}(\vec{\xi})$, $\vec{\xi} \in \mathbb{Z}^m$, be the partition function of the same model, but with b.c.

$$\left.\begin{array}{l}\vec{\phi}_j = \vec{\xi}\ ,\ \text{for}\ j \notin \Lambda\ ,\ j_1 > 0 \\[2em] \vec{\phi}_j = 0\ ,\ \text{for}\ j \notin \Lambda\ ,\ j_1 < 0\end{array}\right\} \tag{33}$$

We set

$$\tau_d(\vec{\xi};\beta) \equiv \lim_{\Lambda\nearrow\mathbb{Z}^d} \log(Z_{\beta,\Lambda}/Z_{\beta,\Lambda}(\vec{\xi})) \tag{34}$$

We note that $\tau_d(\xi = 1;\beta)$ $(m = 1)$ __is expected to behave qualitatively similarly as__ __the step free energy,__ $\sigma(\beta)$, __of the $(d+1)$-dimensional Ising model__. We consider, for simplicity, only the case $m = 1$, assume that $\xi \ne 0$ and that the action is given by (21) or (22). We then have

__Theorem 5.__

1) $\qquad\quad \tau_1(\xi;\beta) = 0$, $\underline{\text{for all}}\ \xi$.

2) $\qquad\quad \tau_2(\xi;\beta) > 0$, $\underline{\text{for large}}\ \beta$

$$\left.\right\}\ \text{for all}\ \xi \ne 0\ .$$

$\qquad\quad \tau_2(\xi;\beta) = 0$, $\underline{\text{for small}}\ \beta$

3) <u>For the models with action given by</u> (22) <u>and</u> $d \geq 3$,

$$\tau_d(\xi;\beta) > 0 , \underline{\text{for all}} \ \beta > 0 , \xi \neq 0 .$$

<u>Remarks.</u>

Part 1) is trivial. The inequality in part 2) is a consequence of a standard low temperature expansion; e.g. [19]. The equation in part 2) follows from the results of Sects. 6 and 7 of [5]. Part 3) follows from the results of Göpfert and Mack [23] $(d = 3)$ and correlation inequalities [18], $(d = 3 \rightarrow d > 3)$. For results related to the ones in [23] but established earlier see also [24].

We believe that Theorem 5 can be extended to all $m \geq 1$ and all actions (21) – (23) , but not all cases have been worked out.

Finally, some recent results in [5,25] suggest that the continuum limits of the <u>two-dimensional</u> models studied in this section are given by massless Gaussian measures, for $\beta < \beta_R$ and for arbitrary $m = 1,2,3,\ldots$. (This is trivial for $d = 1$.) In the next section, we study random surfaces with more complicated continuum limits.

3. <u>Selfavoiding random surfaces and string theories.</u>

In this section we restrict our discussion to two-dimensional random surfaces embedded in a lattice $\mathbb{Z}^d$ (or embedded in $\mathbb{E}^d$) , $d = 3,4,\ldots$. We propose to consider statistical theories of such surfaces which are geometrically more natural than the ones studied in the last section, but which are seemingly almost as simple as the s-o-s models. Our discussion is sketchy; (some details appear elsewhere.)

The models considered in Sect. 2 have a serious defect : All random surfaces admitted in the ensembles introduced in Sect. 2 are required to be graphs of functions. It is natural to study more general ensembles of lattice surfaces and their continuum limits. If one admits lattice surfaces which may pass through each plaquette (unit square) of $\mathbb{Z}^d$ an arbitrary number of times one cannot construct a mathematically meaningful statistical theory : The number of such surfaces of a given area – i.e. containing a given number of plaquettes counted with multiplicites – grows faster than exponentially in the area; see e.g. [26].

There are at least three ensembles of lattice surfaces which are physically natural :

a) Branched random surfaces arising in plaquette percolation models [27].
(They consist of arbitrary connected arrays of "occupied" plaquettes, each plaquette
in $\mathbb{Z}^d$ being either "empty" or "occupied" once. The weight of such a surface, Σ ,
is given by $p^{A(\Sigma)}$, $0 < p < 1$, $A(\Sigma) = \#$ plaquettes belonging to Σ .)

b) Let γ be a closed curve in $\mathbb{Z}^d$, and let $E_\gamma^{s.a.}$ be the class of all
"self-avoiding" connected lattice surfaces bounded by γ , i.e. surfaces, $\Sigma \subset \mathbb{Z}^d$,
with the property that each link $b \in \Sigma$, $b \notin \gamma$, belongs to precisely two plaquettes of Σ and each $b \in \gamma$ to precisely one plaquette of Σ .

c) Let γ be a closed curve in $\mathbb{Z}^d$, and let E_γ be the class of all connected
surfaces bounded by γ which pass through each plaquette of $\mathbb{Z}^d$ at most once.

The ensemble E_γ described in c) occurs naturally in the study of _interfaces_;
(see Sect. 1), while the ensemble $E_\gamma^{s.a.}$ introduced in b) and the one introduced
in a) (which we denote by $E_{perc.}$) arise in models which are limits of _gauge theo-
ries_; see Sect. 4, and [27,28]. For a somewhat detailed discussion of plaquette-
(and general d-cell) percolation see [27] – we limit our review to a discussion of
$E^{s.a.}$ and E_γ , ensembles which are also studied in connection with string theories.

Let $E_\gamma^{\#} = E_\gamma^{s.a.}$, or E_γ . Each surface $\Sigma \in E_\gamma^{\#}$ is assigned the statistical
weight

$$w_{\beta,\mu}^{\gamma}(\Sigma|\gamma) = Z_{\beta,\mu}(\gamma)^{-1} \cdot \exp[-\beta(A(\Sigma)+\mu\chi(\Sigma))] , \qquad (35)$$

where $\chi(\Sigma)$ counts the number of handles of Σ (Euler characteristic), $\beta > 0$,
$\mu \geq 0$; ($A(\Sigma)$, the _area_ of Σ , counts the number of plaquettes in Σ .)

It is an elementary combinatorial exercise to show that for each d and each
γ

$$Z_{\beta,\mu}(\gamma) = \sum_{\Sigma \in E_\gamma^{\#}} \exp[-\beta(A(\Sigma)+\mu\chi(\Sigma))] < \infty , \qquad (36)$$

for β large enough, while

$$Z_{\beta,\mu}(\gamma) \quad \text{diverges,} \qquad (37)$$

for β small enough.

One can argue that there is some value, β_o , of β which only depends on d
(and possibly on μ), but is independent of γ such that (36) holds for all
$\beta > \beta_o$, while (37) holds for all $\beta < \beta_o$.

As a field theorist one is then interested in the question whether β_o is a critical point, in the sense that there is some divergent correlation length, as $\beta \searrow \beta_o$. This question can be investigated by considering, for example, the "<u>string tension</u>"

$$\alpha(\beta,\mu) \equiv - \lim_{d(\gamma)\to\infty} A(\gamma)^{-1} \log Z_{\beta,\mu}(\gamma) \ , \tag{38}$$

where γ is a square loop in a coordinate plane, $A(\gamma)$ the minimal area enclosed by γ , and $d(\gamma)$ its diameter. In a statistical mechanics context, e.g. in the s-o-s model, the string tension is interpreted as the <u>surface tension</u>. The point β_o is a critical point if

$$\alpha(\beta,\mu) \searrow 0 \ , \quad \text{as} \quad \beta \searrow \beta_o \ . \tag{39}$$

More refined methods to analyze the vicinity of β_o would involve the study of "correlations". We sketch one example; (but see [27] for a more detailed discussion): Let γ,γ' be two non-intersecting loops, and define

$$Z_{\beta,\mu}(\gamma,\gamma') \equiv \sum_{\substack{\Sigma \in E^\#_{\gamma\cup\gamma'} \\ \Sigma \text{ connected}}} \exp[-\beta(A(\Sigma)+\mu\chi(\Sigma))]$$

We define a "glue ball mass"

$$m(\beta,\mu) = \lim_{a\to\infty} - \frac{1}{a} \log Z_{\beta,\mu}(\gamma,\gamma'_a) \ , \tag{40}$$

where γ'_a is the loop that corresponds to a translation of γ' in the direction of a lattice axis by a distance a .

If β_o is a critical point in the sense of (39) one expects that

$$m(\beta,\mu) \searrow 0 \ , \quad \text{as} \quad \beta \searrow \beta_o \ . \tag{41}$$

A more subtle question concerns the behaviour of the dimensionless quantity $m(\beta,\mu)^2/\alpha(\beta,\mu)$, as $\beta \searrow \beta_o$. In [23] a model is studied in which the analogue of this quantity tends to 0 , as $\beta \searrow \beta_o$.

Finally, we want to ask whether the three models introduced in this section exhibit <u>roughening</u>. This question can be studied, for example as follows : We choose a square loop, γ , of diameter $d(\gamma)$ lying in a coordinate plane and define the probability

$$P(d|\gamma) = Z_{\beta,\mu}(\gamma)^{-1} \sum_{\substack{\Sigma \subset E^\#_\gamma \\ d(\Sigma,0)\geq d}} \exp[-\beta(A(\Sigma)+\mu\chi(\Sigma))] \ , \tag{42}$$

where $d(\Sigma,0)$ is the maximal distance of the set $\Sigma \cap \pi_o$ from the origin, and π_o is a (d-2)-dimensional plane perpendicular to the curve γ and containing the origin. We consider

$$P(d) = \lim_{d(\gamma)\to\infty} P(d|\gamma) . \qquad (43)$$

It is easy to show that for β sufficiently large

$$P(d) \leq e^{-c(\beta)d} , \qquad (44)$$

for some constant $c(\beta) > 0$.

The question then is whether there exists some value β_R of β , with

$$\beta_R > \beta_o , \qquad (45)$$

such that for $\beta_o < \beta < \beta_R$

$$P(d) = 1 , \quad \text{for all} d < \infty .$$

On the basis of results concerning the Ising model in three dimensions [6,13] and the s-o-s model [5] (see Sects. 1,2) one would conjecture that the percolation model of branched surfaces exhibits a roughening transition.

(If β_o is a critical point it might also be possible that (44) is valid for all $\beta < \beta_o$, with $c(\beta) \searrow 0$, as $\beta \searrow \beta_o$.)

Once all these preliminary questions (see (39) - (45)) - which actually seem to be very hard ones - are out of the way one can address the most interesting one : <u>What are the continuum limits of these lattice models of random surfaces</u> ? So far, there has not been much theoretical progress on these questions.

Until now, there is only one convincing attempt at constructing a continuum theory of random surfaces, the one by Polyakov [29], clarified in [30]. Presumably, this theory, too, can be obtained as a continuum limit of some "lattice theory" : It is essentially the continuum limit of <u>discrete, imaginary-time quantum gravity</u> of piecewise linear, simplicial (two-dimensional) surfaces; a straightforward, functional integral version of Regge calculus. Polyakov's theory is certainly very fascinating, but (as the remark above indicates) its relation to the physics of interfaces in statistical mechanics or to gauge theory is mysterious. In contrast, the other models, a), b) and c), discussed in this section and the s-o-s model discussed in Sect. 2 are related to gauge theory and to the physics of interfaces, respectively, in a definite way : They are obtained as limiting models, as some para- meter tends to 0 or ∞ . See Sects. 2 and 4. Polyakov's theory is intended to represent a correct mathematical formulation of string theories, (dual resonance models.)

4. <u>Lattice gauge theories.</u>

There are (at least) two ways in which random geometrical objects, such as random loops or random surfaces, arise in the analysis of lattice gauge theories (in the imaginary time description.)

1) <u>Sheets of chromo-electric flux.</u>

Lattice gauge theories can be reformulated as gases of random geometrical objects in different ways : The best known such reformulation results from the <u>strong coupling</u> (high temperature) <u>expansion</u> which represents a lattice gauge theory as a gas of closed random surfaces - closed sheets of chromo-electric flux - which interact by hard core exclusion [31]. (For a somewhat different description of chromo-electric flux sheets, see also [32].)

2) <u>Defect gas description of lattice gauge theory</u>

We first consider a pure lattice gauge with a <u>discrete</u> gauge group on a d-dimensional lattice (or a Higgs theory with a non-trivial, discrete unbroken sub-group.) In such a theory, gauge field configurations can be characterized in terms of frustrated plaquettes, i.e. unit squares, where the curvature is non-vanishing. As a consequence of an integral form of the Bianchi identities, frustrated plaquettes form $(d-2)$-dimensional, closed surfaces which one calls (by an abuse of this name) <u>vortices</u>. They are labelled by group elements. The original lattice gauge theory can now be reformulated as a gas of vortices interacting by geometrical constraints. At weak coupling (low temperature) the vortices have small effective activities and form a dilute gas. This observation is the starting point for the low temperature analysis of lattice gauge theories : Vortices play the role of the Peierls contours in the Ising model and can be used to construct an analogue of the Peierls argument (or a contour expansion) which permits one to control the qualitative features of such lattice gauge theories at weak coupling, in three or more dimensions. See [21]. The upshot of this analysis is that gauge theories with discrete gauge groups exhibit deconfining transitions in dimension ≥ 3 .

Clearly, in theories with continuous gauge groups, vortices (as defined above) are not likely to provide us with a useful notion, although vortices of a somewhat different type appear to play an important role in a confinement mechanism in gauge theories with gauge groups containing a non-trivial, <u>discrete</u> center. As an example

of a gauge theory where vortices are not a useful notion we consider the compact
U(1) lattice model (compact QED.) This gauge theory <u>permanently confines</u> electric
charge in two <u>and</u> three dimensions [23], but exhibits a <u>deconfining transition</u> in
<u>four or more</u> dimensions [33,21]. Vortices are not among the "topologically stable
defects" of the U(1) model and cannot be used to explain those facts. (This circumstance is analogous to the one that the interface is unstable in the rotator
model; see Sect. 1.) The topologically stable defects of the U(1) model which are
dilute at weak coupling are its <u>magnetic excitations</u> : Monopoles in three dimensions,
monopole lines (magnetic currents) in four dimensions, etc. In the continuum limit,
such excitations are labelled by first Chern classes of the field configurations at
infinity (identified with $S^2 \times \mathbb{R}^{d-3}$.) Thus they have dimension d-3 and carry an
integer magnetic charge. The corresponding magnetic excitations of the U(1) model
on the <u>lattice</u> can be exhibited by applying a duality transformation (Fourier transformation in the gauge field variables) and a Poisson summation formula, (as explained
at the end of Sect. 1 for the rotator model.) The interactions between different
magnetic excitations have long range. This makes the analysis of these models, at
weak coupling, interesting and mathematically non-trivial; see [21,23,33].

In the U(1) model, confinement breaks down if the magnetic excitations are
bound in <u>finite, neutral clusters</u> which form a <u>dilute gas</u>, thus causing only small
(infrared-irrelevant) corrections to Gaussian "spin wave" theory. This only happens
in four or more dimensions.

In a <u>non-abelian, pure gauge theory</u>, e.g. one with gauge group SU(n) , there
are two kinds of topological excitations, <u>vortices</u>, of co-dimension 2, and <u>instantons</u>,
of co-dimension 4. Vortices are labelled by elements of the center of the gauge
group, instantons by elements of $\pi_3(G)$. One can argue that, in four or more dimensions, it is the statistical mechanics of the instanton gas which determines whether,
at long distances, the theory is in a perturbative or non-perturbative phase. In
<u>four dimensions</u>, it is most likely that the instanton gas is always in a <u>plasma</u>
<u>phase</u>, instantons are <u>not</u> stably bound in neutral clusters, the infrared behaviour
is non-perturbative. However, in <u>five or more dimensions</u>, instantons form closed
surfaces of dimension d-4 , and a simple energy-entropy argument suggests that, at
<u>weak coupling</u>, the effective activity of an instanton decreases <u>exponentially</u> in its
volume (= length for d = 5 ,...). One is thus led to predict that non-abelian models
exhibit a <u>deconfining transition to a perturbative phase at weak coupling</u>, in <u>five</u>
<u>or more</u> dimensions. (In contrast to abelian gauge theories or ones with discrete
gauge group, there are, however, no rigorous results for non-abelian lattice gauge
theories at weak coupling, yet!)

Next, we briefly summarize some recent, rigorous results concerning random
geometrical objects in lattice gauge theories and some limiting models of such theories :

As our lattice we choose $\mathbb{Z}^d$, the gauge group is assumed to be compact and is denoted by G . The gauge field, $g = \{g_{xy}\}$, is a map from oriented pairs of nearest neighbors, xy , in $\mathbb{Z}^d$ to elements, g_{xy} , of G such that

$$g_{yx} = g_{xy}^{-1} \ .$$

Formally

$$g_{xy} = P(e^{\int_x^y A_\mu(\xi)d\xi^\mu}) \ , \text{ for all } xy \ . \tag{46}$$

The Euclidean functional measure (vacuum functional) of a lattice gauge theory is defined by

$$d\mu_\beta(g) = \lim_{\Lambda \nearrow \mathbb{Z}^d} Z_{\beta,\Lambda}^{-1} e^{-\beta A(g_\Lambda)} \prod_{xy \subset \Lambda} dg_{xy} \ , \tag{47}$$

where Λ is a rectangular array of sites, dg_{xy} is the Haar measure on G , for all xy , $A(g_\Lambda)$ is the (Euclidean) lattice action for the model in Λ , $\beta = 1/e^2$ is the inverse square coupling ("inverse temperature"), and $Z_{\beta,\Lambda}$ is the usual partition function (making $d\mu_\beta$ a probability measure.) Given a loop γ in $\mathbb{Z}^d$, we let

$$g_\gamma = \prod_{xy \subset \gamma} {}^{\circlearrowleft} g_{xy} \tag{48}$$

denote the ordered product of gauge fields along γ , (i.e. the holonomy operator associated with γ .) We define the (Wegner-) Wilson loop observable by

$$W_\chi(\gamma) = \chi(g_\gamma) \ , \tag{49}$$

where χ is some (irreducible) character of G .

We shall consider the following examples of lattice actions :

$$(1) \qquad A(g_\Lambda) = - \sum_{p \subset \Lambda} \text{Re } \chi_o(g_{\partial p}) \ , \tag{50}$$

where p ranges over the plaquettes (unit squares) of Λ , ∂p is the oriented boundary of p , and χ_o is a faithful character of G , (e.g. the one of the fundamental representation for $G = SU(n)$.)

$$(2) \qquad A(g_\Lambda) = - \sum_{p \subset \Lambda} \delta_e(g_{\partial p}) \ , \tag{51}$$

where δ_e is the (Kronecker) δ-function on G concentrated at the unit element, e, and G is assumed to be discrete in this example.

Let Σ be a $(d-2)$-dimensional surface in the dual lattice, $(\mathbb{Z}^d)^*$, with boundary $\partial\Sigma$ (a closed, $(d-3)$-dimensional surface.) We define a disorder operator, $D_z(\partial\Sigma)$, where z is an element of the center of G , as follows :

$$D_z(\partial\Sigma) = \exp[-\beta(A(g\cdot z_\Sigma)-A(g))] \quad , \tag{52}$$

where

$$(g\cdot z_\Sigma)_{\partial p} = \begin{cases} g_{\partial p}\cdot z & \text{if } p \text{ is dual to a } d-2 \text{ cell in } \Sigma \text{ ;} \\ \\ g_{\partial p} \, , & \text{otherwise.} \end{cases}$$

In order to analyze the behaviour of chromoelectric flux sheets we study expectation values like

$$< \prod_{j=1}^{n} W_\chi(\gamma_j)D_z(\partial\Sigma)>_\beta \equiv \int \prod_{j=1}^{n} W_\chi(\gamma_j)D_z(\partial\Sigma)d\mu_\beta(g) \, , \tag{53}$$

$n = 1,2,3,\ldots$. In particular, the roughening transition for electric flux sheets can be analyzed in terms of $<W_\chi(\gamma)D_z(\partial\Sigma)>_\beta$.

We also introduce bulk- and surface thermodynamic functions, (see Sect. 1, (6)-(8); Sect. 3, (38) for related definitions) :

(a) The <u>free energy</u> :

$$f(\beta) = \lim_{L,T\to\infty} (TL^{d-1})^{-1}\log Z_{\beta,\Lambda_{L,T}}$$

(b) The <u>string tension</u> :

$$\alpha(\beta,\chi) = \lim_{L\to\infty} -\frac{1}{L^2} \log<W_\chi(\gamma_L)>_\beta \, ,$$

where γ_L is a square loop in the $1-2$ coordinate plane of diameter L . [The string tension corresponds to the surface tension, $\tau(\beta)$, in spin systems. In three-dimensional $\mathbb{Z}_2$ models they are related by a duality transformation.]

We define an expectation

$$<(\cdot)>_\beta^\chi = \lim_{L\to\infty} <(\cdot)W_\chi(\gamma_L)>_\beta/<W_\chi(\gamma_L)>_\beta \, . \tag{54}$$

One can also define the analogue of

(c) The <u>step free energy</u> :

$$\sigma(\beta) = \lim_{L \to \infty} - \frac{1}{L} \log \frac{\langle W_\chi(\gamma_L')\rangle_\beta}{\langle W_\chi(\gamma_L)\rangle_\beta} \quad ,$$

where the loop γ_L' differs from γ_L by a step of height 1 in the middle of two opposite sides. The functions $\alpha(\beta,\chi)$ and $\sigma(\beta)$ serve to describe the thermodynamics of chromo-electric flux. In particular, $\sigma(\beta)$ is of interest in studies of the roughening transition of flux sheets [4] . It is natural to also introduce functions describing the thermodynamics of "magnetic flux" (vortex sheets), or more generally the thermodynamics of the gas of stable ("topological") excitations, like the magnetic excitations in the $U(1)$ model,... . As an example, we define a thermodynamic function for vortex sheets : Let Λ be some rectangular array of sites centered at 0 , and Σ a $(d-2)$-dimensional coordinate plane in $(\mathbb{Z}^d)^*$. We let Ω denote the set of plaquettes on $\partial\Lambda$ which are dual to some $d-2$ cell in Σ . We consider the following boundary conditions on $\partial\Lambda$:

(0 b.c.) $\qquad g_{\partial p} = e$ (the unit element in G) , for all $p \subset \partial\Lambda$;

(twisted b.c.)

$$g_{\partial p} = \begin{cases} e , & p \subset \partial\Lambda , \quad p \notin \Omega \\[2em] z , & p \in \Omega , \quad \text{for some } z \text{ in the center of } G . \end{cases}$$

Let $Z^o_{\beta,\Lambda}$, $\langle(\cdot)\rangle^o_{\beta,\Lambda}$ be the partition function and the expectation with 0 b.c. on $\partial\Lambda$, and $Z^z_{\beta,\Lambda}$, $\langle(\cdot)\rangle^z_{\beta,\Lambda}$ the corresponding quantities with twisted b.c. . We define

(d) The <u>magnetic "string" tension</u>

$$\varphi(\beta,z) = \lim_{L,T \to \infty} - \frac{1}{TL^{d-3}} \log\left(\frac{Z^z_{\beta,\Lambda}}{Z^o_{\beta,\Lambda}}\right) \quad .$$

One will introduce analogous functions associated with other excitations of dimension ≥ 1 , in particular with the stable ones, (like the magnetic current lines in the four-dimensional $U(1)$ model.) Point-like excitations are studied in terms of "topological susceptibilities" and sum rules (like the Stillinger-Lovett sum rule for the gas of magnetic monopoles in the three-dimensional $U(1)$ model.)

These quantities will be studied in more detail elsewhere.

Next, we summarize some recent results.

Theorem 6.

1) [27] Let $G = \mathbb{Z}_n$, let $d\mu_\beta$ be given by eq. (47) with an action $A(g_\Lambda)$ defined as in (51). (This is the n-states Potts lattice gauge theory.) Then there exists an analytic interpolation in n (of thermodynamic functions and correlations) in a neighborhood of the positive real axis with the property that the model corresponding to the limit $n \to 1$ is the plaquette percolation model of branched random surfaces defined in Sect. 3, a), (ensemble $E_{perc.}$)

2) [28] Let $G = SU(n)$, and renormalize the gauge fields such that

$$g_{xy}^{-1} g_{xy} = g_{yx} g_{xy} = n \, \mathbb{1} \ .$$

Then, for β small enough, there exists an analytic interpolation in n with the property that the $n \to 0$ limit yields the model of selfavoiding random surfaces defined in Sect. 3, b) (ensemble $E_\gamma^{s.a.}$) , in particular

$$\lim_{n \to o} n^{-|\gamma|} <W_{\chi_o}(\gamma)>_\beta = Z_{\beta',o}(\gamma) \ ,$$

where $Z_{\beta,\mu}(\gamma)$ is defined in (36).

This result motivates the definition and analysis of the models introduced in Sect. 3.

Next, we discuss some results which are related to the ones in Sect. 1. They are based on the correlation inequalities in [7] which are only known to hold for abelian gauge groups and an action $A(g_\Lambda)$ given by expression (50) , (i.e. the Wilson action.) The analogue of Theorem 3 is the statement that if the free energy $f(\beta)$ is continuously differentiable at some value $\beta = \beta_o$ then there exists only one translation invariant state, $<(\cdot)>_{\beta_o}$, for $\beta = \beta_o$ [7]. Thus non-uniqueness of the vacuum functional in an (abelian) lattice gauge theory only occurs at a first order transition. A result analogous to Theorem 1 is

Theorem 7. [7]

1) If $\alpha(\beta,\chi) = 0$, and $<(\cdot)>_\beta^\chi$ is invariant under translations in the 1-2 plane then

$$<(\cdot)>_\beta^\chi = <(\cdot)>_\beta \ ,$$

(i.e. the electric flux sheet is completely rough.)

2) <u>If</u> $\varphi(\beta,z) = 0$, <u>and</u> ... <u>then</u>

$$<(\cdot)>^z_\beta = <(\cdot)>_\beta \quad ,$$

(i.e. the vortex sheet is rough or "fat".)

3) <u>In the</u> U(1) <u>models</u>,

$$\varphi(\beta,z) = 0 \ , \ \underline{\text{for all}} \ \beta \ ,$$

(i.e. U(1)-vortices are always fat. This is the analogue of the results for the rotator model described in Sect. 1.)

4) [6] <u>In the three-dimensional $\mathbb{Z}_2$ model</u>

$$\alpha(\beta) > 0 \ \Leftrightarrow \ \varphi(\beta) = 0 \ . \qquad\qquad \square$$

Next, we would have to discuss <u>roughening transitions</u> in lattice gauge theories. The electric flux sheet bounded by an (infinitely extended) Wilson loop may, a priori, undergo a roughening transition which does not coincide with a deconfining transition [32,4]. That transition can be described by the following "order parameter" :

$$D(\beta,n) \equiv <D_z(\partial\Sigma)>^\chi_\beta \ , \ z \neq e \ , \tag{55}$$

where Σ is a (d-2)-dimensional, rectangular array of sites with sides of length 2n   which is centered at the origin and is perpendicular to the plane containing the Wilson loop. In the three-dimensional $\mathbb{Z}_2$ model the parameter $D(\beta,n)$ defined in (55) is <u>dual</u> to the parameter $D(\beta,n)$ introduced in Sect. 1, (10). For small β one expects that the phase of $D(\beta,n)$ approaches the value arg z (the phase of the central element z) exponentially fast, as $n \to \infty$. This can presumably be proven by a fairly straightforward extension of the arguments in [8,10]. The behaviour of the function $D(\beta,n)-D(\beta,\infty)$ is a measure for the fluctuations of the infinite flux sheet in directions perpendicular to the plane of the Wilson loop.

The <u>roughening transition</u> is characterized by the circumstance that, for all $\beta > \beta_R$,

$$\arg D(\beta,n) = 0 \ , \ \text{ for all } \ n \ , \tag{56}$$

while, for $\beta < \beta_R$,

$$\lim_{n \to \infty} \arg D(\beta,n) \neq 0 \ . \quad^{1)} \tag{57}$$

(Here $e_R = \sqrt{\beta_R^{-1}}$ is the coupling constant at which the roughening transition occurs.)

It follows from Theorem 7 that, in abelian lattice gauge theories,

$$\alpha(\beta,\chi_o) = 0 \ \Rightarrow \ \arg D(\beta,n) = 0 \ , \ \text{for all} \ \ n \ ,$$

i.e.

$$\beta_c \geq \beta_R \ , \tag{58}$$

where β_c is the point at which the deconfining transition occurs.

It is expected that the roughening transition can also be characterized by the vanishing of the step free energy, $\sigma(\beta)$, i.e.

$$\sigma(\beta) > 0 \ , \quad \text{for} \ \beta < \beta_R \ ,$$
$$\tag{59}$$
$$\sigma(\beta) = 0 \ , \quad \text{for} \ \beta > \beta_R \ .$$

However, there are no rigorous results about roughening transitions known, yet.

Besides chromo-electric flux sheets there can exist other two-dimensional "topological" excitations, like vortex sheets, exhibiting a roughening transition. Such a transition should only occur in a phase characterized by a non-vanishing surface free energy of the excitations in question, (the analogue of the string- or surface tension. Recall that, in the four-dimensional $U(1)$ model , $\varphi(\beta,z) = 0$ implies that $<(\cdot)>_\beta^z = <(\cdot)>_\beta$ is translation invariant!) One expects that in the confinement phase of a (lattice) gauge theory only the string tension is non-vanishing, i.e. only the chromo-electric flux sheet may exhibit a roughening transition, while other two-dimensional defects, e.g. the vortex sheets in four-dimensional theories, are rough or "fat" throughout that phase.

Heuristically, roughening transitions in lattice gauge theories can be described in terms of the models studied in Sect. 2, like the s-o-s model, but there is no rigorous justification of such approximate theories, yet.

---

1) This characterization has been developed in collaboration with E. Seiler.

References.

[1] H.J. Leamy, G.H. Gilmer, K.A. Jackson, in : "Surface Physics of Crystalline Materials", J.M. Blakely (ed.), New York : Academic Press 1976.

H. Müller-Krumbhaar, in : "Crystal Growth and Materials", E. Kaldis, H. Scheel (eds.), Dordrecht : North Holland 1977.

J.D. Weeks, G.H. Gilner and H.J. Leamy, Phys. Rev. Letts. $\underline{31}$, 549 (1973).

H. van Bejeren, Phys. Rev. Letts. 38, 993 (1977).

Refs. 6-10 , below.

[2] G. H. Gilmer and P. Bennema, J. Appl. Phys. $\underline{43}$, 1347 (1972).

[3] Y. Nambu, Phys. Letts. $\underline{80}$ B, 372 (1979); J.-L. Gervais and A. Neveu, Phys. Letts. $\underline{80}$ B, 255 (1979); D. Förster Phys. Letts. $\underline{87}$ B, 87 (1979), T. Eguchi, Phys. Letts. $\underline{87}$ B, 91 (1979); Refs. 29,30,32.

[4] A. Hasenfratz, E. Hasenfratz, and P. Hasenfratz, Nucl. Phys. B $\underline{180}$, 353 (1981).

C. Itzykson, M.E. Peskin and J.-B. Zuber, Phys. Letts. $\underline{95}$ B, 259 (1980).

G. Münster and P. Weisz, Nucl. Phys. B $\underline{180}$, 330 (1981).

Refs. 32 and 22, below.

J. Fröhlich, Physics Rep. $\underline{67}$, 137 (1980).

[5] J. Fröhlich and T. Spencer, Commun. Math. Phys. $\underline{81}$, 527 (1981).

[6] J. Bricmont, J.L. Lebowitz and C.E. Pfister, Ann. N.Y. Acad. Sci. $\underline{337}$, 214 (1980).

J.L. Lebowitz and C.E. Pfister, Phys. Rev. Letts. $\underline{46}$, 1031 (1981).

[7] J. Fröhlich and C.E. Pfister, paper in preparation.

[8] R.L. Dobrushin, Theor. Prob. Appl. $\underline{17}$, 582, (1972) and $\underline{18}$, 252 (1973).

[9] H. van Bejeren, Commun. Math. Phys. $\underline{40}$, 1 (1975).

[10] J. Bricmont, J.L. Lebowitz and C.E. Pfister, J. Stat. Phys. $\underline{26}$, 313 (1981).

[11] M. Aizenman, Commun. math. Phys. $\underline{73}$, 83 (1980).

[12] See ref. 9.

[13] J. Bricmont, J.-R. Fontaine and J.L. Lebowitz, Louvain Preprint UCL-IPT-82-09.

[14] G. Gallavotti, Commun. Math. Phys. $\underline{27}$, 103 (1972).

[15] J.L. Lebowitz and A. Martin-Löf, Commun. Math. Phys. $\underline{25}$, 276 (1972).

[16] J. Bricmont, J.-R. Fontaine and L. Landau, Commun. Math. Phys. $\underline{56}$, 281 (1977).

[17] J.V. José, L.P. Kadanoff, S. Kirkpatrick and D.R. Nelson, Phys. Rev. B $\underline{16}$, 1217 (1977).

[18] J. Fröhlich and Y.M. Park, Commun. Math. Phys. $\underline{59}$, 235 (1978).

[19] R. Brandenberger and C.E. Wayne, Harvard Preprint 1981, to appear in J. Stat. Phys.

[20] J. Fröhlich, B. Simon and T. Spencer, Commun. Math. Phys. $\underline{50}$, 79 (1976).

[21] J. Fröhlich and T. Spencer, Commun. Math. Phys. $\underline{83}$, 411 (1982).

[22] M. Lüscher, DESY Preprint 80/81, Sept. 1980.

[23] M. Göpfert and G. Mack, Commun. Math. Phys. $\underline{82}$, 545 (1982).

[24] D. Brydges and P. Federbush, Commun. Math. Phys. $\underline{73}$, 197 (1980).

[25] K. Gawedzki and A. Kupiainen, Commun. Math. Phys. $\underline{82}$, 407 (1981), $\underline{83}$, 469 (1982) and refs. given there; IHES Preprints 1982.

J. Magnen and R. Sénéor, Ecole Polytechnique Preprint 1982.

[26] D. Weingarten, Phys. Letts. $\underline{90}$ B, 280 (1980).

[27] M. Aizenman and J. Fröhlich, unpublished (1980), and in preparation.

[28] A. Maritan and C. Omero, CERN Preprint, TH. 3182 - CERN, 1981.

[29] A.M. Polyakov, Phys. Letts. $\underline{103}$ B, 207 (1981).

[30] B. Durhuus, P. Olesen and J.L. Petersen, Nordita Preprint, NBI-HE-81-47, 1981.

[31] K. Osterwalder and E. Seiler, Ann. Phys. (NY) $\underline{110}$, 440 (1978).

E. Seiler, Lecture Notes in Physics, Springer-Verlag, to appear, 1982.

[32] B. Durhuus and J. Fröhlich, Commun. Math. Phys. $\underline{75}$, 103 (1980).

[33] A. Guth, Phys. Rev. D $\underline{21}$, 2291 (1980).

A Note on Irreducibility and Ergodicity
of Symmetric Markov Processes

M. Fukushima

College of General Education

Osaka University

Toyonaka, Osaka, Japan

§1 Ergodicity

Let $(X, \mathbf{B}(X), m)$ be a σ-finite measure space and $\{p_t, t > 0\}$ be a transition function over $(X, B(X))$: $p_t = p_t(x,dy)$ is a substochastic kernel on $(X, B(X))$ satisfying $p_t p_s = p_{t+s}$ and $p_0(x,\cdot) = \delta_{\{x\}}(\cdot)$. We call $\{p_t\}$ __conservative__ if $p_t(x,X) = 1$ and m-__symmetric__ if $\int_{B_1} p_t(x,B_2)m(dx) = \int_{B_2} p_t(x,B_1)m(dx)$ for any $B_1, B_2 \in B(X)$. m is said to be __stationary__ with respect to $\{p_t\}$ if $\int_X p_t(x, B)m(dx) = m(B), B \in B(X)$. If $\{p_t\}$ is conservative and m-symmetric, then m is stationary with respect to $\{p_t\}$.

Given a conservative transition function $\{p_t\}$ on $(X,B(X))$, we can construct an associated Markov process $\{\Omega, B^0, X_t, \theta_t, P_x\}$ by
$$\Omega = X^{[0,\infty)}, \quad X_t(\omega) = \omega_t, \quad \omega \in \Omega, \quad B_t^0 = \sigma\{X_s; s \le t\}, \quad B^0 = \bigvee_{t \ge 0} B_t^0,$$
$(\theta_t \omega)_s = \omega_{t+s}$, and, for $0 \le t_1 < t_2 < \cdots \le t_n$, $E_1, \cdots, E_n \in B(X)$,
$$P_x(X_{t_1} \in E_1, \cdots, X_{t_n} \in E_n) = \int_{E_1 \times \cdots \times E_n} p_{t_1}(x,dx_1) p_{t_2-t_1}(x_1,dx_2)$$
$$\cdots p_{t_n-t_{n-1}}(x_{n-1},dx_n).$$
We let $P(\Lambda) = \int_X P_x(\Lambda)m(dx), \Lambda \in B^0$. P is not necessarily a probability measure but a σ-finite measure on (Ω, B^0).

__Lemma 1.__ Suppose that m is stationary with respect to $\{p_t\}$. A P-integrable bounded random variable Z is θ_t-invariant ($Z = Z \circ \theta_t$ P-a.s., $t > 0$) if and only of $Z = g(X_0)$ P-a.s. for some m-integrable bounded function g on X which is p_t-invariant : $p_t g = g$ m-a.e., $t > 0$.

$\underline{\text{Proof}}$ For such a Z, the Markov property yields
$E(Z|B_t^0) = g(X_t)$ P-a.s. with $g(x) = E_x(Z)$.   Hence

$P(|Z - g(X_0)| > \epsilon) = P(|Z \theta_t - g(X_0 \theta_t)| > \epsilon) = P(|Z - g(X_t)| > \epsilon)$

$\underset{t\to\infty}{\longrightarrow} 0$, and $Z = g(X_0)$ P-a.e.   Since $g(X_t) = g(X_0)$ P-a.s.,

$E((g(X_t)-g(X_0))h(X_0))=E((p_t g(X_0)-g(X_0))h(X_0))$ for any bounded h,

which impies $p_t g = g$ m-a.e.   Conversely, given such g as in

Lemma 1, we have $E((g(X_t) - g(X_0))^2 = 0$.   q.e.d.

$\underline{\text{Theorem 1}}$ Let $\{p_t\}$ be an m-symmetric conservative transition

function.   For any $f \in L^p(X;m)$, $p > 1$, the limit $\lim_{t\to\infty} p_t f = g$

exists m-a.e. and in $L^p(X;m)$.   g is p_t-invariant.

$\underline{\text{Proof}}$ This is a consequence of an extension to a σ-finite

measure space of Rota's ergodic theorem(C.Dellacherie and P.A.Meyer

[2]).   In fact using the same notions as before, we have from

symmetry $Y_t = E(f(X_0)|G_t) = p_t f(X_t)$ P-a.s., where $G_t = \sigma\{X_s; s \geq t\}$.

Hence $p_{2t} f(X_0) = E(Y_t|B_0)$.   Since Y_t is an inverse martingale,

the limit $\lim_{t\to\infty} Y_t = Z$ exists and the L^p-martingale inequality yeilds

$\lim_t p_{2t} f(X_0) = E(Z|B_0)$ P-a.s. and in $L^p(\Omega;P)$-sense.   The theorem

follows with $g(x)=E_x(Z)$.   q.e.d.

A conservative transition function $\{p_t\}$ is said to be $\underline{\text{irreduci-}}$

$\underline{\text{ble}}$ if any p_t-invariant bounded m-integrable function is constant m-a.e.

The σ-finite measure P on (Ω, B^0) in Lemma 1 is called $\underline{\text{ergodic}}$

if any θ_t-invariant bounded P-integrable random variable is constant

P-a.s.   The preceding lemma and theorem imply

$\underline{\text{Corollary}}$ Under the situation of Theorem 1, we have
(i) $\{p_t\}$ is irreducible if and only if P is ergodic.

(ii) $\lim_{\to\infty} p_t f(x) = \frac{1}{m(X)} \int_X f(y)m(dx)$ m-a.e. for any bounded

m-intgrable function f.   The right hand side is interpreted to be

zero when $m(X) = \infty$.

§2. Irreducibilty

In this section, we consider a transition function $\{p_t\}$ which is m-symmetric but not necessarily conservative.   Let E be the associated Dirichlet form on $L^2(X;m)$. E is said to be irreducible if $\{p_t\}$ is irreducible.   Irreducibility criteria for E were studied in Albeverio, Fukushima, Karwowsky and Streit [1] in relation to the notion of quantum mechanical tunneling.   The objective of this section is to give some extensions of the results in [1] with a little simplification of the proofs.

As in [1], we assume that X is a locally compact Hausdorff space satisfying the second axiom of countability, m is an everywhere dense positive Radom measure on X and E is a regular and local Dirichlet form on $L^2(X;m)$.   Accordingly we may consider that the associated m-symmetric transition function p_t gives rise to a diffusion process $M = \{\Omega, B, X_t, P_x\}$ on X.

The L^2-semigroup determined by p_t will be denoted by T_t instead of p_t.   A Borel set B is called T_t-<u>invariant</u> if $T_t I_B = I_B T_t$, $t > 0$, namely, $T_t(I_B f) = I_B \cdot T_t f$ for any $f \in L^2(X;m)$. T_t or E is said to be <u>irreducible</u> if any T_t-invariant set B is trivial in the sense that $m(B) = 0$ or $m(X-B) = 0$.   When p_t is conservative, this notion is equivalent to the irreducibility of the preceding section as we can see from Corollary (i).   We call a set B M-invariant if $P_x(\sigma_{X-B} < \infty) = 0$ for any $x \in B$, σ indicating the first hitting time of the set.   We say that an increasing sequence $\{F_n\}$ of closed sets is a nest if $\mathrm{Cap}(X-F_n) \to 0$.   Following LeJan [6], we call a Borel set B quasi-open (resp. quasi-closed) if there esists a nest $\{F_n\}$ such that $B \cap F_n$ is open(resp.closed) in F_n for each n.   For Borel sets B and $\tilde{B}$, $\tilde{B}$ is said to the a modification of B if $m(B \ominus \tilde{B}) = 0$.

<u>Theorem 2</u>   Following conditions are equivalent for a Borel set B.

(i) B is T_t-invariant.

(ii) $u \in D[E] \implies I_B \cdot u \in D[E]$.

(iii) I_B is locally in $D[E]$.

(iv) $I_{\widetilde{B}}$ is quasi-continuous for some modification $\widetilde{B}$ of B.

(v) $\widetilde{B}$ is quasi-open and quasi-closed for some modification $\widetilde{B}$ of B.

(vi)   X can be decomposed as $X = B_1 + B_2 + N$ where B_1 (resp. B_2) is a modification of B(resp. X-B), both B_1 and B_2 are M-invariant and $m(N) = 0$.

Remark 1   Condition (v) is the same as saying that there exists a nest $\{F_n\}$ such that $\widetilde{B} \cap F_n$ and $F_n - \widetilde{B}$ are closed for each n. This simplifies corresponding statements in [1].   The set N in condition (vi) is of zero capacity ([4]).

Proof   (i) $\Longrightarrow$ (ii) : T_t and $D[E]$ are expressible by a resolution of identity $\{E_\lambda, \lambda > 0\}$ on $L^2(x;m)$ as $T_t = \int_0^\infty e^{-\lambda t} dE_\lambda$ and $D[E] = \{u \in L^2 : \int_0^\infty \lambda d(E_\lambda u, u) < \infty\}$.   (i) then implies $E_\lambda I_B u$ $I_B \cdot E_\lambda u$, $u \in L^2$, from which (ii) is immediate.

(ii) $\Longrightarrow$ (iii) : Since E is regular, there exists for any compact K, a function $u \in D[E] \cap C_0(X)$ such that $u = 1$ on K.

(iii) $\Longrightarrow$ (iv) : I_B admits a quasi-continuous version ϕ.   Since $\phi^2 = \phi$ m-a.e., $\phi = 0$ or 1 q.e.   Hence $\phi = I_{\widetilde{B}}$ q.e. for some modification $\widetilde{B}$ of B.

(iv) $\Longrightarrow$ (v) : trivial.

(v) $\Longrightarrow$ (vi) : Let τ_n be the first leaving time from F_n of the sample path.   Since the sample path is continuous almost surely and $B \cap F_n$ is not arcwise connected with $F_n - B$,

$P_x(X_t \in B \cap F_n$ for any $t < \tau_n) = 1$, $x \in B \cap F_n$ (see Remark 1).

The same property holds for the set $F_n - B$.   We then arrive at (vi) by observing that $P_x(\lim \sigma_{X-F_n} < \zeta) = 0$ q.e. ([4]).

(vi) $\Longrightarrow$ (i) : trivial.   q.e.d.

Corollary   Let $m^{(1)}$ and $m^{(2)}$ be mutually absolutely continuous and let $E^{(1)}$ and $E^{(2)}$ be regular local Dirichlet forms on $L^2(X;m^{(1)})$ and $L^2(X;m^{(2)})$ respectively.   Suppose that any quasi-continuous function with respect to $E^{(2)}$ is also quasi-continuous with respect to $E^{(1)}$.   Then the irreducibility of $E^{(1)}$ implies the same property of $E^{(2)}$.

The condition in this Corollary is satisfied if $E^{(2)}$ is locally dominating $E^{(1)}$. To make this statement more precise, let us consider a dense subalgebra D of $C_0(X)$ satisfying the following two properties : (i) for any $\varepsilon > 0$, there exists a function ϕ_ε making E Markovian ([4]) such that $\phi_\varepsilon(u) \in D$ whenever $u \in D$, (ii) for any compaat set K and a relatively compact open set G with $K \subset G$, D contains a function u with $u = 1$ on K and $u = 0$ on $X - G$. We let $D_G = \{u \in D : u = 0$ on $X - G\}$. We say that D is a core of a Dirichlet form E if D is E_1-dense in $D[E]$.

<u>Theorem 3</u> (i) Let $E^{(1)}$ and $E^{(2)}$ be two local regular Dirichlet forms on $L^2(x;m)$ possessing a set D as above as their common cores. Suppose that for any relatively compact open set $E^{(2)}(u,u) \geq \gamma_K E^{(1)}(u,u)$, u D_G, for some constant $\gamma_K > 0$. Then the irreducibility of $E^{(1)}$ implies the same property of $E^{(2)}$. (ii) Let $m^{(1)}$ and $m^{(2)}$ be related as $dm^{(2)} = \rho\, dm^{(1)}$ with $\gamma_G = \inf\limits_{x \in G} \rho(x) > 0$ for any relatively compact open set G. Let $E^{(1)}$ and $E^{(2)}$ be local regular Dirichlet forms on $L^2(X;m^{(1)})$ and $L^2(X;m^{(2)})$ respectively. Suppose that $E^{(1)}$ and $E^{(2)}$ have a common core D of the above type and $E^{(1)} = E^{(2)}$ on $D \times D$. Then the irreducibility of $E^{(1)}$ implies that of $E^{(2)}$.

<u>Proof</u> (i) The restriction of a local regular Dirichlet form to $\{u \in D[E]$; $\tilde{u} = 0$ q.e. on $X-G$ } is denoted by E_G, which is known to be a local regular Dirichlet form on $L^2(G;m)$. A function on G is E_G-quasi-continuous if and only if it is E-quasi-continuous ([4; Th. 4.4.2]). Moreover, a function which is quasi-continuous on any relatively compact open set is (globally) quasi-continuous. Under the assumption of (i), D_G becomes a common core of $E_G^{(1)}$ and $E_G^{(2)}$ ([4; Prob. 3.3.4]). Inequality in (i) then implies an analogous inequality for the relevent capacities. Hence a function is $E_G^{(1)}$-quasi-continuous whenever it is $E_G^{(2)}$-quasi-continuous.

Thus if a function is quasi-continuous with respect to $E^{(2)}$, so it is with respect to $E^{(1)}$. Now Corollary applies.

(ii) We have

$$E_G^{(2)}(u,u) + (u,u)_{m_2} \geqq E_G^{(1)}(u,u) + \gamma_G \cdot (u,u)_{m_1} \;, \quad u \in D_G. \qquad q.e.e.$$

Example 1 Sobolev space of order 1 on $L^2(R^d)$ is irreducible because the associated diffusion is Brownian motion whose transition function is given by

$$p_t(x,dy) = (2\pi t)^{d/2} \exp(-\frac{|x-y|^2}{2t}) dy \;.$$

Consider a non-negative function $\rho \in L^1_{loc}(R^d)$. Then

$$E(u,v) = \frac{1}{2} \int_{R^d} \nabla u \nabla v \; \rho \; dx, \quad D[E] = C_0^1(R^d),$$

defines a Markovian local symmetric form on $L^2(R^d; dx)$. Assume that $\inf_{x \in K} \rho(x) > 0$ for any compact set $K \subset R^d$, then E is closable on $L^2(R^d; \rho dx)$. The closure $\overline{E}$ becomes a local regular Dirichlet form and consequently admits a diffusion process M_ρ on R^d. Now Theorem 3 applies with $D = C_0^1(R^d)$ and we see that the closure $\overline{E}$ is a irreducible local symmetric Dirichlet form on $L^2(R^d; dx)$.

Example 2 Consider the above example for the one dimensional case. More specifically we are concerned with the symmetric form

$$E(u,v) = \frac{1}{2} \int_{R^1} u'(x) v'(x) \rho(x) dx, \quad D[E] = C_0^1(R^1),$$

on $L^2(R^1, \rho dx)$ and we assume that $\rho \in L^1_{loc}$ and $\inf_{a<x<b} \rho(x) > 0$ whenever $0 \notin (a,b)$. ρ may be degenerate at 0 but E is closable ([4; Th.2.1.4]).
The closure $\overline{E}$ is irreducible if and only if

$$\int_{-b}^{b} \frac{d\xi}{\rho(\xi)} < \infty \quad \text{for} \quad b > 0.$$

In view of Theorem 2 and Remark 1, this assertion follows from the next lemma.

Lemma 2 Let $I_d = (0,d)$.

$$\lim_{d \downarrow 0} Cap(I_d) = 0 \Longleftrightarrow \int_0^b \frac{d\xi}{\rho(\xi)} = \infty \;, \quad b > 0.$$

Proof "$\Longrightarrow$" : Suppose that the above integral is finite. Le $G = (-b,b)$, then it suffices to show $\lim_{d \downarrow 0} Cap_G(I_d) > 0$ where Cap_G

is the capacity related to the form E_G (see the proof of Theorem 3).
By [4;Prob.3.3.2], $Cap_G(K)$ for $K = [c,d]$, $0 < c < d$, can be computed as

$$Cap_G(K) = \inf_{\substack{u \in C_0^1, u=1 \text{ on } K \\ u=0 \text{ on } R^1-G}} E_1(u,u) \geq \inf_{\substack{u \in C_0^1 \\ u(d)=1, u(b)=0}} \frac{1}{2}\int_d^b u'(x)^2 \rho(x)dx$$

which is not smaller than $\frac{1}{2}(\int_d^b d\xi/\rho(\xi))^{-1}$ by Schwarz inequality.

Hence we have $Cap(I_d) \geq \frac{1}{2}(\int_0^b d\xi/\rho(\xi))^{-1} > 0$ uniformly in d.

"$\Longleftarrow$" : Suppose that $\int_0^b \frac{d\xi}{\rho(\xi)} = \infty$. Let $0 < c'< c < d < d'< b$,

$K = [c,d]$ and define u_0 by $u_0(x) = (\int_{c'}^c \frac{d\xi}{\rho(\xi)})^{-1}\int_{c'}^x \frac{d\xi}{\rho(\xi)}$,

$c' \leq x \leq c$; $u_0(x) = 1$, $c \leq x \leq d$; $u_0(x) = (\int_d^{d'} \frac{d\xi}{\rho(\xi)})^{-1}\int_x^{d'} \frac{d\xi}{\rho(\xi)}$,

$d \leq x \leq c'$; $u_0(x) = 0$ outside $[c',d']$. Then $u_0 \in D[E] \cap C_0(R^1)$

(see [5]) and

$$Cap(K) \leq E_1(u_0,u_0) = \frac{1}{2}(\int_{c'}^c \frac{d\xi}{\rho(\xi)})^{-1} + \frac{1}{2}(\int_d^{d'} \frac{d\xi}{\rho(\xi)})^{-1} + \int_{c'}^{d'} \rho(\xi)d\xi.$$

Letting $c' \downarrow 0$ and $c \downarrow 0$, we get

$$Cap(I_d) \leq \frac{1}{2}(\int_d^{d'} \frac{d\xi}{\rho(\xi)})^{-1} + \int_0^d \rho(\xi)d\xi,$$ which tends to zero as $d \downarrow 0$

and $d' \downarrow 0$.   q.e.d.

This lemma means that, if for instance $\int_{-b}^0 \frac{d\xi}{\rho(\xi)} = \infty$ and
$\int_0^b \frac{d\xi}{\rho(\xi)} < \infty$, then R^1 is the sum of two invariant sets $(-\infty,0)$ and
$[0,\infty)$. 0 is attainable by the associated sample path from the
right but not from the left in this case.

Example 3   Consider the same example for the two dimensional
case.   Thus
$$E(u,v) = \frac{1}{2}\int_{R^2} (u_x^2 + u_y^2)\rho(x,y)dxdy, \quad D[E] = C_0^1(R^2).$$

Let $C = \{x = 1\}$.   We assume that $\rho \in L_{loc}^1(R^2)$ and $\inf_{(x,y)\in K} \rho(x,y)$
is positive for any compact set K with $K \cap C = \phi$.   We further
assume that the form above is closable on $L^2(R^2; dxdy)$.   By the
same reasoning as in the preceding example, we can coclude that $\overline{E}$
is irreducible if for some $\alpha < \beta$ and $b > 0$

$$\int_\alpha^\beta \left(\int_{-b}^{b} \frac{d\xi}{\rho(\xi,\eta)} \right)^{-1} d\eta > 0 .$$

A sufficient condition for the reducibility can be stated as follows: For a η-interval J, we denote the integral $\int_J \rho(\xi,\eta)d\eta$ by $\bar{\rho}_J$.

If $\{J_k\}_{k=-\infty}^{\infty}$ is an open covering of R^1 and if for each k,

either $\int_0^{b_k} \frac{d\xi}{\bar{\rho}_{J_k}(\xi)} = \infty$ or $\int_{-b_k}^{0} \frac{d\xi}{\bar{\rho}_{J_k}(\xi)} = \infty$ for some $b_k > 0$,

then the left half plane $\{x < 0\}$ and the right half plane $\{x > 0\}$ are not attainable each other.

References

[1] S. Albeverio, M. Fukushima, W. Karwowski and L. Streit, Capacity and quantum mechanical tunneling,Comm. Math. Physics, 81 (1981), 501-513.

[2] C. Dellacherie and P.A. Meyer, Probabilites et potentiel, Theorie des martingales, Hermann, Paris, 1980.

[3] J.L. Doob, Stochastic processes, Wiley, 1953.

[4] M. Fukushima, Dirichlet forms and Markov processes, Kodansha and North Holland, 1980.

[5] M. Fukushima, On a stochastic calculus related to Dirichlet forms and distorted Brownian motions, Physics reports (A review sections of physics letters) 77, No.3, 1981, 255-262.

[6] Y. LeJan, Quasi-continuous functions and Hunt processes, to appear.

MOMENTUM-POSITION COMPLEMENTARITY IN STOCHASTIC MECHANICS

Francesco Guerra

Istituto Matematico "G.Castelnuovo", Università di Roma

and

Laura Morato

Istituto di Elettrotecnica ed Elettronica, Università di Padova

and LADSEB-CNR, Padova.

Stochastic mechanics /1,2/ provides a quantization procedure for classical dynamical systems based on the theory of stochastic differential equations. In the usual treatment of this method the configuration space plays a distinguished role. In fact the classical position variables are promoted to a stochastic process, while the momentum variables appear in the form of mean forward and backward drift fields, very similar to the expression of the type ∇S in the classical Hamilton-Jacobi thery.

On the other hand, in the operator formulation of quantum mechanics, position and momentum appear on the same footing and the representations of Hilbert space, where each of them becomes diagonal, are simply related through the unitary transformation given by the Fourier transform.

As the experience of Euclidean quantum field theory shows /3,4/ and the success of MonteCarlo methods /5,6/ confirms, stochastic methods are very useful for the treatment of quantum dynamical systems, because they provide a kind of path integral very suitable for mathematical estimation and control and for numerical computer investigations.

Recently there has been some relevant progress /7/ in the formulation of path integrals in momentum space, therefore we believe that it is an important problem to investigate and clarify the role of momentum variables in stochastic mechanics, in order to see whether at least a part of the canonical transformation theory formalism of classical and quantum mechanics can be transferred and adapted to the stochastic frame.

An important contribution came from Shuker /8/ who has given a physically reasonable definition of momentum in stochastic mechanics (based on some large time limiting procedure on the stochastic position variable) and has shown the consequent relevance of the Fourier transform of the wave function. Recently de Falco, De Martino and De Siena /9/ have given a convincing derivation, in the frame of stochastic mechanics, of a simple connection between the quantum Heisenberg momentum position uncertainty relations and the stochastic uncertainty relations connected with the non coincidence of the mean forward and backward drifts for a diffusion process (position-osmotic velocity uncertainty principle).

Taking into account a very simple example, here we consider the problem of directly introducing a stochastic process for the momentum variable. Clearly we must face the difficulties coming from the uncertainty principle. There can be no probability space where the position and the momentum are simultaneously random variables. We will show how in principle stochastic mechanics avoids this difficulty in a very tricky way. In fact if we introduce the representation where the position is a random variable then the momentum splits into two different processes, related to the mean forward and backward drifts, while in the representation where the momentum is a random variable it is the position which splits in two different processes, related to the mean forward and backward accelerations. But all momentum processes (and respectively the position processes), considered in themselves, are stochastically equivalent.

Let us give an explicit account of this very basic phenomenon which allows a form of conciliation between the uncertainty principle and the existence of position and momentum processes. We take as examples the coherent states of the harmonic oscillator.

The Hamiltonian is given by

$$(1) \qquad H = y^2/2m + m\omega^2 x^2/2 \ ,$$

y is the momentum, x is the position, $y, x \in R$, $m > 0$, $\omega^2 > 0$.

For the initial condition

$$(2) \qquad \alpha = (q_0 , p_0)$$

the solution α of the classical dynamics is

(3) $\qquad q_c(t) = q_0 \cos\omega t + (p_0/m\omega)\sin\omega t$,

$\qquad\quad p_c(t) = p_0 \cos\omega t - m\omega q_0 \sin\omega t$.

To each classical solution α we can associate a quantum coherent state ψ_α, which reduces to the classical state (3) in the limit $\hbar \to 0$,

(4) $\qquad \psi_\alpha(x, t) =$

$\qquad \quad = (2\pi\sigma)^{-\frac{1}{4}} \exp(-(x-q_c(t))^2/4 + i(x-q_c(t))p_c(t)/\hbar +$

$\qquad\qquad\qquad\qquad + iq_c(t)p_c(t)/2\hbar - i\omega t/2)$,

$\qquad \sigma = \hbar/2m\omega$.

The position density is

(5) $\qquad \rho(x, t) = \left| \psi_\alpha(x, t) \right|^2 = (2\pi\sigma)^{-\frac{1}{2}} \exp(-(x-q_c(t))^2/2\sigma)$.

Following the general methods explained in /1,2/ it is very simple to introduce the process $q(t)$ associated to the quantum state ψ_α. In fact we can write

(6) $\qquad \psi_\alpha(x, t) = \exp(R(x, t) + iS(x, t)/\hbar)$,

$\qquad\quad R(x, t) = -(x-q_c(t))^2/4\sigma - \log(2\pi\sigma)/4$,

$\qquad\quad S(x, t) = (x-q_c(t))p_c(t) + q_c(t)p_c(t)/2 - \hbar\omega t/2$,

and define

(7) $\qquad v(x, t) \equiv (\nabla S)(x, t)/m = p_c(t)/m \equiv v_c(t)$,

$\qquad\quad \delta v(x, t) \equiv \hbar(\nabla R)(x, t)/m = -\omega(x-q_c(t))$,

$\qquad\quad v_{(\pm)}(x, t) = (v \pm \delta v)(x, t) = v_c(t) \mp \omega(x-q_c(t))$.

Then $q(t)$ must have density given by (5) and satisfy the stochastic differential equation

(8) $\qquad dq(t) = v_{(+)}(q(t), t)dt + dw(t)$,

$\qquad\quad E(dw(t)^2) = (\hbar/m)dt$.

The equation (8) is easily solved (see also /10/) in the form

(9) $\qquad q(t) = q_c(t) + q_0(t)$,

where $q_0(t)$ is the process associated to the ground state /11/. The process $q_o(t)$ is Gaussian with mean zero and covariance given by

(10) $\quad E(q_o(t)q_o(t')) = (\hbar/2m\omega)\exp(-\omega|t-t'|)$

It is worth to recall that Ruggiero and Zannetti /12/ have introduced processes of the type (9), with $q_c(t)$ replaced by the Langevin dissipative process, in order to give a stochastic formulation of the quantum dissipative phenomena.

Clearly we have

(11) $\quad E(q(t)) = q_c(t) \quad , \quad (\Delta q)^2 \equiv E((q(t)-q_c(t))^2) = \hbar/2m\omega \quad .$

Having so defined the position process $q(t)$ let us introduce, in the same probability space where $q(t)$ is defined, the two different processes $p_{(\pm)}(t)$ defined by

(12) $\quad p_{(\pm)}(t) \equiv mv_{(\pm)}(q(t),t) = p_c(t) \mp m\omega(q(t)-q_c(t)) \quad .$

Notice that $p_{(\pm)}(t)$ take the common classical value $p_c(t)$ when $q(t)$ is at $q_c(t)$. We have also

(13) $\quad E(p_{(\pm)}(t)) = p_c(t) \quad ,$

(14) $\quad (\Delta\,\delta p)^2 \equiv E(p_{(+)}-p_{(-)}(t))^2/4 = \tfrac{1}{2}\hbar m\omega \quad ,$ where $\delta p = (p_+-p_-)/2$,

so that from (11) and (14) we have

(15) $\quad \Delta q\,\Delta\,\delta p = \hbar/2 \quad ,$

which is a particular case of the stochastic version of the quantum uncertainty principle, as shown in the general analysis of /9/ (recall that here we have minimum spreading due to the coherent character of the states involved).

Therefore we see that in the probability space of $q(t)$ we can define two different momentum processes $p_{(\pm)}(t)$ so that a stochastic version (15) of the uncertainty principle is verified. But let us give a closer look to $p_{(\pm)}(t)$. We see immediately that they are, separately, two different realizations of the same process $p(t)$ defined by beeing Gaussian, with mean and covariance given by

(16) $\quad E(p(t)) = p_c(t) \quad , \quad E(p(t)p(t')) = \tfrac{1}{2}m\hbar\omega\exp(-\omega|t-t'|) \quad .$

In fact (16) follow easily for $p_{(\pm)}(t)$ defined by (12).

Therefore by a separate monitoring of $p_{(+)}$ and $p_{(-)}$ we could never distinguish them and we would identify them with $p(t)$ defined before in (16).

It is important to remark that $p(t)$ is nothing but the process associated by stochastic mechanics to the momentum wave function related to (4) as a Fourier transform

$$(17) \qquad \widetilde{\psi}_\alpha (y, t) =$$

$$= (2\pi\sigma')^{-\frac{1}{4}} \exp(-(y-p_c(t))^2/4\sigma' - i(y-p_c(t))q_c(t)/\hbar - iq_c(t)p_c(t)/2\hbar - i\omega t/2)$$

$$\sigma' = \hbar m \omega/2.$$

In fact, due to the basic symmetry of the Hamiltonian, we can repeat the same chain of reasoning done through (4)-(10) starting from (17). We find that the process $p(t)$ defined by (16) is associated to the state given by (17). Therefore there are two independent definitions of $p(t)$, either as process associated to (17) or as common realization of $p_{(\pm)}(t)$. The two definitions coincide.

In order to close the circle we must investigate how $q(t)$ is related to $p(t)$. First of all let us find the stochastic differential equation satisfied by $p(t)$. This can be done in two ways. We can start from (17) and follow the route analogous to (4)-(8). Then we find

$$(18) \qquad dp(t) = F_{(+)}(q(t), t)dt + dw'(t) \quad ,$$

where

$$(19) \qquad F_{(+)}(y, t) = -\omega(y-p_c(t)) - m\omega^2 q_c(t) \quad ,$$

$$(20) \qquad E(dw'(t)^2) = \hbar m \omega^2 dt \quad .$$

On the other hand we can start from the definitions (12) and find, in the probability space of $q(t)$,

$$(21) \qquad dp_{(\pm)}(t) = F_{(+)}(p_{(\pm)}(t), t)dt \mp dw' \quad ,$$

$$(22) \qquad dw' = m\omega dw \quad ,$$

with dw defined in (8). Since dw' defined in (22) is stochastically equivalent to dw' in (18) defined in (20) and the sign before dw' in (21) is stochastically irrelevant we find immediately that (21) is completely equivalent to (18) provided $p_{(\pm)}(t)$ are separately monitored in order to realize their stochastic equivalence with $p(t)$.

Formula (21) also shows that the difference between $p_{(+)}(t)$ and $p_{(-)}(t)$

in the probability space of q(t), is uniquely due to the different sign of the Brownian disturbance, a feature which could play a relevant role for the generalization of these considerations to more general states and systems.

Therefore, by following two different routes we have found the stochastic differential equation (18) for p(t).

As in (7) we can easily find the backward force associated to $F_{(+)}$ in the form

$$(23) \qquad F_{(-)}(y, t) = \omega (y-p_c(t)) - m\omega^2 q_c(t) \quad .$$

Formulae (19) and (23) should be compared with the last of (7).

Now let us consider the probability space of p(t) and let us introduce in this space the two processes $q_{(\pm)}(t)$ defined through the mean forward and backward forces in the following way

$$(24) \qquad F_{(\pm)}(p(t), t) = -m\omega^2 q_{(\pm)}(t) \quad ,$$

$$(25) \qquad q_{(\pm)}(t) = q_c(t) \pm (p(t) - p_c(t))/m \quad .$$

Even though they are different processes in the probability space where p(t) is defined, it is immediate to recognize that they are two different realizations of the same process q(t), if separately monitored (i.e. without considering joint distributions of $q_{(+)}$ and $q_{(-)}$).

Moreover this common process q(t) coincides, as it is easily checked, with the stochastic process introduced at the beginning in association with the wave function (4).

Therefore we can see that the complementarity principle assumes the following simple interpretation in the stochastic framework. While in classical mechanics we can give precise values to both the momentum p(t) and the position q(t) at each time, this is no longer possible in stochastic mechanics. In fact if we make an ideal monitoring of the system, sampling the position q(t) time after time, then we have that the velocity splits in two dynamical variables $v_{(\pm)}$ and therefore we have two different processes for the momentum $p_{(+)}(t)$, $p_{(-)}(t)$. We can proceed analogously for the momentum p(t). In the very simple case considered by us the processes $p_{(\pm)}(t)$ are given easily in terms of the drifts $v_{(\pm)}$ as in (12), but in more general cases one can not expect such simple expressions, so that the problem of generalizing the ideas sketched in this note is open for the moment.

Very similar considerations can be done in the field case.
For example in /13/ the magnetic field B is promoted to a random proces
while the electric field splits in two different fields $E_{(+)}$, $E_{(-)}$.

It is clear that these are two different realizations of the same process
E . In the probability space of E it is the magnetic field who splits in
two different fields $B_{(+)}$, $B_{(-)}$, each a realization of the previous B .
More general cases can be considered if the splitting is done in a
relativistically covariant way through general families of space-like
surfaces.

In conclusion we would like to point out that the basis of the results
of this note is given by the invariance of the structure of stochastic
mechanics with respect to the change of sign of the Brownian disturbance
dw. In fact putting +dw or -dw in a stochastic differential equation
leads to different processes (this explains the splitting p(t) $\rightarrow$ $p_{(\pm)}$,
q(t) $\rightarrow$ $q_{(\pm)}$(t) necessary for the stochastic complementarity principle),
but these different processes are stochastically equivalent, because dw
enters only through its square in the Fokker-Planck equations.

REFERENCES.

/1/   E.Nelson, <u>Dynamical Theories of Brownian Motion</u>, Princeton
        University Press, Princeton, N.J., 1967.

/2/   F.Guerra, Structural Aspects of Stochastic Mechanics and
        Stochastic Field Theory, Physics Reports , <u>77</u>,263 (1981).

/3/   B.Simon, <u>The P(ϕ)$_2$ Euclidean (Quantum) Field Theory</u>,
        Princeton University Press, Princeton, N.J., 1974.

/4/   J.Glimm and A.Jaffe, <u>Quantum Physics</u>, Springer-Verlag,Berlin,1981.

/5/   M.Creutz, L.Jacobs and C.Rebbi, Experiments with a Gauge Invariant
        Ising Model, Phys.Rev.Lett.,<u>42</u>,1390 (L979).

/6/   E.Marinari,G.Parisi and C.Rebbi, Computer Estimates of Meson
        Masses in SU(2) Lattice Gauge Theory,Phys.Rev.Lett.,<u>47</u>,1795
        (1981), and references quoted there.

/7/   Ph.Combe,R.Rodriguez,R.Høegh-Krohn,M.Sirugue and M.Sirugue-Collin,
        Generalized Poisson Processes in Quantum Mechanics and Field
        Theory,Physics Reports, <u>77</u>, 221 (1981).

/8/   D.S.Shuker, Stochastic Mechanics of Systems with Zero Potential,
        J.Funct.Anal. <u>38</u>, 146 (1980).

/9/   D. de Falco, S.De Martino and S.De Siena, Position-Momentum Un-
        certainty Relations in Stochastic Mechanics,
        University of Salerno Preprint, 1982.

/10/   F.Guerra and M.I.Loffredo, Thermal Mixtures in Stochastic
        Mechanics, Lettere al Nuovo Cimento <u>30</u>, 81 (1981).

/11/   F.Guerra and P.Ruggiero, New Interpretation of the Euclidean-
        Markov Field in the Framework of Physical Space-Time,
        Phys.Rev.Lett. <u>31</u>, 1022 (1973).

/12/   P.Ruggiero and M.Zannetti, Stochastic Description of the Quantum
        Thermal Mixture, Phys.Rev.Lett. <u>48</u>, 963 (1982).

/13/   F.Guerra and M.I.Loffredo, Stochastic Equations for the Maxwell
        Field, Lettere al Nuovo Cimento <u>27</u>, 41 (1980).

NON-COMMUTATIVE MARTINGALES AND STOCHASTIC INTEGRALS IN FOCK SPACE

by

R.L. Hudson, Mathematics Department, University of Nottingham,
University Park, Nottingham NG7 2RD, England

and

R.F. Streater, Mathematics Department, Bedford College,
University of London, Regent's Park, London NW1 4NS, England.

Abstract

Boson Fock space is used to construct some non-commutative martingales, and a definition of stochastic integrals based on exponential vectors is given. It is shown that Itô's formula reduces to the chain rule with Wick ordering.

§1. An algebra of operators in Fock space

Let h be a Hilbert space and let

$$\Gamma(h) = \bigoplus_{n=0}^{\infty} \left(\bigotimes_{j=1}^{n} h \right)_s$$

be its symmetric Fock space. The _exponential vector_ $e(f)$ corresponding to $f \in h$ is defined to be the element of $\Gamma(h)$

$$e(f) = (1, f, (2!)^{-\frac{1}{2}} f \otimes f, (3!)^{-\frac{1}{2}} f \otimes f \otimes f, \ldots).$$

The _vacuum_, denoted by Ω or $\Omega(h)$ is $e(0)$. The exponential vectors form a total linearly independent set [3].

If $h = h_1 \oplus h_2$ is a direct sum there is a natural identification

$$\Gamma(h) = \Gamma(h_1) \otimes \Gamma(h_2) \tag{1.1}$$

between $\Gamma(h)$ and the Hilbert space tensor product of the Fock spaces of the summands in which each exponential vector is a product vector

$$e(f) = e(f_1) \otimes e(f_2) \quad (f = (f_1, f_2) \in h = h_1 \oplus h_2). \tag{1.2}$$

We denote by $\mathcal{E}$ or $\mathcal{E}(h)$ the dense subspace of $\Gamma(h)$ comprising all finite linear combinations of exponential vectors and their (strong) derivatives

$$\frac{\partial}{\partial x_1} \cdots \frac{\partial}{\partial x_n} e\left(f + \sum_{j=1}^{n} x_j f_j \right)\Bigg|_{x_1 = \ldots = x_n = 0} \quad (f, f_1, \ldots, f_n \in h).$$

From (1.2) it is clear that under the identification (1.1)

$$\mathcal{E}(h) = \mathcal{E}(h_1) \,\hat{\otimes}\, \mathcal{E}(h_2) \tag{1.3}$$

where $\hat{\otimes}$ denotes the algebraic tensor product.

We define operators denoted $e^{a^\dagger(f)}$, $e^{a(f)}$, $f \in h$ by their actions on exponential vectors

$$e^{a^\dagger(f)} e(g) = e(g+f) \tag{1.4}$$

$$e^{a(f)} e(g) = e^{\langle f,g \rangle} e(g). \tag{1.5}$$

These operators extend uniquely by commutation with partial differentiations to mutually adjoint operators on all of $\mathcal{E}(h)$. The operators $a^\dagger(f)$ and $a(f)$ can be defined by (strong) differentiation on $\mathcal{E}(h)$

$$a^\dagger(f) = \frac{d}{dx} e^{a^\dagger(xf)}\Big|_{x=0} , \quad a(f) = \frac{d}{dx} e^{a(xf)}\Big|_{x=0} \tag{1.6}$$

$\mathcal{E}(h)$ is then a common invariant domain for the operators $e^{a^\dagger(f)}$, $e^{a(f)}$, $a(f)$ and $a^\dagger(f)$ on which the relations

$$e^{a^\dagger(f)} e^{a^\dagger(g)} = e^{a^\dagger(f+g)} \tag{1.7}$$

$$e^{a(f)} e^{a^\dagger(g)} = e^{\langle f,g \rangle} e^{a^\dagger(g)} e^{a(f)} \tag{1.8}$$

$$a^\dagger(\alpha f+g) = \alpha a^\dagger(f) + a^\dagger(g) \tag{1.9}$$

$$a^\dagger(f) a^\dagger(g) = a^\dagger(g) a^\dagger(f) \tag{1.10}$$

$$a(f) a^\dagger(g) = a^\dagger(g) a(f) + \langle f,g \rangle \tag{1.11}$$

$$a^\dagger(f) e^{a^\dagger(g)} = e^{a^\dagger(g)} a^\dagger(f) \tag{1.12}$$

$$a(f) e^{a^\dagger(g)} = e^{a^\dagger(g)} (a(f) + \langle f,g \rangle) \tag{1.13}$$

hold together with relations obtained from these by the adjunction $a(f) \to a^\dagger(f)$, $e^{a(f)} \to e^{a^\dagger(f)}$. We denote by $\mathcal{M}$ or $\mathcal{M}(h)$ the involutive algebra with identity generated by these operators on $\mathcal{E}(h)$. consists of linear combinations of <u>Wick monomials</u>

$$W_{m,n}(f_0;f_1,\ldots,f_m;g_0;g_1,\ldots,g_n) = a^\dagger(f_1)\ldots a^\dagger(f_m) e^{a^\dagger(f_0)} e^{a(g_0)} a(g_n)\ldots a(g_1). \tag{1.14}$$

Under the identification (1.1), corresponding to (1.3) we have

$$\mathcal{M}(h) = \mathcal{M}(h_1) \hat{\otimes} \mathcal{M}(h_2)$$

where $\hat{\otimes}$ now denotes the algebraic tensor product algebra.

We write $\mathbb{E}[T]$ to denote the vacuum expectation

$$\mathbb{E}[T] = \langle \Omega, T\Omega \rangle$$

of any operator in $\Gamma(h)$, and notice that

$$\mathbb{E}[W_{m,n}(f_0;f_1,\ldots,f_m;g_0;g_1,\ldots,g_n)] = \delta_{m,0}\, \delta_{n,0}. \tag{1.15}$$

Let K be a sub Hilbert space of h. Corresponding to the canonical decomposition $h = K \oplus K^\perp$ write

$$\mathcal{M}(h) = \mathcal{M}(K) \hat{\otimes} \mathcal{M}(K^\perp)$$

and denote by $\mathcal{M}^K$ or $\mathcal{M}^K(h)$ the subalgebra

$$\mathcal{M}^K = \mathcal{M}(K) \,\hat{\otimes}\, 1 .$$

For each $T \in \mathcal{M}$ there exists a unique element $\mathbb{E}[T|\mathcal{M}^K]$ of $\mathcal{M}^K$, the <u>vacuum conditional expectation given</u> $\mathcal{M}^K$ of T, having the property that, for arbitrary $S_1, S_2 \in \mathcal{M}^K$

$$\mathbb{E}[S_1 T S_2] = \mathbb{E}[S_1 \mathbb{E}[T|\mathcal{M}^K] S_2].$$

The map $\mathbb{E}[\ |\mathcal{M}^K]$ is defined either by its action on product elements $B \otimes C$ of $\mathcal{M}(h) = \mathcal{M}(K) \otimes \mathcal{M}(K^{\perp})$,

$$\mathbb{E}[B \otimes C|\mathcal{M}^K] = \mathbb{E}[C]B \otimes 1 ,$$

or by its action on Wick monomials

$$\mathbb{E}[W(f_0;\ldots,g_n)|\mathcal{M}^K] = W(Ef_0;\ldots,Eg_n) \tag{1.16}$$

where E is the projector onto K in h. A similar construction for the CAR algebra has been given in [2].

§2. <u>Quantum Brownian motion</u>

Now let $h = L^2([0,\infty[)$ and define operators $A_t^\dagger$, A_t in $\mathcal{M}$ by

$$A_t^\dagger = a^\dagger(\chi_{[0,t]}), \qquad A_t = a(\chi_{[0,t]}), \qquad t \geq 0 \tag{2.1}$$

where χ_C denotes the indicator of the set C. The operators

$$Q_t = A_t + A_t^\dagger, \qquad t \geq 0$$

are essentially self-adjoint on $\mathcal{E}(h)$ and commute with each other in the sense that their spectral projectors commute. Equipped with the probabilities determined by the vacuum state we may therefore regard $(Q_t : t \geq 0)$ as a classical stochastic process; that this process is essentially Brownian motion follows from the computations

$$\mathbb{E}\left[e^{i \sum_{j=1}^{n} u_j Q_{t_j}}\right] = \exp\{-\tfrac{1}{2} \text{ quadratic in } u_1,\ldots,u_n\},$$

showing that the process is Gaussian and

$$\mathbb{E}[Q_s Q_t] = s \wedge t,$$

the covariance characteristic among all Gaussian processes of Brownian motion. However exactly the same thing is true of

$$Q_t^\theta = e^{-i\theta} A_t + e^{i\theta} A_t^\dagger, \qquad t \geq 0$$

for each fixed $\theta \in [0,2\pi[$. Because for different values of θ the operators Q_t^θ do not commute with each other, the corresponding Brownian motions cannot be simultaneously realised on one and the same classical probability space. We regard them collectively as <u>quantum mechanical Brownian motion</u>, noting that the pair $(2^{-\frac{1}{2}}Q^{\frac{\pi}{2}}, 2^{-\frac{1}{2}}Q)$ is a canonical Wiener process in the sense of [1]. However it is more convenient to regard the family of nonnormal operators $(A_t : t \geq 0)$ as defining quantum Brownian motion.

§3. Martingales

Now let K_t denote the subspace of $L^2([0,\infty[)$ comprising functions which vanish in $[t,\infty[$, let E_t be the projector onto K_t and write $\mathfrak{M}^t$ for $\mathfrak{M}^{K_t}$. A function $f: [0,\infty[\to \mathbb{C}$ is $\underline{\text{locally square-integrable}}$ if for each $t \geq 0$ $f^t = f\chi_{[0,t]} \in K_t$. For locally square-integrable $f_0, f_1, \ldots, f_m, g_0, g_1, \ldots, g_n$ we can form the family of Wick monomials

$$M_t = W_{mn}(f_0^t; f_1^t, \ldots f_m^t; g_0^t; g_1^t, \ldots g_n^t) \in \mathfrak{M}^t.$$

Since, for $s \leq t$, $E_s f^t = f^s$ we see from (1.16) that

$$\mathbb{E}[M_t | \mathfrak{M}^s] = M_s,$$

that is the operators $(M_t : t \geq 0)$ form a $\underline{\text{martingale}}$.

Among the non-commutative martingales formed in this way we note those of $\underline{\text{polynomial type}}$

$$W(0, f_1^t \ldots, f_m^t; 0, g_1^t \ldots g_n^t) = a^+(f_1^t) \ldots a^+(f_m^t) a(g_n^t) \ldots a(g_1^t)$$

and those of $\underline{\text{exponential type}}$

$$W(f^t; g^t) = e^{a^+(f^t)} e^{a(g^t)};$$

the former can be expressed as (strong) limits of linear combinations of exponential martingales by partial differentiation.

§4. Stochastic integrals

From (2.1) (1.6) and (1.5) we have, for $f \in L^2([0,\infty[)$

$$A_t e(f) = \langle \chi_{[0,t]}, f \rangle e(f)$$

$$= \int_0^t f(s) \, ds \, e(f).$$

This suggests introduction of an operator $\dfrac{dA_t}{dt}$ whose action is

$$\frac{dA_t}{dt} e(f) = f(t) e(f)$$

on exponential vectors corresponding to sufficiently smooth $f \in L^2([0,\infty[)$. Stochastic integrals of the form $\int_0^t G_s \, dA_s$ should then be given by

$$\int_0^t G_s \, dA_s \, e(f) = \int_0^t G_s f(s) e(f) \, ds$$

and integrals of form $\int dA_s^+ F_s$ by formal adjunction,

$$\int_0^t dA_s^\dagger F_s = \left[\int_0^t F_s^\dagger dA_s \right]^\dagger ;$$

more generally a stochastic integral of form $\int_0^t (dA_s^\dagger F_s + G_s dA_s + H_s ds)$ should be given by specifying its matrix elements between elements of the total family of exponential vectors;

$$\langle e(f), \int_0^t (dA_s^\dagger F_s + G_s dA_s + H_s ds) e(g) \rangle = \int_0^t \langle e(f), (\bar{f}(s) F_s + G_s g(s) + H_s) e(g) \rangle \, ds ;$$

we adopt this as our formal definition.

A family $(M_t = t \geq 0)$ of elements of $\mathcal{M}(h)$ is <u>adapted</u> if, for each $t \geq 0$, $M_t \in \mathcal{M}^t$, and to be a <u>stochastic integral</u>, written as

$$M_t = \int_0^t (dA_s^\dagger F_s + G_s dA_s + H_s ds)$$

or as

$$dM = dA^\dagger F + GdA + Hdt$$

if there exist adapted families $(F_t = t \geq 0)$, $(G_t = t \geq 0)$, $(H_t = t \geq 0)$ such that, for arbitrary $f, g \in h$ and $t \geq 0$, the Lebesque integral

$$\int_0^t \langle e(f), (\bar{f}(s) F_s + G_s g(s) + H_s) e(g) \rangle \, ds$$

exists and is equal to $\langle e(f), M_t e(g) \rangle$.

Since the action on $\mathcal{E}(h)$ of an element of $\mathcal{M}(h)$ is determined through partial differentiation by its action on exponential vectors, the stochastic integral is unique.

Simple calculations show that each martingale of polynomial type is a stochastic integral of integrands of the same type;

$$a_t^\dagger(f_1^t) \ldots a^\dagger(f_m^t) a(g_n^t) \ldots a(g_1^t)$$

$$= \int_0^t \left\{ dA_s^\dagger \sum_{j=1}^m f_j(s) a^\dagger(f_1^s) \ldots \wedge \ldots a^\dagger(f_m^s) a(g_n^s) \ldots a(g_1^s) + \sum_{k=1}^n \overline{g_k(s)} a^\dagger(f_1^s) \ldots a^\dagger(f_m^s) \right.$$

$$\left. \overset{k}{a(g_n^s)} \ldots \wedge \ldots a(g_1^s) dA_s \right\},$$

and that the exponential martingale

$$W_t = W(f^t; g^t) = e^{a^\dagger(f^t)} e^{a(g^t)}$$

is the solution of the stochastic differential equation

$$dW = dA^\dagger f(t) W + W \overline{g(t)} dA, \quad W_0 = 1.$$

More generally we have

$$W_{mn}(f_0^t; f_1^t, \ldots, f_m^t; g_0^t; g_1^t, \ldots g_n^t) - \delta_{m0,n0}$$

$$= \int_0^t \left[dA_s^\dagger \left\{ \sum_{j=1}^m f_j(s) W_{m-1,n}(f_0^s; f_1^s, \ldots \overset{j}{\wedge} \ldots f_m^s; g_0^s; g_1^s, \ldots g_n^s) + f_0(s) W_{mn}(f_0^s; \ldots, g_n^s) \right\} \right.$$

$$\left. + \left\{ \sum_{k=1}^n g_k(s) W_{m,n-1}(f_0^s; f_1^s, \ldots f_m^s; g_0^s; g_1^s, \ldots \overset{j}{\wedge} \ldots g_n^s) + \overline{g_0(s)} W_{mn}(f_0^s; \ldots, g_n^s) \right\} dA_s \right].$$

§5. Itô's formula

We consider, as a generalisation of a functional of Brownian motion depending only on the position at time t together with an explicit t-dependence, a family of elements of $\mathcal{M}$ of the form

$$M_t = \sum_j F_j(A_t^\dagger, t) e^{\alpha_j A_t^\dagger} e^{\beta_j A_t} G_j(A_t, t) \tag{5.1}$$

where the F_j and G_j are polynomials in $A_t^\dagger$ and A_t respectively whose coefficients are smooth functions of t. The <u>Wick ordered partial derivatives</u> of M_t are then defined as

$$\frac{\partial M_t}{\partial A_t^\dagger} = \sum_j \left[\partial_1 F_j(A_t^\dagger, t) + \alpha_j F_j(A_t^\dagger, t) \right] e^{\alpha_j A_t^\dagger} e^{\beta_j A_t} G_j(A_t, t)$$

$$\frac{\partial M_t}{\partial A_t} = \sum_j F_j(A_t^\dagger, t) e^{\alpha_j A_t^\dagger} e^{\beta_j A_t} \left[\partial_1 G_j(A, t) + \beta_j G_j(A_t, t) \right],$$

while

$$\frac{\partial M}{\partial t} = \sum_j \left\{ \partial_2 F_j e^{\alpha_j A_t^\dagger} e^{\beta_j A_t} G_j + F_j e^{\alpha_j A_t^\dagger} e^{\beta_j A_t} \partial_2 G_j \right\}.$$

Differentiating with respect to t the identity

$$\langle e(f), M_t e(g) \rangle = \sum_j \overline{F}_j\left(\int_0^t \overline{f}, t \right) e^{\overline{\alpha}_j \int_0^t \overline{f}} e^{\beta_j \int_0^t g} G_j\left(\int_0^t g, t \right)$$

we see that

$$dM_t = dA^\dagger \frac{\partial M}{\partial A_t^\dagger} + \frac{\partial M}{\partial A_t} dA^\dagger + \frac{\partial M}{\partial t} dt. \tag{5.2}$$

In other words, <u>Itô's formula is the chain rule with Wick ordering</u>.

The formula (5.2) implies the classical Itô formula

$$d[F(Q_t, t)] = \partial_1 F \, dQ + (\partial_2 F + \tfrac{1}{2} \partial_1^2 F) \, dt \tag{5.3}$$

[4] for a smooth functional F of Brownian motion which can be expressed in the form (5.1) by the substitution $Q_t = A_t + A_t^\dagger$. To see this, consider first the case

$$F(Q_t, t) = \beta(t) e^{\alpha Q_t}$$
$$= \beta(t) e^{\alpha(A_t + A_t^\dagger)}$$
$$= \beta(t) e^{\frac{1}{2}\alpha^2 t} e^{\alpha A_t^\dagger} e^{\alpha A_t}.$$

According to (5.2)

$$d[F(Q_t, t)] = \alpha(dA^\dagger F + FdA) + (\beta(t) + \tfrac{1}{2}\alpha^2\beta) e^{\frac{1}{2}\alpha^2 t} e^{\alpha A_t^\dagger} e^{\alpha A_t} = \partial_1 FdQ + (\partial_2 F + \tfrac{1}{2}\partial_1^2 F)\, dt$$

as required. The general case now follows by parametric differentiation.

References

1. Cockroft, A.M. and Hudson R.L., Quantum mechanical Wiener processes, J. Multivariate Anal. $\underline{7}$, 107-124 (1977).

2. Evans, D.E., Completely positive quasi-free maps on the CAR algebra, Commun. Math. Phys. $\underline{70}$, 53-68 (1979).

3. Guichardet, A., Symmetric Hilbert spaces and related topics, Springer lecture notes in mathematics $\underline{261}$, Springer, Berlin (1972).

4. Lipster, R.S. and Shirayev, A.N., Statistics of random processes I, General Theory, Springer, Berlin (1977).

SOME RESULTS ON THE SPECTRA OF RANDOM SCHRÖDINGER

OPERATORS AND THEIR APPLICATION TO RANDOM POINT

INTERACTION MODELS IN ONE AND THREE DIMENSIONS

by

W.Kirsch   and   F.Martinelli [+]

Institut für Mathematik
Ruhr-Universität Bochum
D 4630 Bochum, FRG

I Introduction

Schrödinger operators with a random potential occur naturally in the theory of alloys,
liquids,solids with impurities and amorphus materials.In this paper we give several
results concerning the spectra of random Schrödinger operators H_ω of the form:
$H_\omega = H_o + V_\omega$,on $L^2(\mathbb{R}^d)$ $H_o = - \Delta$ being the Laplacian and V_ω a function or a quadratic form which depends on ω ,a point in a probability space $(\Omega ,\mathcal{F}, P)$.In most
considerations that will follow V_ω will be taken as the total potential acting on the
quantum particle generated by "heavy" particles of random "charges" $q_i(\omega)$ located at
the random positions $\xi_i(\omega)$;thus V_ω is of the form:

$$V_\omega(x) = \sum_{i\in Z^d} q_i(\omega)f(x-\xi_i(\omega)) \tag{1}$$

for a suitable f.

Let us now go through the sections of the paper.In section II.1 we derive weak conditions under which V_ω is well defined,i.e. the sum in (1) converges,and moreover the
operators $H_\omega = H_o + V_\omega$ are essentially selfadjoint on $C_o^\infty(\mathbb{R}^d)$ in the case when f is
actually a function.If f is a quadratic form rather than a function we give conditions
under which V_ω is well defined as a quadratic form that is form bounded with respect
to H_o with relative bound less than one; in this case the operator H_ω defined as form
sum is selfadjoint.In section II.2 we state an ergodicity assumption on V_ω which ensures that the spectrum of H_ω is for P-almost all ω a fixed non random set Σ .In II.3
we continue the discussion of II.1 ,II.2 and specialize to potentials of the form (1).
A method that connects the spectrum Σ of H_ω,with V_ω given by (1),with the spectra
of certain periodic potentials is discussed in the next paragraph where in particular
it is also proved that Σ has a band structure.In the third section we concern ourselves to random point interaction Hamiltonians (random Kronig-Penney models).In III.1
and III.2 we restrict ourselves to the case when the "heavy" particles producing the

[+] On leave of absence from Istituto di Fisica Università di Roma, G.N.F.M. - C.N.R.

potential V_ω have non random positions $\{x_i\}$ which furthermore form a lattice in $\mathbb{R}^d$. This potential serves as the model of an alloy with randomly mixed ions of different metals at the positions of the lattice $\{x_i\}$. In the one dimensional case d=1 it is shown that under reasonable conditions on the random field $\{q_i(\omega)\}$ that models the coupling constants ("charges") the spectrum Σ of H_ω has infinitely many gaps. Moreover Σ can be computed rather explicitly from the solutions of a certain algebraic inequality. In the three dimensional case we show that -as for the non random case- one gap occur in the spectrum Σ and that under weak additional assumptions on the random field $\{q_i(\omega)\}$ Σ can be computed explicitly. If in the above model one puts away the particle in x_i with some probability $1 > p > 0$, and this for all $i \in Z^d$, one arrives at a model with random point defects. Theorem 9 in III.3 tells us that throwing away particles in a random way does not effect the spectrum Σ , i.e. the spectrum for this case is the same as for the previous model. In the final section III.4 we then consider a model in one dimension where the positions of the heavy particles are allowed to be random; here we prove that if the fluctuations around the equilibrium positions x_i are small enough then a finite number of gaps appears and that this number can be taken arbitrarily large provided the fluctuations are sufficiently small.

This paper summarizes partially works by the authors [1],[2],[3],by Albeverio, Høegh-Krohn and the authors [4] as well some material from Kirsch [5].At the same time it contains new recent results and the extension of older ones to more general physically interesting situations.

Section 2   Some general results on the spectra of random Schrödinger operators.

§ 1   <u>Definitions and Selfadjointness</u>.

In this section we introduce the random Schrödinger operators we will be concerned with, and state some rather weak conditions under which these operators turn out to be essentially selfadjoint on $C_o^\infty(\mathbb{R}^d)$, the space of infinitely differentiable functions with compact support.

Let $(\Omega, \mathcal{F}, P)$ be a probability space and $V(\omega, x) = V_\omega(x)$ a jointly measurable (realvalued) function on $\Omega \times \mathbb{R}^d$, i.e. a measurable random field. We shall investigate Schrö - dinger operators H_ω of the form:

$$H_\omega = H_o + V_\omega \tag{1}$$

where $H_o = -\Delta$ denotes the Laplacian on $L^2(\mathbb{R}^d)$.

Since we are interested in applications to solid state physics, it is natural to assume that the random field V_ω satisfies some weak symmetry property with respect to spatial shifts, namely we assume that there is a d-dimensional lattice $\{x_i\}_{i \in \mathbb{Z}^d}$ in $\mathbb{R}^d$ such that the finite dimensional distributions of the processes $V_\omega(\cdot)$ and $V_\omega(\cdot - x_i)$ are identical for all $i \in \mathbb{Z}^d$. We will call such random fields homogeneous with respect to the lattice $\{x_i\}$. A special case is a homogeneous random field (with respect to $\mathbb{R}^d$) i.e. a random field V_ω for which $V_\omega(\cdot)$ and $V_\omega(\cdot - y)$ have the same distributions for all $y \in \mathbb{R}^d$, which of course is homogeneous with respect to any lattice.

In order to have the action of the operator H_ω well defined on $C_o^\infty(\mathbb{R}^d)$, at least for almost all ω, we assume that V_ω is almost surely square integrable on any compact subset of $\mathbb{R}^d$, briefly $V_\omega \in L^2_{loc}(\mathbb{R}^d)$ a.s.. It is easy to see that for this it is suffi - cient that:

$$E \left(\int_{C_o} |V_\omega(x)|^2 \, dx \right) < +\infty \tag{2}$$

where E denotes the expectation with respect to P and C_o is the unit cell of the lattice $\{x_i\}$. If V_ω is homogeneous with respect to $\mathbb{R}^d$ then (2) is implied by:

$$E \left(|V_\omega(o)|^2 \right) < +\infty \tag{3}$$

While condition (2) implies that H_ω is well defined on $C_o^\infty(\mathbb{R}^d)$ it does not ensure in general the essential selfadjointness of H_ω on that domain. However a condition of the same type but slightly stronger guarantees the essential selfadjointness on $C_o^\infty(\mathbb{R}^d)$.

<u>Theorem 1</u>   Let $p > p_o := \max(2, \frac{d}{2})$ where d is the dimension and assume:

$$E\left(\left| \int_{C_o} |V_\omega(x)|^p dx \right|^k \right) < +\infty \tag{4}$$

for some $k > k_o := \dfrac{(p_o+1)p - p_o}{p - p_o}$, $k \in \mathbb{N}$. Then $H_\omega = H_o + V_\omega$ is essentially self-adjoint on $C_o^\infty(\mathbb{R}^d)$ for almost all ω.

This theorem is proved in [1] where also several examples are given (see also [5]).
Theorem 1 is strong enough to prove the essential selfadjointness of virtually all
interesting random Schrödinger operators,as long as they can be defined as operators
sum .

In the examples we are going to discuss in section 3 the potential V_ω actually will
not be a function but rather a quadratic form so that (1) can be defined only in the
sense of forms sum.In this case for all $\omega \in \Omega$ we are given a quadratic form V_ω on $Q(H_o)$,
the form domain of H_o,such that $\omega \to V_\omega(\phi,\psi)$ is measurable for any $\phi ,\psi \in Q(H_o)$,and
define $H_\omega = H_o + V_\omega$ as the sum of quadratic forms.This form sum is (the form of)
a selfadjoint operator if V_ω is form bounded with respect to H_o with relative constant
less that one (see e.g. Reed-Simon II $[6]$). In section 2 §4 we will give examples
of such random quadratic forms.Finally in the sequel we will set for any quadratic
form ß and y in $\mathbb{R}^d$:

$$\beta_y(\phi,\psi) = \beta(\phi_y,\psi_y) \tag{5}$$

where for any $\phi \in L^2(\mathbb{R}^d)$: $\phi_y(x) = \phi(x+y)$.Sometimes we will also write $\beta(\cdot - y)$ or e-
ven $\beta(x-y)$ instead of β_y in order to stress the analogy between potentials that are
functions and those that are quadratic forms.

§2   Ergodicity

In §1 we have assumed the random field $V_\omega(x)$ to be homogeneous either with respect to
a lattice $\{x_i\}$ or with respect to $\mathbb{R}^d$. By passing to an equivalent process,if necessary,we can assume that we are given for any lattice point $\{x_i\}$ (respectively any $y \in \mathbb{R}^d$)
a measure preserving transformation T_i (respectively T_y) on $(\Omega , \mathcal{F},P)$ such that :

$$V_{T_i\omega}(x) = V_\omega(x - x_i) \tag{6}$$

respectively:

$$V_{T_y\omega}(x) = V_\omega(x- y) \tag{6'}$$

As a further assumption we now suppose that the transformations $\{T_i\}_{i \in Z^d}$ (respecti-
vely $\{T_y\}_{y \in \mathbb{R}^d}$) are <u>ergodic</u> in the sense that any invariant event $A \in \mathcal{F}$,i.e. $T_i^{-1}A=A$
for all $i \in Z^d$ (respectively $T_y^{-1}A = A$ for all $y \in \mathbb{R}^d$), has probability one or zero.
In this case we call V <u>metrically transitive</u> with respect to the lattice $\{x_i\}$(resp.
to $\mathbb{R}^d$).This notion extends in an obvious way to random quadratic forms.Examples of
metrically transitive random potentials will be given in §4 of this section (see also
$[2]$ and $[5]$).

Let us now denote by $\sigma(H_\omega)$ ($\sigma_{dis}(H_\omega)$, $\sigma_{ess}(H_\omega)$) the spectrum (discrete spectrum,
essential spectrum) of H_ω . $\sigma_{dis}(H_\omega)$ is the set of isolated eigenvalues of finite mul-
tiplicity and $\sigma_{ess}(H_\omega) = \sigma(H_\omega)\backslash\sigma_{dis}(H_\omega)$.Furthermore we denote by $\sigma_{pp}(H_\omega)$, $\sigma_c(H_\omega)$,
$\sigma_{ac}(H_\omega)$ the pure point spectrum,continuous spectrum,absolutely continuous spectrum of
H_ω .We remark that $\sigma_{pp}(H_\omega)$ is the set of all eigenvalues of H_ω,no matter whether they
are isolated or have finite multiplicity.

<u>Theorem 2</u>   Let V_ω be a measurable random field or a random quadratic form,metrically transitive with respect to a lattice $\{x_i\}$ (or to R^d).Suppose in addition that the operator $H_\omega = H_o + V_\omega$ is essentially selfadjoint on $C_o^\infty(\mathbb{R}^d)$ or that V_ω is form bounded with respect to H_o with relative bound less than one. Then there exist non-random sets Σ , Σ_{pp} , Σ_c , $\Sigma_{ac} \subset \mathbb{R}$ such that:

   i) $\Sigma = \sigma(H_\omega)$

   ii) $\Sigma_{pp} = \overline{\sigma_{pp}(H_\omega)}$

   iii) $\Sigma_c = \sigma_c(H_\omega)$

   iv) $\Sigma_{ac} = \sigma_{ac}(H_\omega)$

   v) $\sigma_{dis}(H_\omega) = \emptyset$

This theorem partially goes back to Pastur [7] who proved i) and v) for random finite difference operators.See also Kunz and Souillard for an extension of Pastur's work. In the case of Schrödinger operators on $L^2(\mathbb{R}^d)$ and also more general situations Theorem 2 is proved in [2] and [5] .As pointed out by Simon [9] this theorem also applies to almost periodic Schrödinger operators (see also [5]).

While the discrete part of the spectrum ,$\sigma_{dis}(H_\omega)$,is almost surely empty the pure point spectrum $\sigma_{pp}(H_\omega)$ may be rather thick. For example it was proved by Gol'dsheid, Molchanov and Pastur [10]that for some one dimensional situations $\overline{\sigma_{pp}(H_\omega)} = \sigma(H_\omega) = [0,\infty[$ almost surely.In their case it can be shown that a given $\lambda \in \mathbb{R}$ belongs to $\sigma_{pp}(H_\omega)$ with probability zero.Thus as ω varies $\sigma_{pp}(H_\omega)$ must vary rather strongly; nevertheless its closure does not vary on a set of probability one.

§3   A special model

We now turn to more specific random potentials V_ω,at the same time giving examples for the general theorem above and investigating the spectrum of $H_\omega = H_o + V_\omega$ in more details.

Let us start with a rather simple model.We assume that in each lattice point x_i of a (ideal) crystal we are given a particle generating a potential $q_i f(x-x_i)$ around its position.If the coupling constant q_i (or "charges" in analogy to electrostatic potentials) are independent of $i \in Z^d$ we are concerned with an ideal crystal with periodic total potential :

$$V(x) = \sum_{i \, Z^d} q_i f(x-x_i) \tag{7}$$

Of course,in order to have the infinite sum in (7) well defined we need some assumptions on the decay of the function $f(x)$ at infinity and moreover to have $V \in L^2_{loc}(\mathbb{R}^d)$ or even $H = H_o + V$ essentially selfadjoint on $C_o^\infty(\mathbb{R}^d)$ we need some control on the local singularities of $f(x)$.

For both these questions it is convenient to introduce the following Banach spaces (see also Simon [11] for their use in scattering theory).We denote by $\ell^1(L^p)$ the space of measurable functions $f : \mathbb{R}^d \to \mathbb{C}$ such that:

$$\|f\|_{1,p} = \sum_{i \in \mathbb{Z}^d} \left(\int_{C_0 + x_i} |f(x)|^p dx \right)^{1/p} < \infty \qquad (8)$$

As it can be easily checked a function $f \in L^p_{loc}$ which decays at infinity like $|x|^{-d-\varepsilon}$ belongs to $\ell^1(L^p)$.

It is not difficult to show that V defined by (7) is an L^2_{loc}-function and furthermore $H = H_0 + V$ is essentially selfadjoint on $C_0^\infty(\mathbb{R}^d)$ if $f \in \ell^1(L^p)$ for p = 2 if d < 3, p > 2 if d = 4, $p = \frac{d}{2}$ if d > 5.

Now we let the "charges" q_i to be random. The total potential then is given by :

$$V_\omega = \sum_{i \, \mathbb{Z}^d} q_i(\omega) f(x-x_i) \qquad (9)$$

<u>Proposition 1</u>   Let $\{q_i\}_{i \, \mathbb{Z}^d}$ be a homogeneous random field on $\mathbb{Z}^d$,let the function f belong to $\ell^1(L^p)$ with $p > \max(2,\frac{d}{2})$ and assume $E(|q_0|) < \infty$. Then:

i) For P-almost all ω the infinite sum (9) converges absolutely for Lebesgue-almost all $x \in \mathbb{R}^d$ and V_ω is locally integrable.a.s.

ii) The random field V_ω is homogeneous with respect to the lattice $\{x_i\}$ .If $\{q_i\}$ is metrically transitive V_ω is metrically transitive with respect to $\{x_i\}$.

iii) If $E(|q_0|^k) < \infty$ for $k > k_0$, k_0 as in theorem 1, then the operator : $H_\omega = H_0 + V_\omega$ is essentially selfadjoint on $C_0^\infty(\mathbb{R}^d)$ for almost all $\omega \in \Omega$.

Proposition 1 tells us in particular that theorem 2 applies to the operator H_ω .
We now generalize this model in order to include also randomness of the position of the particles.This means that we are interested in potentials of the form:

$$V_\omega = \sum_{i \, \mathbb{Z}^d} q_i(\omega) f(x-\xi_i(\omega)) \qquad (10)$$

In order to define and to investigate such random potentials it is convenient to introduce the notion of random point measures.This means that for any $\omega \in \Omega$ we are given a measure μ_ω on $\mathbb{R}^d$ of the form $\mu_\omega = \sum q_i(\omega)\delta_{\xi_i(\omega)}$ where δ_y is the Dirac measure concentrated at the point y,such that for any bounded Borel set $A \subset \mathbb{R}^d$ $\mu_\omega(A)$ is a random variable.(For the definition and the properties of random measures we refer to the book by Kallenberg [12]).We will call a random measure μ_ω homogeneous with respect to the lattice $\{x_i\}$ (resp. to $\mathbb{R}^d$) if for any bounded Borel set A and any point x_i in the lattice (resp. $y \in \mathbb{R}^d$) the random variables $\mu_\omega(A)$ and $\mu_\omega(A+x_i)$ (resp. $\mu_\omega(A+y)$) have the same distributions.Again we find measure preserving transformations $\{T_i\}_{i \in \mathbb{Z}^d}$ (resp. $\{T_y\}_{y \in \mathbb{R}^d}$) on such that:

$$\mu_{T_i\omega}(A) = \mu_\omega(A-x_i) \qquad (\text{resp.} \quad \mu_{T_y\omega}(A) = \mu_\omega(A-y))$$

We will call a random measure metrically transitive with respect to the lattice $\{x_i\}$ (resp. to $\mathbb{R}^d$) if the corresponding measure preserving transformations are ergodic. Now we can define more rigorously the potential given by (9) :

$$V_\omega(x) \;=\; \int f(x+y)\mu_\omega(dy) \;=\; f * \mu_\omega(x) \tag{11}$$

<u>Proposition 2</u>   Let μ_ω be a random point measure homogeneous with respect to the lattice $\{x_i\}$ whose unit cell we denote by C_o, and such that $E(|\mu_\omega|(C_o)) < \infty$ .Let f belong to $\ell^1(L^p)$ with $p > \max(2,\frac{d}{2})$. Then:

i) For P-almost all $\omega \in \Omega$ the integral in (11) is absolutely convergent for Lebesgue almost all x and V_ω is P-almost surely locally integrable.

ii) The random field V_ω is homogeneous with respect to $\{x_i\}$(resp. to $\mathbb{R}^d$) if μ_ω is homogeneous with respect to $\{x_i\}$ $(\mathbb{R}^d)$ and the same holds for the metrically transitivity property.

iii) If $E(||\mu_\omega(C_o)|^k) < \infty$ for $k > k_o$,k_o as in theorem 1,then $H_\omega = \dot{H}_o + V_\omega$ is essentially selfadjoint on $C_o^\infty(\mathbb{R}^d)$ for almost all ω .

We now define random quadratic forms analogous to potentials of the form (11). We call a quadratic form β defined on $Q(H_o)$,the form domain of H_o,<u>distributive</u> if $\beta(\chi\phi,\psi) = \beta(\phi,\overline{\chi}\psi)$ for any $\phi,\psi \in Q(H_o)$ and $\chi \in C_o^\infty(\mathbb{R}^d)$ .Furthermore β is said to be <u>summably bounded</u> with respect to H_o if β is form bounded with respect to H_o and for any $i \in \mathbb{Z}^d$ there are constants a_i , $b_i > 0$ such that $\sum_i a_i < 1$, $\sum_i b_i < \infty$ and
$$|\beta(\phi,\phi)| \leqslant a_i\langle H_o\phi,\phi\rangle + b_i\langle\phi,\phi\rangle \qquad \text{whenever } \phi \in Q(H_o) \text{ and supp}\phi \subset x_i + C_o$$

For a given random point measure μ_ω we will define:

$$V_\omega(\phi,\psi) \;=\; \int\beta_y(\phi,\psi)\mu_\omega(dy) \tag{12}$$

or for short:
$$V_\omega \;=\; \int\beta_y\,\mu_\omega(dy)$$

<u>Proposition 3</u>   Let μ_ω be a random point measure homogeneous with respect to a lattice $\{x_i\}$ such that $|\mu_\omega(C_o)| < c < \infty$ for a ω -independent constant c and almost all ω ,and let β be a distributive quadratic form summably bounded with respect to H_o . Then:

i) For P-almost all ω the integral (12) is absolutely convergent for any ϕ ,$\psi \in Q(H_o)$ so that V_ω is a quadratic form well defined on $Q(H_o)$.

ii) The random quadratic form V_ω is homogeneous with respect to $\{x_i\}$.If μ_ω is metrically transitive with respect to $\{x_i\}$ the same is true for V_ω .

iii) V_ω is form bounded with respect to H_o with relative constant not exceeding $c \sum_i a_i$.

The proof of this result is given in [5] using techniques due to Morgan [13]   .
In the sequel we will call a homogeneous random measure μ_ω satisfying the condition

$|\mu_\omega|(C_o) < C < \infty$ for all ω uniformly <u>locally bounded</u>. We remark that for $d = 1$ any Borel measure μ defines a distributive summably bounded quadratic form β_μ via: $\beta_\mu(\phi,\psi) = \int \phi(x)\psi(x)\,\mu(dx)$. Moreover the constants a_i can be chosen in such a way that $\Sigma_i\, a_i$ is arbitrarily small. Also a function $f \in \ell^1(L^p)$ defines a distributive summably bounded quadratic form provided $p > 1$ for $d = 2$, $p > \frac{d}{2}$ for $d > 3$; also in this case Σa_i can be chosen arbitrarily small. Thus for uniformly locally bounded random measures the potentials considered in Proposition 2 are a special case of those in Proposition 3. However if μ_ω fails to be uniformly locally bounded the potentials considered in Proposition 2 are in general no longer form bounded.

§ 4   <u>Admissible potentials and spectra of random operators</u>

In this paragraph we recall and extend some of the results from [1] and [5] concerning the spectrum of the operators $H_\omega = H_o + V_\omega$ with V_ω defined by (11) or (12). We will call a measure ν a realisation of the measure μ_ω if for any compact set $K \subset \mathbb{R}^d$ and any $\varepsilon > 0$:

$$P(\{\omega;\ |\nu-\mu_\omega|\ (K) < \varepsilon\}) > 0 \tag{13}$$

where $|.|$ denotes the total variation.

First let us assume that the random measure μ_ω is uniformly locally bounded (see § 3). Let furthermore β be a summably bounded distributive quadratic form and let V_ω be given by: $V_\omega = \int \beta_y \mu_\omega(dy)$. Since μ_ω is uniformly locally bounded this covers also the case of potentials given by (11). Finally we will call a quadratic form W : $W = \int \beta_y \nu(dy)$ on $Q(H_o)$ an <u>admissible potential</u> with respect to V_ω if the measure ν is a realisation of the measure μ_ω.

<u>Theorem 3</u>   Let the random measure μ_ω be metrically transitive with respect to a lattice $\{x_i\}$ or to $\mathbb{R}^d$ and let μ_ω be uniformly locally bounded. Let furthermore β be a summably bounded distributive quadratic form and define $V_\omega = \int \beta_y \mu_\omega(dy)$. Denote by Σ the almost surely constant spectrum of $H_\omega = H_o + V_\omega$. Then for any admissible potential W one has:

$$\sigma(H_o + W) \subseteq \Sigma$$

We shall sketch very briefly the idea of the proof of the above theorem.
Let us fix a sequence of compact sets $K_n \subseteq \mathbb{R}^d$. Then for any $n \in \mathbb{N}$ we can find a set $\Omega_n \subseteq \Omega$ of positive probability such that:

$$|V_\omega(\phi,\phi) - W(\phi,\phi)| < \frac{1}{n}\, \|(H + 1)^{1/2}\,\phi\| \tag{14}$$

for any $\phi \in Q(H_o)$ with supp $\phi \subset K_n$ and for any $\omega \in \Omega_n$. Furthermore in Ω_n we can always find a point ω_n such that $\sigma(H_o + V_{\omega_n}) = \Sigma$ since Ω_n has positive probability. But (14) implies that $H_o + V_{\omega_n}$ converges to $H_o + W$ as $n \to \infty$ in in the strong resolvent sense and this implies (see e.g. Reed-Simon [14])that:

$$\sigma(H_0 + W) \subset \lim_{n \to \infty} \sigma(H_0 + V_{\omega_n}) = \Sigma \quad .$$

Call now $\mathcal{A}$ the class of all admissible potentials; then:

<u>Corollary</u>   Under the same assumptions of theorem 3 we have:

$$\Sigma = \bigcup_{W \in \mathcal{A}} \sigma(H_0 + W)$$

The Corollary follows immediately from theorem 3 since V_ω is an admissible potential
for almost all ω .

The above corollary tells us in particular that the spectrum Σ of the random operator
H_ω depends only on the supports of the finite dimensional distributions $P_{A_1 .. A_n}$.
As a concrete example consider a random mixture of particles with two possible charges
λ_1 and λ_2 the potential of which is given by: $V_\omega(x) = \Sigma q_i(\omega)f(x-x_i)$ where q_i are
independent random variables with $P(q_0 = \lambda_1) = p$ and $P(q_0 = \lambda_2) = 1-p$, and $\{x_i\}$ is some lattice.Then the spectrum of the corresponding random Hamiltonian is almost sure-
ly independent of p as long as $0 < p < 1$.

In order to extend theorem 3 to the case of random point measures not necessarily uniformly locally bounded (see proposition 2) we need some further assumptions which allow us to have some control on the potential "at infinity" (a precise meaning of this
will be given later on).For simplicity we state these assumptions in a way stronger
than necessary,as far as next theorem is concerned,since we will need them in order
to prove our main theorem (theorem 5 of this section).

For any Borel set A let $\mathcal{F}_A$ denote the σ-algebra generated by the random variables
$\{\mu_\omega(B); B \subset A, B \in \mathcal{B}(\mathbf{R}^d)\}$ .We will call a random measure positive correlated with
respect to the lattice $\{x_i\}$ if for any $\Omega_1 \in \mathcal{F}_{A_1}$, $\Omega_2 \in \mathcal{F}_{A_2}$ with $P(\Omega_1) > 0$ and
$P(\Omega_2) > 0$, also $P(\Omega_1 \cap \Omega_2) > 0$ whenever for any x_i or $A_1 \cap C_0 + x_i = \phi$ or $A_2 \cap C_0 + x_i = \phi$.

<u>Example</u>: If the $\{q_i\}$ are independent random variables and $\{x_i\}$ is a lattice, then
$\mu_\omega = \Sigma q_i(\omega)\delta_{x_i}$ is positive correlated with respect to $\{x_i\}$.

If the random measure μ_ω is not necessarily uniformly locally bounded we call a
function $W(x) = \int f(x+y) \, \nu(dy)$ an <u>admissible potential</u> with respect to
$V_\omega(x) = \int f(x+y) \, \mu_\omega(dy)$ if ν <u>is</u> a realisation of μ_ω and

    (i)   W is locally square integrable

    (ii) $H_\omega = H_0 + W$ is essentially selfadjoint on $C_0^\infty(\mathbf{R}^d)$.

Since for uniformly locally bounded μ_ω the conditions (i) and (ii) are satisfied by
any realisation ν of μ_ω the two definitions of admissible potentials agree in the
cases where they are both applicable.

__Theorem 4__   Let the random measure μ_ω be metrically transitive with respect to a lattice $\{x_i\}$ (resp. $\mathbf{R}^d$) and let μ_ω be positive correlated with respect to a lattice $\{y_i\} \subset \{x_i\}$. Let furthermore $f \in \ell^1(L^p)$, $p > \max(2,\frac{d}{2})$, and assume $E(||\mu_\omega|(C)|^k) < \infty$ for a $k > k_o$, k_o as in theorem 1; then for any potential W admissible with respect to $V_\omega(\cdot) = \int f_y(\cdot)\mu_\omega(dy)$ we have:

$$\sigma(H_o + W) \subset \Sigma$$

__Corollary__   Under the assumptions of Theorem 4 we have:

$$\Sigma = \bigcup_{W \in \mathcal{A}} \sigma(H_o + W)$$

Theorems 3 and 4 are as they stand of limited usefulness for practical computations of the spectrum Σ of the random Hamiltonian since the class $\mathcal{A}$ of admissible potentials is in general very large and furthermore for most admissible potentials the determination of the spectrum of the corresponding Hamiltonian is as difficult as the computation of Σ. Fortunately in several situations a smaller class of nicer admissible potentials is sufficient to determine the spectrum Σ. Let us define $\mathcal{P}$ to be the class of all admissible potentials which are periodic (with arbitrary period). then the following result holds:

__Theorem 5__ Let the random measure μ_ω be metrically transitive with respect to a lattice $\{x_i\}$ (resp. to $\mathbf{R}^d$) and assumed μ_ω to be positive correlated with respect to a lattice $\{y_i\} \subset \{x_i\}$. If either β is a summably bounded, distributive quadratic form, μ_ω is uniformly locally bounded and V_ω is given by: $V_\omega = \int \beta_y \mu_\omega(dy)$, or f is a $\ell^1(L^p)$-function with $p > \max(2,\frac{d}{2})$, $E(||\mu_\omega|(C_o)|^k) < \infty$ for $k > k_o$ (k_o as in theorem 1) and V_ω is given by: $V_\omega = \int f_y \mu_\omega(dy)$, then:

$$\Sigma = \overline{\bigcup_{W \in \mathcal{P}} \sigma(H_o + W)}$$

__Sketch of the proof__: By theorems 3 and 4 $\sigma(H_o + W) \subset \Sigma$ for any $W \in \mathcal{P}$ hence also

$\overline{\bigcup_{W \in \mathcal{P}} \sigma(H_o + W)} \subset \Sigma$ since Σ is a closed set. Thus it remains to prove the converse inclusion. For this one shows that for almost all ω the Hamiltonian H_ω can be approximated in the strong resolvent sense by Hamiltonian of the form: $H_n = H_o + W_n$, $n \in \mathbb{N}$ with $W_n \in \mathcal{P}$ for any n. Hence: $\sigma(H_\omega) = \Sigma \subset \overline{\bigcup_{n \in \mathbb{N}} \sigma(H_o + W_n)} \subset \overline{\bigcup_{W \in \mathcal{P}} \sigma(H_o + W)}$.

Theorem 5 has some remarkable consequences:

1) For $d \leq 3$ (and under additional mild assumptions also for $d \geq 4$) the spectrum of $H_o + W$ for $W \in \mathcal{P}$ has a band structure that is $\sigma(H_o + W) = \cup [a_i , b_i]$, $a_i < b_i$. Hence also Σ has a band structure; in particular Σ contains no isolated points.

2) Σ is the closure of an open set. This is in contrast to some almost periodic potentials considered by Avron and Simon [15] who proved that the spectrum of generic limit of periodic potentials is nowhere dense (see also Moser [16]). While these potentials can be looked upon as random potentials,they are of course not positive correlated.

3) For d=2 Skriganov [17] proved that the spectrum of a periodic Schrödinger operator contains the infinite interval $[L,+\infty[$ for some L depending on the operator,provided that the periodicity lattice satisfies a rationality condition (see [17]).Thus Theorem 5 gives us the same result for the random Hamiltonians we are concerned with. Very recently Skriganov generalized his theorem to arbitrary dimension $d \geq 2$.

Theorem 5 leaves open the question whether actually there are gaps in the spectrum since it may happen that the energy bands of the periodic admissible potentials may overlap in such a way that $\Sigma = [a,+\infty[$ for some a .

Although one can derive from Theorem 5 some general criteria for the existence of gaps in Σ (see [1] and [5]),we will not discuss them here but in the next section in connection with some concrete models.

III Random point interactions

In this chapter we consider point interactions in one and three dimensions as examples
and illustrations of the general theorems of the second chapter. The two dimensional
case actually is completely analogous and is omitted here. The material of this
chapter may be found in Kirsch and Martinelli [1], [2] for section III. 1 and
III. 2, in Albeverio, Høegh-Krohn, Kirsch and Martinelli [4] for section III. 3,
while the material in section III. 4 so far is unpublished.

§ 1 Definitions

Throughout this chapter let μ_ω be a uniformally locally bounded random measure
metrically transitive with respect to a lattice $\{x_i\}$ or to $\mathbb{R}^d$. Let δ denote the quadratic form $\delta(\phi,\psi) = \phi(0)\overline{\psi(0)}$. In the one dimensional case this quadratic form is
infinitesimally form bounded with respect to $H_o = -\dfrac{d^2}{dx^2}$ and moreover, as can easily
been seen, is summably bounded and distributive. Thus following Proposition 3 the quadratic form:

$$V_\omega = \int \delta_y \mu_\omega (dy) \tag{1}$$

is well defined and infinitesimally form bounded with respect to H_o, hence:

$$H_\omega = H_o + V_\omega \tag{2}$$

defined as a form sum is selfadjoint on $L^2(\mathbb{R})$ with form domain $Q(H_o)$.
The resolvent of H_ω can be computed explicitly. For let $\mu_\omega = \Sigma\, q_i(\omega)\delta(\cdot-\xi_i(\omega))$ and denote by G_k the kernel of the resolvent $(H_o-k^2)^{-1}$. G_k (the Green function) is given by:

$$G_k(x,y) = G_k(x-y) = \frac{1}{2ik}\, e^{ik|x-y|}. \text{Let us furthermore denote by } H_{\lambda_j,x_j} \text{ the Hamiltonian:}$$

$$H_{\lambda_j,x_j} = -\frac{d^2}{dx^2} + \sum_{j\,\in\, Z} \lambda_j \delta(\cdot -x_j) \tag{3}$$

which is well defined as form sum if there exist constants $C,D>0$ such that $|\lambda_j| \leq C$
for all $j \in Z$ and $|x_j - x_k| \geq D$ for all $j,k \in Z$ with $j \neq k$.
With this notation the kernel of the resolvent of H_{λ_j,x_j} is given by:

$$(H_{\lambda_j,x_j} - k^2)^{-1}(x,y) = G_k(x-y) + \sum_{i,j\,\in\, Z} (T_k^{-1})_{ij}\, G_k(x-x_i)G_k(y-x_j) \tag{4}$$

where T_k is the operator in $\ell^2(Z)$ with matrix elements $(T_k)_{ij} = \dfrac{1}{\lambda_j}\, \delta_{ij} + G_k(x_i-x_j)$ (5)

(δ_{ij} being the Kronecker symbol).
For a periodic point interaction, i.e. for $x_i = ip$ for some $p>0$ and $\lambda_i=\lambda$ the spectrum of $H_{\lambda,ip} = H_\lambda$ was computed already by Kronig and Penney [18] in 1931. As usual
for periodic Hamiltonians the spectrum consists of " bands " separated by " gaps "
(forbidden zones). In general let $H = H_o + V$ be a periodic Hamiltonian with periodicity lattice $\{x_i\}_{i\in Z^d}$ and unit cell C_o. Under weak assumptions on V (see e.g. Reed-

Simon IV [19],Avron,Grossman and Rodriguez [20] , Thomas [21]) the operator H can be looked upon as the direct integral over the reduced Hamiltonians H(θ):

$$H \simeq \int_B^{\oplus} H(\theta)d\theta \ , \ B = [0,2\pi[^d \tag{6}$$

The reduced Hamiltonians H(θ) are the restrictions of H to the unit cell C_o with θ-depending boundary conditions,namely: $\psi(x+a_j) = e^{i\theta}\psi(x)$, $\frac{\partial}{\partial x_j}\psi(x+a_j) = e^{i\theta}\frac{\partial}{\partial x_j}\psi(x)$ for all $x \in \overline{C_o}$, such that also $x+a_j \in \overline{C_o}$

Furthermore H(θ) has purely discrete spectrum for all $\theta \in B$;we will call $E_n(\theta)$ the n-th eigenvalue of H(θ) (counting multiplicity).Each $E_n(\theta)$ as a function of θ are continuous in $\theta \in B$ and moreover $B_n = \{E_n(\theta); \ \theta \in B\}$ is a compact interval of the real line that never reduces to a single point.The spectrum of H is then given by:

$$\sigma(H) = \bigcup_n B_n = \{E_n(\theta); \ \theta \in B, \ n \in \mathbb{N}\} \tag{7}$$

For this reason B_n is usually called the n-th band of H and $\Gamma_n = \{x;y<x<z,\forall y \in B_n$, $\forall z \in B_{n+1}\}$ the n-th gap although it can be empty if B_n and B_{n+1} overlap. In the case of $H_\lambda = -\frac{d^2}{dx^2} + \lambda\Sigma_i \delta(\cdot - ip)$ it is possible to compute the bands B_n

rather explicitely.If we set $B_n = [a_n(\lambda),b_n(\lambda)]$,then for $\lambda \neq 0$ $a_n(\lambda) > b_{n-1}(\lambda)$ so that there is no overlapping between the bands and all (infinitely many) gaps are open,i.e. $\Gamma_n \neq \emptyset$.

For $\lambda > 0$ the whole spectrum is positive,i.e. $a_1(\lambda) > 0$ and $b_n(\lambda) = (2n\pi)^2 p^{-2}$ independent of λ ,while $a_n(\lambda)$ is a strictly monotone increasing continuous function of λ and $a_n(\lambda) \to b_n(\lambda)$ as $\lambda \to 0$ .For $\lambda < 0$ the band $[a_1(\lambda),b_1(\lambda)]$ is at least partially negative,i.e. $a_1(\lambda) < 0$, while $b_1(\lambda)$ is a strictly monotone increasing continuous function such that $b_1(\lambda_o) = 0$ for some $\lambda_o < 0$.The $a_n(\lambda)$ for $n \geq 2$ are given by: $a_n(\lambda) = (2\pi(n-1))^2 p^{-2}$ while the $b_n(\lambda)$ are strictly monotone increasing continuous functions. We remark here that both a_n and b_n can be expresses as the zeros of a certain trascendntal equation (see e.g. Kronig,Penney [18] ,Saxon,Hunter [22] , Flügge [23]).

In dimension $d \geq 2$ the point measure δ unfortunately does not define a quadratic form that is form bounded with respect to H_o.Thus the above procedure breaks down for $d \geq 2$. However Hamiltonians with point interactions for $d = 2$ and $d = 3$ can be still defined while for $d \geq 4$ there is definitely no point interaction Hamiltonian since in this case H_o is essentially selfadjoint on $C_o^\infty(\mathbb{R}^d \setminus \{0\})$. In the sequel we will restrict ourselves to $d = 3$ although all our results after slight modifications extend to $d = 2$ too.

There are several methods to define three dimensional point interaction,for example by approximation with smooth shrincking potentials (see Albeverio,Høegh-Krohn [24] , Albeverio,Gesztesy and Høegh-Krohn [25]) and by the methods of non standard analysis (see Albeverio,Fenstad and Høegh-Krohn [26] ,Albeverio,Fenstad,Høegh-Krohn and Lindstrøm [27]).Perhaps the most intuitive approach is the one provided by non standard methods.There the point interactions are defined as a non standard step function with infinitesimal support and infinite height.In this paper we will define the point interaction Hamiltonian just by giving its resolvent and refering to the work by Grossmann,Høegh-Krohn and Mebkhout [28] for a proof that this Hamiltonian is the one obtained by smooth approximations.We will denote by the symbol:

$$H_{\lambda_i,x_i} = H_o + \sum_i \lambda_i \delta(\cdot - x_i) \tag{8}$$

that has only a formal meaning in three dimensions,the unique selfadjoint operator on $L^2(\mathbb{R}^3)$ whose resolvent is given by:

$$(H_{\lambda_i,x_i} - k^2)^{-1}(x,y) = G_k(x-y) + \sum_{i,j \in Z^3} (T_k)^{-1}_{ij} G_k(x-x_i)G_k(y-x_j) \tag{9}$$

where now $G_k(x-y) = \dfrac{1}{4\pi|x-y|} e^{ik|x-y|}$ is the Green function in three dimensions and T_k is

the operator on $\ell^2(Z^3)$ given by: $(T_k)_{ij} = (\lambda_i^{-1} - \dfrac{ik}{4})\delta_{ij} - G_k(x_i - x_j)$ $\tag{10}$

For the points x_i we have again to assume that $|x_i - x_j| > D$ for $i \neq j$ and some positive constant D,and on the λ_i we assume $-C \leq \lambda \leq -B < 0$ for some $C,B > 0$.We can restrict ourselves to negative coupling constants λ_i since for positive λ_i and d=3 the Hamiltonian H_{λ_i,x_i} is equivalent to H_o.For the periodic case i.e. $\lambda_i = \lambda$ for all $i \in Z^3$ and $\{x_i\}$ a lattice in $\mathbb{R}^3$,the spectrum of $H_\lambda = H_{\lambda,x_i} = H_o + \lambda\sum\delta(\cdot - x_i)$ $\tag{11}$ was computed by Grossmann,Høegh-Krohn and Mebkhout [28] and their result is the following: there exist two strictly monotonically increasing continuous functions of λ , $a_o(\lambda)$ and $b_o(\lambda)$,such that:

$$\sigma(H_\lambda) = [a_o(\lambda),b_o(\lambda)] \cup [0,\infty[\tag{12}$$

Thus there is at most one gap in the spectrum of H_λ .Actually $a_o(\lambda) < 0$ for $\lambda < 0$ while $b_o(\lambda) < 0$ for $\lambda < \lambda_o$ and $b_o(\lambda)$ if $\lambda > \lambda_o$ for some $\lambda_o < 0$.
Although the three dimensional δ-interaction is neither a function nor a quadratic form it is still possible to prove the results of the previous section for Hamiltonians given by (9) (see Kirsch,Martinelli [1] ,Albeverio,Høegh-Krohn,Kirsch,Martinelli [4])
In particular for a uniformly locally bounded metrically transitive random measure $\mu_\omega = \sum q_i(\omega)\delta_{\xi_i(\omega)}$ positive correlated,the random Hamiltonian H_ω formally given by:

$$H_\omega = H_o + \int \delta_y \mu_\omega(dy) = H_o + \sum q_i(\omega)\delta(\cdot - \xi_i(\omega)) \tag{13}$$

satisfies:

<u>Proposition 4</u> (i) There exists a subset Σ of the real line such that $\sigma(H_\omega) = \Sigma$ a.s.
(ii) The discrete part of $\sigma(H_\omega)$ is empty.
(iii) If υ is a realization of the measure μ_ω ,then the spectrum of $H_\upsilon = H_o + \int \delta_y \upsilon(dy)$ is contained in Σ .
(iv) $\Sigma = \overline{\cup \sigma(H_\upsilon)}$ where the union is extended over the class of all realizations

of the random measure μ_ω which are periodic (with arbitrary period).

§ 2 Models with random charges (alloys)

Here we investigate the case of point interactions of random strengths located at the points of a lattice $\{x_i\}_{i \in Z^d}$ (d=1,2,3). The random Hamiltonian is thus formally given by:

$$H_\omega = H_o + \sum_{i \in Z^d} q_i(\omega) \delta(\cdot - x_i) \tag{14}$$

where $\{q_i(\omega)\}_{i \in Z^d}$ is a metrically transitive, positive correlated random field on Z^d, with $|q_i(\omega)| < C$ for all $i \in Z^d$ and all $\omega \in \Omega$. Such Hamiltonians may serve as models of alloys where atoms or ions of different kind randomly distributed, over the lattice $\{x_i\}$ interact with the electrons via a point interaction of strength $q_i(\omega)$, $q_i(\omega)$ being the strength parameter of the atom located at the point x_i. We will that under some additional assumptions on the random field $\{q_i\}$ the spectrum Σ of H_ω can be computed rather explicitely; more precisely we assume that either $q_i(\omega) \geq c_1 > 0$ for all $i \in Z^d$ and $\omega \in \Omega$, or $q_i(\omega) \leq c_2 < 0$ for all $i \in Z^d$ and $\omega \in \Omega$. Let us set $\lambda_{min} = $ inf suppP_{q_o} and $\lambda_{max} = $ sup suppP_{q_o} where P_{q_o} is the distribution function of the random variable q_o and suppP_{q_o} its support. By theorem 5 (or Proposition 4 in the three dimensional case) we have:

$$\Sigma = \overline{\bigcup \sigma(H_o + \sum \lambda_i \delta(\cdot - x_i))} \tag{15}$$

where the union is taken over all sequences $\{\lambda_i\}$ periodic in $i \in Z^d$ and with $\lambda_i \in$ suppP_{q_o}. Let analyze the spectra of the periodic Hamiltonians $H_o + \sum \lambda_i \delta(\cdot - x_i)$ (we remark that for d=3 the following arguments are only almost rigorous; for a detailed discussion see Kirsch, Martinelli [1]). Using the direct integral decomposition (see § 1 of this section and references therein) the spectrum of $H_{\{\lambda_i\}} = H_o + \sum \lambda_i \delta(\cdot - x_i)$ consists of the eigenvalues $E_n(\theta, \{\lambda_i\})$ of the reduced Hamiltonians $H_{\{\lambda_i\}}(\theta)$ associated to $H_{\{\lambda_i\}}$. Let $C_{\{\lambda_i\}}$ be the unit cell of the periodicity lattice of $H_{\{\lambda_i\}}$. Since this lattice is a sublattice of $\{x_i\}$, $C_o \subseteq C_{\{\lambda_i\}}$ and any Hamiltonian $H_{\{\lambda\}}$ with period C_o can be decomposed into a direct integral with respect to this sub-lattice. Thus the corresponding reduced Hamiltonians $\tilde{H}_{\{\lambda\}}(\theta)$ act on $L^2(C_{\{\lambda_i\}})$ as the $H_{\{\lambda_i\}}(\theta)$. The eigenvalues of $\tilde{H}_{\{\lambda\}}(\theta)$, $\tilde{E}_n(\theta, \lambda)$ are clearly different from the eigenvalues $E_n(\theta, \lambda)$ given by the direct integral decomposition with respect to the lattice $\{x_i\}$, nevertheless the set : $\{E_n(\theta, \lambda); n \in \mathbb{N}, \theta \in [0, 2[^d\} = \{E_n(\theta, \lambda); n \in \mathbb{N}, \theta \in [0, 2[^d\}$ since both sets agree with $\sigma(H_\lambda)$. Now since $\tilde{H}_{\lambda_{min}}(\theta) \leq H_{\{\lambda_i\}}(\theta) \leq \tilde{H}_{\lambda_{max}}(\theta)$ by the min-max principle we also have $\tilde{E}_n(\theta, \lambda_{min}) \leq E_n(\theta, \{\lambda_i\}) \leq \tilde{E}_n(\theta, \lambda_{max})$ and since the functions $\tilde{E}_n(\theta, \lambda)$ are continuous in λ we can find a $\tilde{\lambda}$ between λ_{min} and λ_{max} such that : $\tilde{E}_n(\theta, \tilde{\lambda}) = E_n(\theta, \{\lambda_i\})$ which implies that $E_n(\theta, \{\lambda_i\})$ belongs to $\sigma(H_{\tilde{\lambda}})$. Thus we have proven:

<u>Theorem 6</u> $\qquad \Sigma \subseteq \bigcup_{\lambda_{min} \leq \lambda \leq \lambda_{max}} \sigma(H_\lambda)$

From this and theorem 5 (or proposition 4) we get immediately:

<u>Corollary</u> If $\text{supp}P_{q_0}$ is connected (i.e. $\text{supp}P_{q_0} = [\lambda_{min}, \lambda_{max}]$

$$\Sigma = \bigcup_{\lambda \in \text{supp}P_{q_0}} \sigma(H_\lambda)$$

We remark that so far our discussion was independent of the special nature of the interaction and actually Theorem 6 and its Corollary are true for any Hamiltonian $H_\omega = H_o + \Sigma q_i(\omega)f(x-x_i)$ whenever H_ω is almost surely essentially selfadjoint on $C_o^\infty(\mathbb{R}^d)$ or the potential is form bounded with respect to H_o, if $q_i(\omega)$ are e.g. positive i.i.d random variables and f is positive (or q_i and f both negative) (see Kirsch,Martinelli [1]). For the point interactions we can go further.First let $d=1$ and $\lambda_{min} > 0$.Then,with the notations of § 1 of this section,we have:

$$\sigma(H_\lambda) = \bigcup_{n=1}^{\infty} [a_n(\lambda), b_n(\lambda)] = \bigcup_{n=1}^{\infty} [a_n(\lambda), (\frac{2n\pi}{p})^2]$$

Since furthermore $a_n(\lambda)$ is monotonically increasing in λ and $b_n(\lambda)$ is independent of λ we get: $\sigma(H_\lambda) \subset \sigma(H_{\lambda_{min}})$ for any $\lambda > \lambda_{min}$ .Thus by Theorem 6 : $\Sigma \subset \sigma(H_{\lambda_{min}})$.On the other hand by Theorem 4 : $\sigma(H_{\lambda_{min}}) \subset \Sigma$,hence

$$\Sigma = \sigma(H_{\lambda_{min}}).$$

Let now $\lambda_{max} < 0$ (but still d=1);then $\sigma(H_\lambda) = [a_1(\lambda), b_1(\lambda)] \cup \bigcup_{n=2}^{\infty}[\frac{(2\pi(n-1))^2}{p^2}, b_n(\lambda)]$

Here we have: $a_1(\lambda_{min}) \leq a_1(\lambda)$, $b_n(\lambda) \leq b_n(\lambda_{max})$ for any λ between λ_{min} and λ_{max} and any $n \in \mathbb{N}$ .Thus:

$$\sigma(H_\lambda) \quad [a_1(\lambda_{min}), b_1(\lambda_{max})] \cup \bigcup_{n=2}^{\infty} [\frac{(2\pi(n-1))^2}{p^2}, b_n(\lambda_{max})]$$

for any $\lambda \in [\lambda_{min}, \lambda_{max}]$, so that using Theorem 6 we have:

$$\Sigma \subset [a_1(\lambda_{min}), b_1(\lambda_{max})] \cup \bigcup_{n=2}^{\infty} [\frac{(2\pi(n-1))^2}{p^2}, b_n(\lambda_{max})]$$

Hence the positive part of $\sigma(H_\omega)$ agrees with the positive part of $\sigma(H_{\lambda_{max}})$.Moreover if $b_1(\lambda_{min}) \geq a_1(\lambda_{max})$ we have: $\Sigma = [a_1(\lambda_{min}), b_1(\lambda_{max})] \cup \bigcup_{n=2}^{\infty}[\frac{(2\pi(n-1))^2}{p^2}, b_n(\lambda_{max})]$

since both $\Sigma \lambda_{min} \delta(\cdot -x_i)$ and $\Sigma \lambda_{max} \delta(\cdot -x_i)$ are admissible potentials.Let us now assume $\lambda_{max} < 0$ and d=3.Then the same type of considerations yield:

$$\Sigma \subset [a_1(\lambda_{min}), b_1(\lambda_{max})] \cup [0, +\infty[$$

and if $b_1(\lambda_{min}) \geq a_1(\lambda_{max})$:

$$\Sigma = [a_1(\lambda_{min}), b_1(\lambda_{max})] \cup [0, +\infty[$$

Let us summarize these results in a theorem:

<u>Theorem 7</u>   Let $\{q_i\}_{i \in Z^d}$ be a metrically transitive, positive correlated random field on Z^d and assume that the distribution P_{q_o} of q_o has compact support and denote λ_{min} = inf supp P_{q_o}, λ_{max} = sup supp P_{q_o}. Let $\{x_i\}_{i \in Z^d}$ be a lattice in R^d. Denote by H_ω the Hamiltonians formally given by: $H_\omega = H_o + \Sigma q_i(\omega)\delta(\cdot - x_i)$,and by H_λ the periodic Hamiltonians: $H_\lambda = H_o + \lambda\Sigma\delta(\cdot - x_i)$ (see section III.1 for the rigorous definition). Finally denote by Σ the almost surely constant spectrum of H_ω . Then:

i) If d=1 and $\lambda_{min} > 0$, $\Sigma = \sigma(H_{\lambda_{max}}) = \bigcup_{n=1}^{\infty} [a_n(\lambda_{max}),b_n(\lambda_{max})]$, where $[a_n(\lambda),b_n(\lambda)]$ is the n-th band of the operator H_λ^{max}.

ii) If d=1 and $\lambda_{max} < 0$, and if for any $E \in [a_1(\lambda_{min}),b_1(\lambda_{max})]$ there exists a $\lambda \in$ suppP_{q_o} with $a_1(\lambda) \leq E \leq b_1(\lambda)$ (which is the case e.g. if supp$P_{q_o} = [\lambda_{min},\lambda_{max}]$ or if $b_1(\lambda_{min}) \geq a_1(\lambda_{max})$) $\Sigma = [a_1(\lambda_{min}),b_1(\lambda_{max})] \cup \bigcup_{n=2}^{\infty} [a_n(\lambda_{max}),b_n(\lambda_{max})]$

iii) If d=3 and if the assumptions of (ii) are satisfied $\Sigma = [a_1(\lambda_{min}),b_1(\lambda_{max})] \cup [0,+\infty[$

In the case d=1 the above theorem tells us that infinitely many gaps occur in the spectrum of H_ω. It is worthwhile to remark at this point that already in 1949 Saxon and Hutner [22]   conjectured that in a random mixture of one dimensional point interaction Hamiltonians with only two types of charges, i.e. $q_i(\omega) = \lambda_1$ or $q_i(\omega) = \lambda_2$,any common gap for H_{λ_1} and H_{λ_2} is also a gap for H_ω. Such a conjecture was proved by Luttinger in 1951 [29]   ,but it turned out to be false in general for interactions different from the point interactions (see Kerner [30] ,James and Ginzbarg [31]). Theorem 7 of course contains the Saxon-Hutner conjecture for rather general distribution function P_{q_o} thus extending Luttinger's work. From the proof of the theorem it is also clear the result is peculiar for the point interactions since only in this case one of the band edges is independent of the coupling constant λ .Let us consider this aspect in more details. Let $H_\omega = H_o + \Sigma q_i(\omega)f(x-x_i)$ and assume $f \geq 0, \lambda_{min} =$ inf supp$P_{q_o} \geq 0$. The same kind of considerations made before show that:

$$\sigma(H_\omega) \subset \bigcup_{n=1}^{\infty} [a_n(\lambda_{min}),b_n(\lambda_{max})] \tag{16}$$

where $\sigma(H_\lambda) = \sigma(H_o + \lambda \Sigma f(x-x_i)) = \bigcup_{n=1}^{\infty} [a_n(\lambda),b_n(\lambda)]$

Let now for example (as in the Saxon-Hutner conjecture) supp$P_{q_o} = \{\lambda_{min},\lambda_{max}\}$ and let Γ be a common gap for $H_{\lambda_{min}}$ and $H_{\lambda_{max}}$, e.g. $\Gamma \cap [b_n(\lambda_{min}),a_{n+1}(\lambda_{min})] = \emptyset$ and

$\Gamma \cap [b_m(\lambda_{max}),a_{m+1}(\lambda_{min})] = \emptyset$; then from (16) we cannot conclude that $\Gamma \cap \sigma(H_\omega) = \emptyset$ and actually there are examples where $\Gamma \cap \sigma(H_\omega) \neq \emptyset$ (Kerner [30]). But if we know that n = m i.e. Γ is a gap for both $H_{\lambda_{min}}$ and $H_{\lambda_{max}}$ with the same counting, or in the notations of sect.III.1 $\Gamma \subset [\Gamma_n(\lambda_{min}) \cap \Gamma_n(\lambda_{max})]$ then $\Gamma \cap \bigcup_{n=1}^{\infty} [a_n(\lambda_{min}),b_n(\lambda_{max})] = \emptyset$

240

and thus from (16) $\Gamma \cap \sigma(H_\omega) = \emptyset$. We have thus shown a modified Saxon-Hutner conjecture that we state in a theorem:

Theorem 8 (modified Saxon-Hutner conjecture)

Let $\{q_i(\omega)\}_{i \in Z}d$ be a metrically transitive,positive correlated random field on Z^d and denote by λ_{min} (resp. λ_{max}) the inf (resp. sup) of $suppP_{q_o}$ .Assume $0 \le \lambda_{min} \le \lambda_{max} \le \infty$ or $-\infty \le \lambda_{min} \le \lambda_{max} \le 0$.Let furthermore $\{x_i\}_{i \in Z}d$ be a lattice in $\mathbb{R}^d$ and $f \in \ell^1(L^p)$ p=2 for $d \le 3$,p > 2 for d=4 and $p=\frac{d}{2}$ for $d \ge 5$,$f \ge 0$.Denote by $\Gamma_n(\lambda)$ the n-th gap of the operator H_λ (according to the definition in sect.III.1).Then if an open set Γ satisfies: $\Gamma \subset \Gamma_n(\lambda_{min}) \cap \Gamma_n(\lambda_{max})$ for some n $\in \mathbb{N}$,then $\qquad \underline{\Gamma \cap \Sigma = \emptyset}$

Several extensions of this theorem can be found in Kirsch [5]   .

§ 3   Models with random point defects

In this section we consider a point interaction Hamiltonian with random charges at positions of a lattice $\{x_i\}$ but with only a random subset of $\{x_i\}$ occupied by charges. More precisely let as before $\{q_i(\omega)\}_{i \in Z}d$ (d=1,2,3) be a metrically transitive positive correlated random field on Z^d and let $\{\chi_i(\omega)\}_{i \in Z}d$ be independent identically distributed random variables,that are also independent of the field $\{q_i(\omega)\}$,with values in $\{0,1\}$ and such that $0 < P(\chi_o(\omega) = 0) < 1$ .We then consider the operator H_ω formally given by:

$$H_\omega = H_o + \sum_{i \in Z}d \chi_i(\omega)q_i(\omega)\delta(\cdot - x_i) \qquad\qquad (17)$$

While this operator has a direct meaning as sum of quadratic forms for d=1,for d=2,3 it has to be interpreted as a Hamiltonian with point interactions of strngths $q_i(\omega)$ at the points of the random set $\Gamma(\omega) = \{ x_i; \chi_i(\omega)=1\}$ according to the formula (9). As above we denote by λ_{min} and λ_{max} the inf and sup of $suppP_{q_o}$ and assume: $-\infty < \lambda_{min} \le \lambda_{max} < 0$. Let furthermore $a_1(\lambda)$ and $b_1(\lambda)$ be the lower and upper edge respectively of the negative band of the operator $H_\lambda = H_o + \sum_i \lambda \delta(\cdot - x_i)$.

Then one has (see Albeverio,Høegh-Krohn,Kirsch and Martinelli [4]):

Theorem 9   Let H_ω be given by (17) and assume that λ_{min} and λ_{max} are such that $b_1(\lambda_{min}) \le a_1(\lambda_{max})$.Then:

$$\sigma(H_\omega) = [a_1(\lambda_{min}) ,b_1(\lambda_{max})] \cup [0 , +\infty[\qquad a.s.$$

As a special case Theorem 9 contains the following:

Corollary   Let $\tilde{H}_\omega$ be given by: $\tilde{H}_\omega = H_o + \lambda \sum \chi_i(\omega)\delta(\cdot -x_i)$ where $\{\chi_i(\omega)\}_{i \in Z}d$ are as above.   Then for almost all ω :

$$\sigma(\tilde{H}_\omega) = \sigma(H_o + \lambda \sum_i \delta(\cdot - x_i)$$

It is worthwhile to remark that,using Theorem 7 and the above result,the spectrum of H is almost surely equal to the spectrum of the random point interactions Hamiltonian with centers at the <u>all</u> points of the lattice Z^d of strength $q_i(\omega)$;this amounts of saying that switching out some of the centers in a random way does not change the spectrum.

§ 4 <u>Models with random positions</u> [1]

In this final section we investigate the spectra of $H_\omega = H_o + \Sigma q_i(\omega)\delta(\cdot -x_i+\xi_i(\omega)) = H_o + \int \delta_y \mu_\omega(dy)$,where both $\{q_i(\omega)\}$ and $\{\xi_i(\omega)\}$ are metrically transitive,positive correlatated random fields independent of each other.Here we restrict ourselves to the one dimensional case although similar considerations are possible for d=3.Thus we can write the lattice $\{x_i\}$ as: $x_i = ip$ for some p >0.We will assume that in each unit cell $C_i = \{x\in\mathbb{R} ; ip-\frac{p}{2} \leq x < ip+\frac{p}{2} \}$ there exists one and only one particle;this is ensured by the condition[2]:

$$|\xi_i(\omega)| < \frac{a}{2} < \frac{p}{2} \tag{18}$$

Hence the distance between $x_i + \xi_i(\omega)$ and $x_j + \xi_j(\omega)$ for $i\neq j$ is at least p-a > 0.Furthermore : $|x_i+ \xi_i(\omega) -x_{i+1} - \xi_{i+1}(\omega)| < p+a$ (19)

According to theorem   the operators H_ω have almost surely constant spectrum Σ given by: $\Sigma = \overline{\underset{W \in \mathcal{P}}{\cup} \sigma(H_o + W)}$ where $\mathcal{P}$ denotes the class of periodic admissible potentials. Our aim now is to estimate the bands of an arbitrary periodic potentials in terms of the bands of periodic potentials with period p in a way similar to that one of section III.2. For this let first assume $q_i(\omega)> 0$ and let $W = \Sigma\lambda_i\delta(\cdot - y_i)$ be an admissible potential of period Np.It is not difficult to show that one can construct a periodic C^∞ function T on $\mathbb{R}$ such that: $Ty_i = x_i$, $T'y_o = 1$ and $(1+\frac{a}{p})^{-1} \leq T' \leq (1-\frac{a}{p})^{-1}$

Now consider the eigenvalues $E_n(\theta)$ of the reduced Hamiltonian $H_o(\theta) + W$, $\theta\in [0,2\pi[$. which are given by the min–max principle applied to the form $\langle H_o(\theta)\varphi,\varphi\rangle + \langle W\varphi,\varphi\rangle$, with $\varphi \in L^2([y_o,y_N])$ in the domain of the operator $H_o(\theta) + W$ (see Reed–Simon IV [19]). The function T generates a transformation from $L^2([0,Np])$ onto $L^2([y_o,y_N])$ via the formula: $(\tau f)(x) = f(Tx)$.Using this transformation the quantity:

$$\frac{\langle H_o\varphi,\varphi\rangle}{\langle\varphi,\varphi\rangle} + \frac{\langle W\varphi,\varphi\rangle}{\langle\varphi,\varphi\rangle} = \frac{\langle\varphi',\varphi'\rangle}{\langle\varphi,\varphi\rangle} + \frac{\Sigma_i \lambda_i|\varphi(y_i)|^2}{\langle\varphi,\varphi\rangle} \tag{*}$$

can be estimated by:

$$(*) \geq (1 + \frac{a}{p})^{-2} \frac{[\ \langle(\varphi\circ T^{-1})',(\varphi\circ T^{-1})'\rangle + (1 + \frac{a}{p})\lambda_{min}\Sigma|(\varphi\circ T^{-1})(x_i)|^2\ ]}{\langle(\varphi\circ T^{-1}),(\varphi\circ T^{-1})\rangle}$$

---

[1]

   The material of this section is part of a forthcoming paper by the authors.

$$(*) \quad \leq (1 - \tfrac{a}{p})^{-2} \frac{[\ <(\varphi \circ T^{-1})', (\varphi \circ T^{-1})'> \ + \ (1 - \tfrac{a}{p})\lambda_{max} \ \Sigma \ (\varphi \circ T^{-1})(\mathbf{x}_i)\,]}{<(\varphi \circ T^{-1}) , (\varphi \circ T^{-1})>}$$

From this and the min-max principle we conclude:

$$\Sigma \subset \bigcup_{n=1}^{\infty} \ [\ (1 + \tfrac{a}{p})^{-2} a_n ((1 + \tfrac{a}{p})\lambda_{min}), (1 - \tfrac{a}{p})^{-2} b_n ((1 - \tfrac{a}{p})\lambda_{max}]$$

Hence there is a gap for H between $\beta_n = (1 - \tfrac{a}{p})^{-2} b_n ((1 - \tfrac{a}{p})\lambda_{max})$ and

$\alpha_{n+1} = (1 + \tfrac{a}{p})^{-2} a_{n+1} ((1 + \tfrac{a}{p})\lambda_{min})$ whenever the latter is larger than the former. Since $b_n(\lambda) = (\pi n)^2 p^{-2}$ for any λ, the above condition reads:

$$a_{n+1}((1 + \tfrac{a}{p})\lambda_{min}) \ > (p-a)^2 (p+a)^2 (\pi n)^2 \ p^{-2} \tag{19}$$

On the other hand we know (see e.g Flügge [23]) that $a_{n+1}(\lambda) > (\pi n)^2 \ p^{-2} + C$, for some positive constant $C = C(\lambda, p)$ <u>independent</u> of n . Since $\lim_{a \ 0} (p-a)^2 (p+a)^{-2} = 1$ we have that (19) is satisfied for all $a < a_o$, where a_o depends on λ_{min} ,p and n.

<u>Theorem 10:</u> Let $\{q_i(\omega)\}$ be a metrically transitive positive correlated random field. Let λ_{min} and λ_{max} be the inf and sup of $\mathrm{supp}P_{q_o}$ and assume $0 < \lambda_{min} \leq \lambda_{max} < \infty$ .Let furthermore $\{\xi_i(\omega)\}$ be a metrically transitive positive correlated random field independent of $\{q_i\}$ and assume $\mathrm{supp}P_{\xi_o} = [-\tfrac{a}{2} , + \tfrac{a}{2}]$ .For any p >a denote by H_ω the random Hamiltonian :

$$H_\omega = -\frac{d^2}{dx^2} + \Sigma \ q_i(\omega) \ \delta(x - ip - \xi_i(\omega)) \tag{20}$$

Then there are (at least) n gaps open in the spectrum of H_ω if a is taken less than some a_n .

References:

[1] Kirsch W., Martinelli F.: On the spectrum of Schrödinger operators with a random potential; to appear in Comm. Math. Phys.

[2] Kirsch W., Martinelli F.: On the ergodic properties of the spectrum of general random operators; to appear in Journal für die reine und angewandte Mathematik

[3] Kirsch W., Martinelli F.: On the essential selfadjointness of stochastic Schrödinger operators; Preprint Bochum (1981)

[4] Albeverio S., Høegh-Krohn R., Kirsch W., Martinelli F.: The spectrum of the three-dimensional Kronig-Penney model with random point defects; to appear in Adv. in Applied Mathematics

[5] Kirsch W.: Über Spektren stochastischer Schrödingeroperatoren; Thesis, Bochum (1981)

[6]   Reed M., Simon B.: Methods of modern Mathematical Physics Vol. II Academic Press,
      New York 1975

[7]   Pastur L.A.: Spectral properties of disordered systems in the one-body approxi-
      mation; Comm. Math. Phys. 75, 179-196 (1980)

[8]   Kunz H., Souillard B.: Sur le spectre des opérateurs aux differences finies
      aléatoires; Comm. Math. Phys. 78, 201-246 (1980)

[9]   Simon B.: Almost Periodic Schrödinger operators: A Review; Adv. Appl. Math.
      to appear

[10] Gol'dsheid I.J., Molchanov S.A., Pastur L.A.: A pure point spectrum of the
      stochastic one-dimensional Schrödinger operator; Funct. Anal. Appl. 11, 1-10
      (1977)

[11] Simon B.: Trace ideals and their applications, Cambridge University Press, 1979

[12] Kallenberg O.: Random Measures, Akademie Verlag Berlin, 1976

[13] Morgan J.D.: Schrödinger operators whose potentials have separated singularities;
      J. Oper. Theory 1, 109-115 (1979)

[14] Reed M., Simon B.: Methods of Modern Mathematical Physics, Vol. I; Academic Press,
      New York 1972

[15] Avron J., Simon B.: Almost periodic Schrödinger operators: I Limit periodic
      potentials; Comm. Math. Phys.

[16] Moser J.: An example of a Schrödinger equations with almost periodic potential
      and nowhere dense spectrum; Preprint Zürich (1981)

[17] Skriganov M.M.: Proof of the Bethe-Sommerfeld conjecture in dimension two;
      Soviet Math. Dokl. 20, 956-959 (1979)

[18] Kronig R.L., Penney W.G.: Quantum mechanics of electrons in crystal lattices;
      Proc. Roy. Soc. (London) A130, 499-513; reprinted in [32]

[19] Reed M., Simon B.: Methods of modern Mathematical Physics Vol. IV; Academic
      Press, New York 1978

[20] Avron J., Grossmann A., Rodriguez R.: Spectral properties of reduced Bloch
      Hamiltonians; Ann. Phys. 103, 47-63 (1977)

[21] Thomas L.E.: Time dependent approach to scattering from impurities in a crystal;
      Comm. Math. Phys. 33, 335-343 (1973)

[22] Saxom D.S., Hutner R.A.: Some electronic properties of a one-dimensional
      crystal model; Philips Res. Rep. 4, 81-122 (1949)

[23] Flügge S.: Rechenmethoden der Quantenmechanik; Springer, Berlin 1975

[24] Albeverio S., Høegh-Krohn R.: Point interaction as limit of short range inter-
      actions; J. Op. Theory 6, 313-339 (1981)

[25] Albeverio S., Gesztesy F., Høegh-Krohn R.: The low energy expansion in non
      relativistic scattering theory; to appear in Ann. Inst. H. Poinc.

[26] Albeverio S., Fenstad J.E., Høegh-Krohn R.: Singular perturbation and non-
      standard analysis; Trans. Amer. Math. Soc. 252, 275-295 (1979)

[27] Albeverio S., Fenstad J. E., Høegh-Krohn R., Lindstrøm T.: <u>Nonstandard Analysis</u>; book in preparation

[28] Grossmann A., Høegh-Krohn R., Mebkhout M.: The one particle theory of point interactions; Comm. Math. Phys. <u>77</u>, 87-110 (1980)

[29] Luttinger J.M.: Wave propagation in one-dimensional structure; Philips Res. Rep. <u>6</u>, 303-310 (1951) reprinted in [32]

[30] Kerner E.H.: Periodic impurities in a periodic lattice; Phys. Rev. <u>55</u>, 687-689 (1954)

[31] James H.M., Ginzbarg A.S.: Band structure in disordered alloys and impurity semiconductors; J. Phys. Chem. <u>57</u>, 840-848 (1953)

[32] Lieb E.H., Mattis D.C.: <u>Mathematical Physics in one dimension</u>; Academic Press, New York 1966

Wiener Measures for Quantum Mechanical Path Integrals

John R. Klauder, Bell Laboratories, Murray Hill, N. J. 07974, USA,

Ingrid Daubechies[*], Physics Department, Princeton University, Princeton, N.J.

08544 USA.

Our purpose here is to show that it is possible to represent certain quantum mechanical evolution operators by path integrals with mathematically well-defined measures. For the Hamiltonians we consider here, this measure will be a Wiener measure on phase space. To achieve this goal we exploit the overcompleteness of the coherent states.

Coherent states are defined as

$$|z> \equiv |p,q> \equiv e^{i(pQ-qP)}|0>$$
$$\equiv e^{-\frac{1}{2}|z|^2} e^{z A^\dagger}|0> \quad ,$$

where $|0>$ is the harmonic oscillator ground state, and where $z \equiv (q+ip)/\sqrt{2}$.
In the Schrödinger representation one has

$$<x|z> = \pi^{-\frac{1}{4}} \exp\ [ip(x-q/2) - (x-q)^2/2]. \tag{1}$$

The overcompleteness of the coherent states is illustrated by the overlap function

$$<z|z'> = \exp[-\frac{1}{2}|z|^2 - \frac{1}{2}|z'|^2 + z^*z']$$

which shows that these states are never mutually orthogonal. Nevertheless, one of the most useful properties of the coherent states is the "resolution of the identity"

$$I = \int |z><z|\ d\mu \tag{2}$$

where $d\mu \equiv \frac{1}{2\pi}\ dp\ dq$, integrated over $\mathbb{R}^2$.

Equation (2) can be used to represent operators by means of their matrix elements between coherent states. If B is a (bounded) operator, we can always write B as

$$B = \int |z''><z''|B|z'><z'|\ d\mu''d\mu' \quad . \tag{3}$$

Due to the overcompleteness of the coherent states, many other functions can be found to replace $<z''|B|z'>$ without invalidating (3). These functions form an equivalence class (labelled by B), a generic element of which we denote by $<z''|B|z'>_{E.C.}$, and therefore

$$B = \int |z''><z''|B|z'>_{E.C.}\ <z'|\ d\mu''d\mu' \quad . \tag{4}$$

---

[*] On leave from Dienst voor Theoretische Natuurkunde, Vrije Universiteit Brussel, Belgium, and Interuniversitair Instituut voor Kernwetenschappen, Belgium.

The class of dynamical systems we shall consider here is given by the Hamiltonians

$$H(t) = \frac{1}{2} (P^2 + Q^2 - 1) + s(t)Q \quad .$$

The coherent state matrix element of the evolution operator can be written as a formal coherent state path integral[1] ,

$$<z'',t''|z',t'> \equiv <z''|T \exp[-i\int H(t)dt]|z'>$$

$$= \mathcal{N} \int \exp(i\int \{ \frac{1}{2} [p(t)\dot{q}(t)-q(t)\dot{p}(t)]$$

$$- \frac{1}{2} [p^2(t) + q^2(t)]-s(t)q(t)\}dt) \prod dp(t)dq(t) \quad . \tag{5}$$

The "volume element" $\prod dp(t)dq(t)$ does not define a genuine measure, and the right hand side of (5) should be understood as a limit[1],

$$<z'',t''|z',t'> = \lim_{\varepsilon \to 0} \int \ldots \int \exp(\sum \{\frac{i}{2} [p_k(q_{k+1}-q_k)-q_k(p_{k+1}-p_k)]$$

$$- \frac{1}{4}[(p_{k+1}-p_k)^2+(q_{k+1}-q_k)^2]- \frac{i}{2} \varepsilon(q_{k+1}-ip_{k+1})(q_k+ip_k)$$

$$- \frac{i}{2}\varepsilon \, s_k[q_{k+1}+q_k+i(p_{k+1}-p_k)]\}) \prod (dp_k dq_k / 2\pi)$$

We claim that it is however possible to write a path integral over a genuine measure for a member in the same equivalence class as $<z'',t''|z',t'>$ in the fashion

$$<z'',t''|z',t'>_{E.C.} = 2\pi \int \exp(\int \{ (\frac{i}{2} + \frac{1}{\nu})[p(t)\dot{q}(t)-q(t)\dot{p}(t)]$$

$$- \frac{1}{2} (i+ \frac{1}{\nu})[p^2(t)+ q^2(t)] - \frac{1}{2\nu} [s^2(t)- \nu^2]$$

$$- s(t)[(i+ \frac{1}{\nu})q(t)+ \frac{1}{\nu} \dot{p}(t)]\}dt)d\rho_w(p) \, d\rho_w(q) \quad , \tag{6}$$

where each ρ_w is a Wiener measure, and ν can be arbitrarily chosen in $\mathbb{R}_+^*$ (every $\nu>0$ leads to a member of the equivalence class). For p the measure is pinned at the initial point p' and at the final point p'', and is normalised such that

$$\int d\rho_w(p) = [2\pi\nu(t''-t')]^{-\frac{1}{2}} \exp\{-(p''-p')^2/[2\nu(t''-t')]\} \quad ;$$

the Wiener measure for q is defined entirely analogously. The integrand in (6) is a well-defined stochastic integral. For example, the term $\int(\dot{p}q-\dot{q}p)dt$ should be understood as $\int pdq-\int qdp$; since p and q are independent stochastic variables all procedures to define these integrals are equivalent. The construction[2] of (6) exploits well-known properties of the harmonic oscillator eigenstates and of the Weyl operators. Note that while there are similarities between the integrands in (5) and (6), some coefficients are different: we have a coefficient $(\frac{i}{2} + \frac{1}{\nu})$ for $(p\dot{q}-q\dot{p})$ in (6) instead of $\frac{i}{2}$ in (5), and $- \frac{1}{2}(i + \frac{1}{\nu})$ for (p^2+q^2) instead of $- \frac{i}{2}$. Moreover an extra term $- \frac{1}{2\nu}\int dt[s^2(t)-\nu^2]$ appears in (6) (we assume s to be square integrable).

Some further remarks regarding this construction are:

1. One can incorporate the term $-\frac{1}{2\nu}(p^2+q^2)$ in the integrand in (6) into the Wiener measures, and so write $<z",t"|z',t'>$ as an integral over an associated Ornstein-Uhlenbeck measure.

2. For $t"-t' \to \infty$, with ν fixed, the equivalence class function $<z",t"|z',t'>_{E.C.}$ converges to the matrix element $<z",t"|z',t'>$; this happens whatever the fixed value is assigned to ν . Hence (6) gives a genuine path integral expression for the matrix element $<z",t"|z',t'>$ in the limit when the time interval diverges.

3. Alternatively in the limit $\nu \to \infty$, with $t"-t'$ fixed, the equivalence class function $<z",t"|z',t'>_{E.C.}$ converges to the matrix element $<z",t"|z',t'>$. This result may be heuristically seen since (6) formally converges to (5) in the limit $\nu \to \infty$.

4. One can use (1) and (4) to write the matrix element $<x",t"|x',t'>$ as a well-defined path integral,

$$<x",t"|x',t'> = \int <x"|z"><z",t"|z',t'>_{E.C.} <z'|x'>d\mu"d\mu' \ .$$

Carrying out the integral over p, $<x",t"|x',t'>$ is then expressed as a path integral over a Wiener measure in q alone; the integrand, though very complicated, is still a well-defined stochastic integral.

5. An analogous construction holds for other kinematical systems as well. In particular, for spin systems, special spin coherent state matrix elements in the equivalence class of the evolution operator for certain systems may be expressed as a path integral involving Wiener measure on the surface of the unit sphere.[3]

<u>References</u>

1. See e.g., J. R. Klauder, Acta Physica Austriaca, Suppl. XXII, 3 (1980); in <u>Path Integrals</u>, ed. by G. J. Papadopoulos and J. T. Devreese (Plenum, 1978), p. 5.
2. I. Daubechies and J. R. Klauder, "Constructing Measures for Path Integrals", to be published in J. Math. Phys.
3. J. R. Klauder, "Constructing Measures for Spin-Variable Path Integrals", submitted for publication.

<u>EXISTENCE OF A FIRST-ORDER PHASE TRANSITION FOR THE POTTS MODEL</u>

R.Kotecký
Dept. Mathematical Physics
Charles University
Povltavská 1
Praha 8, Czechoslovakia

S.B.Shlosman
Institute for Problems of
Information Transmission
Acad. of Sciences
Moscow, USSR

I. <u>The results</u>

Our aim in this lecture is to sketch a proof that for many-state Potts model an expected discontinuity of the average energy as a function of temperature really occurs. We base the presentation on our recent paper [1] though we give an alternative proof of one part of the statement.

We consider a q-state Potts model [2] with a spin variable $\sigma(i)$ attached to each lattice site i of a ν-dimensional lattice Z^ν and taking values in the set $\{1,2,\ldots q\}$. The nearest neighbour interaction favours configurations with neighbouring spins in the same state; (formal) Hamiltonian is

$$H(\sigma) = - \sum_{\langle i,j \rangle} \delta_{\sigma(i),\sigma(j)}$$

with $\delta_{\alpha,\beta}$ a Kronecker delta. Now, it is anticipated that average energy develops a discontinuity in temperature whenever q is larger than a critical value which depends on the dimension ν. (See e.g. [3] for a recent review about the Potts model.) This anticipation will be met (at least for q large) once we prove the following

<u>Theorem</u>: A ν-dimensional q-state Potts model undergoes a first-order phase transition in temperature whenever $\nu \geq 2$ and q is large enough. Namely, for each $\nu \geq 2$ there is a number $q(\nu)$ such that whenever $q \geq q(\nu)$ there exist an inverse temperature $\beta_c = \beta_c(q,\nu)$ and two different translation invariant Gibbs states $\langle \rangle_{\beta_c}^=$, $\langle \rangle_{\beta_c}^{\neq}$ of the q-state Potts model at the inverse temperature β_c such that $\langle \delta_{\sigma(i),\sigma(j)} \rangle_{\beta_c}^= > \frac{1}{2}$ and $\langle \delta_{\sigma(i),\sigma(j)} \rangle_{\beta_c}^{\neq} > \frac{1}{2}$ for each pair of nearest neighbour sites $i,j \in Z^\nu$.

Before sketching the proof we shall mention two other results from [1]. One concerns a more accurate description of the phase diagram of

the Potts model. At low temperatures there are q different translatio-
nally invariant Gibbs states (ordered phases). This is easy to show by
standard Peierls argument. Using the technical means that led to the
proof of the above Theorem we show in [1] that those q pure states sur-
vive up to the critical temperature, including the critical temperature
itself. Thus at the critical temperature we have all together (at least)
q+1 different Gibbs states: q of them contained in $\langle \ \rangle^{=}_{\beta_c}$ and an additional one - the chaotic state $\langle \ \rangle^{\neq}_{\beta_c}$.

The second remark concerns lattice gauge theory. The correspondence
Ising model $\longleftrightarrow$ lattice Z_2 gauge model may be extended to Potts model
by introducing a lattice Potts gauge model [4]. It turns out that the
Theorem may be reformulated and proven also for the latter model providing that the gauge group is large enough. The phase transition again
reveals itself in a jump of average energy (represented by average plaquette frustration).

II. The proof

Consider $\langle \ \rangle_{\beta}$ a Gibbs state of the Potts model resulting by thermodynamic limit from finite volume Gibbs states $\langle \ \rangle_{\beta,\Lambda}$ at inverse temperature β , with periodic boundary conditions. To prove that average
energy jumps at certain β_c amounts to proving that for any bond l
the expectation $\langle P^{=}_l \rangle_{\beta}$ jumps. Here $P^{=}_l$ denotes an indicator (cha-
racteristic function) of the event $\{\sigma \,|\, \sigma(i)=\sigma(j)\}$ with l a bond of
neighbouring lattice sites i,j $(l=\langle i,j \rangle)$. It is easy to show that
at low temperatures there is complete order: $\lim_{\beta \to \infty} \langle P^{=}_l \rangle_{\beta} = 1$; while at
high temperatures, complete chaos: $\lim_{\beta \to 0} \langle P^{=}_l \rangle_{\beta} = \frac{1}{q}$. *) If we knew that
ordered and chaotical regions (i.e. regions with different elementary
contributions to energy) do not like to coexist in typical configurations - we shall exemplify this by showing that for every two bonds
l_1, l_2 and every temperature β the bound

$$\langle P^{=}_{l_1}(1-P^{=}_{l_2}) \rangle_{\beta} < \frac{1}{5} \tag{1}$$

is valid - then the expectation $\langle P^{=}_l \rangle_{\beta}$ would inevitably jump.

---

*) Actually it will be clear later on that evaluating the temperatures
for which $\langle P^{=}_l \rangle_{\beta}$, a function monotonous in β , climbs above and below $\frac{1}{2}$ one gets rough bounds on the critical temperature. In this way
we have got $\dfrac{1}{\nu 2^{\nu-1}} \log q \lesssim \beta_c(q) \lesssim 2^{\nu-1} \log q$ for q large.

There are several ways how to reach the conclusion from the bound (1
In [1] we used a strategy of Dobrushin and Shlosman [5]. Other possibility is to show that

$$\langle P_1^= \rangle_\beta \, (1 - \langle P_1^= \rangle_\beta) \leq \tfrac{1}{4} - \delta \tag{2}$$

for a dense set of β's, which forces $\langle P_1^= \rangle_\beta$ to keep out of a forbidden interval around $\tfrac{1}{2}$ for the same set of β's. Taking into account
the monotonicity of $\langle P_1^{\underline{=}} \rangle_\beta$ and the fact that $\langle P_1^= \rangle_\beta$ takes a value
below or above forbidden strip for β respectively small or large, one
concludes that at certain β_c the expectation $\langle P_1^= \rangle_\beta$ jumps over forbidden interval allowing thus to introduce $\langle \; \rangle_{\beta_c}^=$ ($\langle \; \rangle_{\beta_c}^{\neq}$) limiting
by $\langle \; \rangle_\beta$ from above (below) of β_c.

The fact that (1) implies (2) may be shown using a variant of an
argument of Guerra [6]. Actually a slight modification of the version
from [7] will do. Here we have chosen yet another way [*]. Consider
$I = [\beta_1, \beta_2]$ an arbitrary interval. We will show that there is $\beta \in I$ such
that (2) holds. Choose l_0 a bond and $\varepsilon > 0$ and consider the set
$M_\Lambda = \{\beta \in I \mid$ there is $l \in \Lambda$ such that $\langle P_{l_0}^= ; P_1^= \rangle_{\beta,\Lambda} < \varepsilon \}$.
Here $\langle P_{l_0}^= ; P_1^= \rangle_{\beta,\Lambda}$ denotes a truncated correlation

$\langle P_{l_0}^= P_1^= \rangle_{\beta,\Lambda} - \langle P_{l_0}^= \rangle_{\beta,\Lambda} \langle P_1^= \rangle_{\beta,\Lambda}$. If λ is the Lebesgue measure on I
one has $\lambda(M_\Lambda) > \dfrac{\varepsilon}{1+\varepsilon} \dfrac{\lambda(I)}{2}$. Indeed, if it were $\lambda(M_\Lambda) \leq \dfrac{\varepsilon}{1+\varepsilon} \dfrac{\lambda(I)}{2}$
then taking into account that $\langle P_{l_0}^= ; P_1^= \rangle_{\beta,\Lambda} \geq -2$ one would infer

$$\langle P_{l_0}^= \rangle_{\beta_2,\Lambda} - \langle P_{l_0}^= \rangle_{\beta_1,\Lambda} = \int_{\beta_1}^{\beta_2} \langle P_{l_0}^= ; \sum_{l \in \Lambda} P_1^= \rangle_{\beta,\Lambda} \, d\beta \geq$$

$$\geq |\Lambda| \left[\varepsilon \left(\lambda(I) - \dfrac{\varepsilon}{1+\varepsilon} \dfrac{\lambda(I)}{2} \right) - 2 \dfrac{\varepsilon}{1+\varepsilon} \dfrac{\lambda(I)}{2} \right] = |\Lambda| \lambda(I) \dfrac{\varepsilon^2}{2(1+\varepsilon)}$$

which for Λ large contradicts the fact that $\langle P_{l_0}^= \rangle_{\beta_2,\Lambda} - \langle P_{l_0}^= \rangle_{\beta_1,\Lambda}$
can never be larger than 2. Denoting χ_{M_Λ} the characteristic function
of M_Λ one has

$$\int \limsup_{\Lambda \to \infty} \chi_{M_\Lambda} \geq \dfrac{\varepsilon}{1+\varepsilon} \dfrac{\lambda(I)}{2}$$

by Fatou's lemma and there is $\beta \in I$ belonging to M_Λ for infinite num-
ber of Λ's. For this β and those Λ one has $\langle P_{l_0}^= ; P_1^= \rangle_{\beta,\Lambda} < \varepsilon$ for some
$l \in \Lambda$ which together with $\langle P_{l_0}^= \rangle_{\beta,\Lambda}^2 - \langle P_{l_0}^= P_1^= \rangle_{\beta,\Lambda} < \tfrac{1}{5}$ implies

$$\langle P_{l_0}^= \rangle_{\beta,\Lambda} - \langle P_{l_0}^= \rangle_{\beta,\Lambda}^2 = \langle P_{l_0}^= \rangle_{\beta,\Lambda} - \langle P_{l_0}^= \rangle_{\beta,\Lambda} \langle P_1^= \rangle_{\beta,\Lambda} < \tfrac{1}{5} + \varepsilon$$

and hence (2). Note that we used (1) for finite volume Gibbs state

---

[*] We would like to thank D.Preiss whose suggestions were decisive
for the following argument.

$\langle \ \rangle_{\beta,\Lambda}$ but this is what we shall actually prove.

Thus what remains is to show that for Λ large and l_1, l_2 bonds from Λ one has

$$\langle P_{l_1}^= (1 - P_{l_2}^=) \rangle_{\beta,\Lambda} < \frac{1}{5} \tag{3}$$

for each inverse temperature β . We shall accomplish this using a variant of Peierls argument. One proves that there is a jump in magnetization of Ising model by estimating a probability of contours between regions with different elementary magnetizations. The fact that it is a jump in energy we search for suggests to consider contours separating regions of different elementary contributions to energy. Namely, given a configuration σ, we introduce its <u>contours</u> as follows. First (restricting ourselves to the case $\nu = 2$ for the sake of simplicity) we define <u>precontours</u> as connected lines in the dual lattice separating regions of excited ($P_l^=(\sigma)=0$) and nonexcited ($P_l^=(\sigma)=1$) bonds. More precisely one first introduces <u>islands</u> as maximal connected (in the sense of connection by a sequence of nearest neighbours) sets of vertices on which the configuration σ is constant and which contain at least two vertices. An island, say Q, is then bordered by precontours introduced exactly in the same way as contours for Ising model with the Ising configuration constant on all vertices of Q and of opposite sign outside of Q. A precontour Γ^* is thus a closed line on the dual lattice crossing solely excited bonds of the configuration σ. But what will turn out important is that it intersects certain number $N_a(\Gamma^*)$ of elementary plaquettes called <u>acceptable</u> ones distinguished by the fact that each of them contains at least one nonexcited bond from the island Q. Finally a contour Γ is introduced as a precontour Γ^* together with the excitation pattern on all bonds of all its acceptable plaquettes.

Let us follow the wisdom of Peierls argument and estimate the probability $\text{Prob}_{\beta,\Lambda}\{\Gamma\}$ of the event $\{\sigma | \Gamma \text{ is a contour of } \sigma\}$ in the Gibbs state $\langle \ \rangle_{\beta,\Lambda}$. Using chessboard estimates [8] one shows that

$$\text{Prob}_{\beta,\Lambda}\{\Gamma\} \leqslant \left\langle \prod_{\square \text{ is } \Gamma^* \text{ accept.}} P_\square \right\rangle_{\beta,\Lambda} \leqslant \left\{ \min_r \langle P_\Lambda^r \rangle_{\beta,\Lambda}^{\frac{1}{|\Lambda|}} \right\}^{\frac{|\Gamma^*|}{10}}$$

where: $P_\square$ is the indicator of the event

$$\left\{ \sigma \Big| \begin{array}{l} \text{the excitation pattern of the plaquette } \square \text{ coincides} \\ \text{with the excitation pattern of that plaquette in } \Gamma \end{array} \right\}$$

and P_Λ^r are indicators of "universal contours" resulting from reflections (with respect to the planes perpendicular to the lattice axes and going through lattice vertices) of the excitation pattern of an acceptable plaquette. Since there is finite number of possible excitation patterns on acceptable plaquettes, there is also a finite number of universal contours labeled by r. In the last inequality we used the observation that there are not too many unacceptable plaquettes crossed by Γ^* (they sit only on convex corners of Γ^*) which yields an estimate $N_a(\Gamma^*) \geqslant \frac{|\Gamma^*|}{10}$. Universal contours may be listed and the expectations $\langle P_\Lambda^r \rangle_{\beta,\Lambda}$ may be estimated one by one. (Though in [1] we also give an argument allowing to estimate $\langle P_\Lambda^r \rangle_{\beta,\Lambda}$ without listing all universal contours but using instead only the fact that it was an acceptable plaquette that gave rise to the universal contour.) We will illustrate the method of evaluation of $\langle P_\Lambda^r \rangle_{\beta,\Lambda}$ on one particular case (which actually turns out to yield $\min\limits_r \langle P_\Lambda^r \rangle_{\beta,\Lambda}$). Let thus P^1 be the indicator corresponding to the excitation pattern

Here wavy lines picture excited bonds and straight lines unexcited ones. This excitation pattern resulted e.g. from reflecting a plaquette . Denoting $|\{\ \}|$ the number of elements of the set described within brackets we have

$$\langle P_\Lambda^1 \rangle_{\beta,\Lambda} = \frac{\sum\limits_{\sigma \text{ consistent with } P_\Lambda^1} e^{-\beta H_\Lambda(\sigma)}}{\sum e^{-\beta H_\Lambda(\sigma)}} \leqslant \frac{e^{\frac{|\Lambda|}{2}}|\{\sigma\,|\,\sigma \text{ consistent with } P_\Lambda^1\}|}{e^{\frac{|\Lambda|}{2}}|\{\sigma\,|\,\sigma \text{ such that } H_\Lambda(\sigma)=-\frac{|\Lambda|}{2}\}|}$$

Realizing that $|\{\sigma\,|\,\sigma \text{ consistent with } P_\Lambda^1\}| \leqslant q^{\frac{|\Lambda|}{2}+\frac{\sqrt{|\Lambda|}}{2}}$ and

$$|\{\sigma\,|\,H_\Lambda(\sigma)=-\frac{|\Lambda|}{2}\}| \geqslant \left[q(q-4)\right]^{\frac{3}{8}}$$

(one may check the last inequality considering only configurations with $H_\Lambda(\sigma)=-\frac{|\Lambda|}{2}$ such that all excited bonds are jammed into one corner of Λ) we conclude that

$$\langle P_\Lambda^1 \rangle_{\beta,\Lambda} \leqslant \left[\frac{q}{(q-4)^3}\right]^{\frac{|\Lambda|}{8}}$$

Hence we have

$$\text{Prob}_{\beta,\Lambda}(\Gamma) \leqslant \left[\frac{q}{(q-4)^3}\right]^{\frac{|\Gamma^*|}{80}} \tag{4}$$

for all β . Note that for q large is the probability of long contours small <u>for every</u> β . This is possible since our contours play a double role: for low temperatures the space is filled with an ordered phase with rare contours confining a disorder inside of them, while at high temperatures there is overwhelming chaos everywhere with only few contours around islands of ordered phase.

Having established an estimate (4) the proof of our crucial
inequality (3) is standard:

$$\left\langle P^=_{l_1}(1-P^=_{l_2})\right\rangle_{\beta\Lambda} \leqslant \sum_{\Gamma \text{ surrounds } l_1} \text{Prob}(\Gamma) + \sum_{\Gamma \text{ surrounds } l_2} \text{Prob}(\Gamma) + \sum_{\substack{\Gamma \text{ wrapped} \\ \text{around } \Lambda}} \text{Prob}(\Gamma) \leqslant$$

$$\leqslant 2 \sum_{k=2} 4^{2k}\, 2^k \left[\frac{q}{(q-4)^3}\right]^{\frac{k}{40}} .$$

We used a three way argument to estimate the number of precontours of
given length and the observation that given a precontour Γ^* the num-
ber of corresponding contours is bounded by $2^{|\Gamma^*|}$.

References

[1] Kotecký,R.,Shlosman,S.B.:Commun.Math.Phys.83,493-515(1982)

[2] Potts,R.B.:Proc.Camb.Philos.Soc.48,106-109(1952)

[3] Wu,F.Y.:Rev.Mod.Phys.54,235-268(1982)

[4] Kogut,J.B.:Phys.Rev.D21,2316-2326(1980)

[5] Dobrushin,R.L.,Shlosman,S.B.:Phases corresponding to the local
energy minima (in print)

[6] Guerra,F.:in Mathematical Methods of Quantum Field Theory,
CNRS Marseille (1976)

[7] Gawedzki,K.:Commun.Math.Phys.59,117-142(1978)

[8] Glimm,J.,Jaffe,A.,Spencer,T.:Commun.Math.Phys.45,203-216(1975)
Fröhlich,J.,Lieb,E.:Commun.Math.Phys.60,233-267(1978)
Fröhlich,J.,Israel,R.,Lieb,E.,Simon,B.:Commun.Math.Phys.62,1-37(78)

LAGRANGIANS WITH ANTICOMMUTING

ARGUMENTS FOR DIRAC FIELDS

by

P. KREE

Lagrangians of classical fields are usually written using functions
with commuting arguments. For example, Lagrangians arising in some
gauge field theories are sometimes defined by differential forms on
the Minkovski space ; but the coefficients of these forms are usual
functions (i.e. functions with commuting arguments) of the field. After
quantization, this commutativity is well adapted to the commutativity
of Bosons components of the field and suggests sometimes usefull functional methods. But this commutativity is not well adapted to the anti-
commutativity of Fermions components of quantum fields. For example,
Noether's theorem applied to the classical Dirac field, or to the classical neutrino field gives errors of signs in the expressions of energy
charge, spin etc... These errors of signs disappear after quantization, assuming the existence of a quantum and anticommutative extension
of Noether theory : see for example $\begin{bmatrix} 1 \end{bmatrix}$ p 125. In conclusion the errors
of signs arising applying the usual Noether theorem come from the com-
mutativity of the arguments in usual Euler-Lagrange (E.L.) and Noether
theories. In this optic, these theories must be reformulated using
Lagrangians with anticommuting arguments for pure Fermions fields, and
Lagrangians with mixed commutativity for Fermions-Bosons fields. The
possibility of such extension of classical Lagrangian theory is also
suggested in $\begin{bmatrix} 7 \end{bmatrix}$. The goal of present work is to begin this extension
using similar formalisms for variables of the first kind ($\varepsilon = -$) and
variables of the second kind ($\varepsilon = +$). Therefore a free variable ε
with two possible values + and - is introduced and the following conventions are used bellow :

	$\varepsilon = -$	$\varepsilon = +$
ε - symmetric	symmetric	antisymmetric
ε commutator $[A,B]_\varepsilon$ = $AB + \varepsilon BA$ of two operators A and B	commutator $[A,B]$	anticommutator $[A,B]_\varepsilon$
ε -commuting	commuting	anticommuting

For example, the algebra of all ε-symmetric tensors on a vector space Y is denoted $T^\varepsilon(Y) = \bigoplus_{k \geqslant 0} T^\varepsilon_k(Y)$ where $T^\varepsilon_k(Y)$ denotes the space of ε-symmetric tensors homogeneous of degree $k = 0,1...$; and $T^+(Y) = \wedge Y = \bigoplus \wedge_k Y$ is called the Grassman algebra, or the exterior algebra of Y. Using these conventions a theory containing a free index ε is proposed. This theory coincides for $\varepsilon = -$ with usual Euler-Lagrange and Noether theories, and gives for $\varepsilon = +$ an anticommutative reformulation of these theories. The results are then applied to Dirac fields ; see [6] for applications to neutrino fields and connection with anticommuting functional integration. These results was announced in [5] [6] .

1 - <u>CALCULUS WITH FUNCTIONALS WITH ε - COMMUTING ARGUMENTS</u>

(1.1) <u>Motivations</u>

The contravariant coordinates in a Minkovski space are denoted $x^o = t, x^1,...x^s$ with $s = 0,1,2$ or 3 ; hence $\vec{x} = (x^1,...x^s)$ denotes the set of all spatial coordinates. In the usual E.L. theory a field map is defined by a C^1 map

$$(1.2) \qquad u = u(x) = (u_j(x)) \; : \; \mathbb{R}^{s+1} \longrightarrow \mathbb{C}^d$$

where d denotes the number of field components. For arbitrary $x \in \mathbb{R}^{s+1}$, the derivative $Du(x)$ is a linear map $\mathbb{C}^{s+1} \longrightarrow \mathbb{C}^d$, hence we have a linear mapping

$$(1.3) \qquad Y = C^1(\mathbb{R}^{s+1}, \mathbb{C}^d) \xrightarrow{\ \alpha(x)\ } \mathbb{C}^M = \mathbb{C}^d \oplus (\mathbb{C}^{s+1} \otimes \mathbb{C}^d)$$

$$u \longmapsto (u(x); Du(x))$$

A polynomial Lagrangian is defined by a function $x \longrightarrow \mathcal{L}(\bullet,\bullet;x)$ on $\mathbb{R}^{s+1}$ with values in the space $F^-_{sta}(\mathbb{C}^M)$ of all polynomial functions on $\mathbb{C}^M$. The space $F^-_{sta}(\mathbb{C}^M)$ in duality with $T^-(\mathbb{C}^M)$, is endoved with the weak topology. We assume that the map $x \longrightarrow \mathcal{L}(\bullet,\bullet;x)$ is locally and weakly integrable. The inverse image of $\mathcal{L}(\bullet,\bullet;x)$ by $\alpha(x)$ is a polynomial functional on Y denoted $Q_x : u \longmapsto \mathcal{L}(u(x), Du(x), x)$. A subset B of $\mathbb{R}^{s+1}$ is called smooth if B is connected, bounded, closed and if ∂B is an orientable and piecewise differentiable sub-manifold of $\mathbb{R}^{s+1}$. For arbitrary smooth subset B of $\mathbb{R}^{s+1}$, the action functional S_B is the following element of $F^-_{STA}(Y)$ = the set of all polynomial functional S on Y :

$$(1.4) \qquad u \longmapsto S_B(u) = \int_B \mathcal{L}(u(x), (Du)(x), x)\ dx$$

Euler-Lagrange and Noether theories use these action functionals with commuting arguments u_j and $\partial_k u_j$. Therefore an anticommuting reformulation of these theories needs first a concept of functionals with anticommuting arguments on a vector space X. If X is finite dimensional, according to F.A. Berezin [1], elements of $\Lambda(X^*)$ can be used. These elements are linear combinations of wedge products of coordinates forms $x^1, x^2 \ldots$ in X. This definition is not convenient for (dim X) infinite since an element of $\Lambda(X^*)$ defines a cylindrical functional on X and since non cylindrical functionals are used in physics. Therefore the definitions and results of [3] will be used.

(1.5) <u>Functionals with ε commuting arguments</u>

The space $F^\varepsilon(X)$ of all ε symmetric forms on X is defined by

$$F^{\varepsilon}(X) \;=\; \prod_{k=0}^{\infty} \; F_k^{\varepsilon}(X)$$

where $F_k^{\varepsilon}(X) = T_k^{\varepsilon}(X)^{*}$ is the space of all ε symmetric forms homogeneous of degree k on X. An element of $F^{\varepsilon}(X)$ is written :

$$f(x) \;=\; \sum_{j=0}^{\infty} \; f_j(x) \;\; ; \quad f_j \in F_j^{\varepsilon}(X)$$

Hence $F^{\varepsilon}(X)$ can also be viewed as the space of all functionals with ε-commuting arguments on X. For ex. $F^{-}(X)$ is the space of all formal power series on X, and subspaces of converging power series can easely be defined ; if $f_j = 0$ for j big enough, $f = \sum f_j$ belongs to the subspace $F_{sta}^{-}(X)$ of all polynomial functionals on X. The subspace

$$\left\{ \; f = f_1 + f_2 \;\; ; \quad f_1 \in F_1^{+}(X) \quad \text{and} \quad f_2 \in F_2^{+}(X) \; \right\}$$

of $F^{+}(X)$ is denoted $F_{STA}^{+}(X)$ for reasons given bellow :

(1.6) Stationary point of a functional $f \in F_{STA}^{\varepsilon}(X)$

For an arbitrary fixed $u \in T^{\varepsilon}(X)$, the left directional derivation $f \longrightarrow \partial_u f$ in $F^{\varepsilon}(X)$ is defined by transposition of the right product by u in $T^{\varepsilon}(X)$: see point (1.20) of [3] . For an arbitrary given element $f \in F^{\varepsilon}(X)$, the global left derivative $D^k f$ is defined by the linear map $u \longrightarrow \partial_u f \; : \; T_k^{\varepsilon}(X) \longrightarrow F^{\varepsilon}(X)$. For example, for an homogeneous form f_m of degree $m \geqslant 1$, the left derivative Df_m is the linear map $\delta\varphi \longrightarrow (Df_m)(\delta\varphi) \; : \; X \longrightarrow F_{m-1}^{\varepsilon}(X)$ such that for arbitrary u_2, $u_3 \ldots u_m$ in X.

$$(1.7) \quad (Df_m)(\delta\varphi)(u_2, u_3 \ldots, u_m) \;=\; f_m(\delta\varphi, u_2, u_3 \ldots, u_m)$$

Moreover $Df_0 = 0$ for arbitrary $f_0 \in \mathbb{K}$. Hence by linearity, the derivative of $f = f_0 + f_1 + \ldots \in F^{\varepsilon}(X)$ can be computed assuming $f_0 = 0$. Let X_0 be a vector subspace of X and let $f =$

$f_1 + \ldots + f_M \in F_{STA}^{\varepsilon}(X)$. Then $f \in F_{STA}^{-}(X)$ is called stationary in some point $u \in X$ with respect to variations in X_o if Df restricted to X_o is vanishing in u i.e.

$$(1.8) \qquad \forall \; \delta\varphi \in X_o \qquad (Df)(u) = \sum_1^M f_m(\delta\varphi, u^{m-1}) = 0$$

A similar definition is not possible for $\varepsilon = +$ since $f_m(\delta\varphi, u^{m-1}) = 0$ for $m > 2$. Therefore a point $u \in X$ is called a stationary point of $f = f_1 + f_2 \in F_{STA}^{+}(X)$ with respect to variations in X_o if

$$(1.9) \qquad \forall \delta\varphi \in X_o \qquad (Df)(u) = f_1(\delta\varphi) + f_2(\delta\varphi, u) = 0$$

(1.10) <u>Inverse image by a given linear map $1 : X \longrightarrow Y$</u>.

If $g = g_m \in F_m^{\varepsilon}(Y)$ is homogeneous of degree m on Y, the inverse image $g_m \circ 1 \in F_m^2(X)$ is defined by $(g \circ \ell)(x_1, x_2, \ldots x_m) = g(\ell x_1, \ldots, \ell x_m)$. Hence the map $g \longrightarrow g \circ \ell : F^{\varepsilon}(Y) \longrightarrow F^{\varepsilon}(X)$ is defined by linear extension.

(1.11) <u>Integral of functionals with ε- commuting arguments</u>

The algebraic dual V^* of a vector space V is endoved with the weak topology. A functional $f : \mathbb{R}^{s+1} \longrightarrow V^*$ is called weakly and locally integrable if for arbitrary $v \in V$, the scalar function $\langle f, v \rangle \in L_{loc}^1(\mathbb{R}^{s+1}, dx)$; for an arbitrary smooth subset B of $\mathbb{R}^{s+1}$, the weak integral $\int_B f \, dx \in V^*$ is defined by the linear form $v \longrightarrow \int_B \langle f, v \rangle \, dx$ on V. For example $V = T^-(Y)$ and using the définitions of (1.1), the weak integral $\int_B f(x) dx$ coïncides with the action functional S_B. Hence the notation Q_x is conveniently replaced by $\mathcal{L}(u(x), D\,u(x), x)$.

2 - ε COMMUTING EXTENSION OF EULER-LAGRANGE EQUATIONS

A theory with a free index $\varepsilon = \pm$ is formulated bellow. This theory agrees with usual E.L. theory for $\varepsilon = -$ and gives an anticommuting extension for $\varepsilon = +$

(2.1) Definition of a class of fields

A class of fields is defined by :

- a subset $\mathcal{U}$ of $Y = C^1(\mathbb{R}^{s+1}, \mathbb{C}^d)$ and by an equivalence relation $\mathcal{R}$ on $\mathcal{U}$

- a real Lie group G and a morphism of groups

$$(2.2) \qquad G \longrightarrow (\text{Diff } \mathbb{R}^{s+1}) \times (\text{Diff } \mathbb{C}^d)$$

$$g \longmapsto (g' \ ; \ g'')$$

An element $u \in \mathcal{U}$ is called a field map ; G is called the symmetry group of the given class of fields. The equivalence relation $\mathcal{R}$ is called the physical equivalence of field maps. A covariant action of G in Y is defined by $v \longmapsto gv$ with

$$(2.3) \qquad (gv)(x) = g''(v(g'^{-1}x))$$

(2.4) Remark

If $\mathcal{R}$ is trivial i.e. if any equivalence class in $\mathcal{U}$ is a point, the field is called "without gauge transforms". In general, G contains the Poincaré group P or its covering group $\tilde{P}$. For arbitrary $g \in P$, g' can be interpreted as an affine transformation of $\mathbb{R}^{s+1}$, or as a transformation of coordinates.

(2.5) Lagrangian field

Using the notations (1.5), a Lagrangian $\mathcal{L}$ is defined by a locally and weakly integrable function $x \longrightarrow \mathcal{L}(.,\ .;x) : \mathbb{R}^{s+1} \longrightarrow F^{\varepsilon}_{STA}(\mathbb{C}^M)$. For an arbitrary smooth subset B of $\mathbb{R}^{s+1}$, the corresponding action functional is defined by the following weak integral

$$(2.6) \qquad S_B(u) = \int_B \mathcal{L}(u(x),\ Du(x),x)\ dx \in F^{\varepsilon}_{STA}(Y)$$

where $\mathcal{L}(u(x), Du(x),x)$ denotes the inverse image of $\mathcal{L}(\bullet, \bullet;x)$ by the map (1.3). A field map $u \in Y$ admits the Lagrangian $\mathcal{L}$ if for arbitrary smooth subset B of $\mathbb{R}^{s+1}$, u is a stationary point of $S_B(u)$ with respect to variations in $Y_o = \mathcal{D}(\mathbb{R}^{s+1}, \mathbb{C}^d)$. In order to extend E.L. equations to the anticommuting case, the coordinates in $\mathbb{C}^d$ and $\mathbb{C}^d \otimes \mathbb{C}^{s+1}$ are resp. denoted u_j and $\partial_k u_j$, $j = 1 \ldots d$ and $k = 0, \ldots s$ Hence the left directional derivatives of the form $\mathcal{L}(\bullet, \bullet;x)$ on $\mathbb{C}^M$ are resp. denoted $\partial\mathcal{L}/\partial u_j$ and $\partial\mathcal{L}/\partial(\partial_k u_j)$. For arbitrary $u \in Y$ and $x \in \mathbb{R}^{s+1}$

$$I_j(x) = (\partial\mathcal{L}/\partial u_j)(u(x),Du(x),x) \quad ; \quad II_{jk}(x) = (\partial\mathcal{L}/\partial_{0_k} u_j)(u(x),Du(x),x)$$

denote resp. the inverse image by $\alpha(x)$ of the forms $\partial\mathcal{L}/\partial u_j(\bullet,\bullet;x)$ and $(\partial\mathcal{L}/\partial(\partial_k u_j))(\bullet,\bullet;x)$ on $\mathbb{C}^M$

(2.7) Extension of Euler–Lagrange equations

Let $\varepsilon = \pm$. A field map $u \in Y$ admits the Lagrangian density $x \longrightarrow \mathcal{L}(\bullet,\bullet;x) \in F_{STA}^{\varepsilon}(\mathbb{C}^M)$ iff the following equalities of distributions hold for $j = 1,2\ldots d$

$$(2.8) \qquad (\partial\mathcal{L}/\partial u_j)(u(x),Du(x),x) - \sum_k \partial_k (\partial\mathcal{L}/\partial(\partial_k u_j))(u(x),Du(x),x) = 0$$

Proof

We have $\mathcal{L} = \mathcal{L}_1 + \mathcal{L}_2$. For arbitrary $x \in M$, the forms $\mathcal{L}_1(\bullet;\bullet;x)$ and $\mathcal{L}_2(\circ,\circ;x)$ on Y are resp denoted $\mathcal{L}_1(t,t';x)$ and $\mathcal{L}_2(t,t';t_1,t'_1;x)$ with t and $t_1 \in \mathbb{C}^d$; t' and $t'_1 \in \mathbb{C}^d \otimes \mathbb{C}^{s+1}$. The canonical basis of $\mathbb{C}^M$ is denoted $\left\{ (\varepsilon_j; \varepsilon_{kj'}), j \text{ and } j' = 1\ldots d ; k = 0,1\ldots s \right\}$ Hence u is Lagrangian iff for any smooth open subset B of M and any $\varphi = \sum_{j=1\ldots d} \varphi_j \varepsilon_j \in \mathcal{D}(B,\mathbb{C}^d)$

$$(2.9) \quad 0 = 2 \int_B \mathcal{L}_2(\varphi(x),D\varphi(x);u(x),Du(x),x)dx + \int_B \mathcal{L}_1(\varphi(x),D\varphi(x),x)dx$$

By linearity :

$$2 \mathcal{L}_2(\varphi(x), D\varphi(x) \; ; \; u(x), D u(x); x)$$

$$= 2 \sum_j \varphi_j(x) \mathcal{L}_2(\varepsilon_j, 0 \; ; \; u(x), D u(x), x)$$

$$+ 2 \sum_{kj} \partial_k \varphi_j(x) \mathcal{L}_2 (0, \varepsilon_{jk} \; ; \; u(x), D u(x), x)$$

$$= \sum_j \varphi_j(x) (\partial \mathcal{L}_2 / \partial u_j) (u(x), D u(x), x)$$

$$+ \sum_{kj} \partial_k \varphi_j(x) (\partial \mathcal{L}_2 / \partial (\partial_k u_j))(u(x), D u(x), x)$$

In the same way :

$$\mathcal{L}_1(\varphi(x), D\varphi(x), x) = \sum \varphi_j(x) \partial \mathcal{L} / \partial u_j + \sum_{kj} \partial_k \varphi_j(x)(\partial \mathcal{L} / \partial (\partial_k u_j))$$

Hence substituing these expressions of $\mathcal{L}_2$ and $\mathcal{L}_1$ in (2.9) and using the duality brackets of distribution theory:

$$0 = \sum_j < \varphi_j, \; \partial \mathcal{L} / \partial u_j > + \sum_{kj} < \partial_k u_j, \; \partial \mathcal{L} / \partial (\partial_k u_j) >$$

$$= \sum_j < \varphi_j, \; \partial \mathcal{L} / \partial u_j - \sum_k \partial_k (\partial \mathcal{L} / \partial (\partial_k u_j)) >$$

for all $\varphi_1, \ldots \varphi_d \in \mathcal{D}(\mathbb{R}^{s+1})$. This proves (2.8). In particular if $(\partial \mathcal{L} / \partial (\partial_k u_j))(u(x), D u(x), x)$ is C^1, (2.8) is in fact an equality of functions.

(2.10) <u>Definition of a class of Lagrangian fields</u>

> A class of Lagrangian fields is defined by
> - a Lagrangian $x \longrightarrow \mathcal{L}(\bullet, \bullet; x) \in F_{STA}^{\varepsilon}(\mathbb{C}^M)$
> - the subset $Y_{\mathcal{L}}$ consisting of all fields maps $u \in Y$ with Lagrangian $\mathcal{L}$; an equivalence relation on $Y_{\mathcal{L}}$
> - a real Lie group G and a morphism (2.2) of groups.

(2.11) <u>Symmetry group preserving the Lagrangian</u>

> For a given class of Lagrangian fields, we says that some sub-group G_{SUB} of G preserves the Lagrangian if for an

(2.12)   arbitrary smooth subset B of $\mathbb{R}^{s+1}$, the following function :

$$g \longrightarrow I_u(g) = S_{g'B}(gu) = \int_{g'B} \mathcal{L}(gu)(x), D(gu)(x), x) \, dx$$

is constant on G_{SUB}

(2.13) <u>Proposition</u>

A given class of Lagrangian fields is considered. If some subgroup G_{SUB} of G preserves the Lagrangian, and if g'' acts linearely in $\mathbb{C}^d$ for arbitrary $g \in G_{SUB}$, then the action (2.3) of G_{SUB} in Y preserves the set $Y_{\mathcal{L}}$ of Lagrangian fields

In fact let $u \in Y_{\mathcal{L}}$. For arbitrary λ real, B smooth and $\varphi \in \mathcal{D}(\mathbb{R}^{s+1}, \mathbb{C}^d)$, $g \in G_{SUB}$, we have since g'' is linear :

$$g(u + \lambda \varphi) = g'' \left[(u + \lambda \varphi')(g^{-1} x) \right] = gu + \lambda (g\varphi)$$

and since G_{SUB} preserves $\mathcal{L}$:

$$\int_B \mathcal{L}(u + \lambda \varphi, D(u + \lambda \varphi'), x) \, dx = \int_{g^{-1}B} \mathcal{L}(gu + \lambda g\varphi, D(gu + \lambda g\varphi), x) \, dx$$

Since $u \in Y_{\mathcal{L}}$, the L.H.S. is stationary for $\lambda = 0$ and arbitrary $\varphi \in \mathcal{D}(\mathbb{R}^{s+1}, \mathbb{C}^d)$. Hence the R.H.S. is stationary in the same conditions, i.e. gu admits the Lagrangian $\mathcal{L}$.

3 - <u>THE ACTION THEOREM</u>

(3.1) <u>The problem</u>

In the present paragraph, a class of Lagrangian fields and a subgroup G_{SUB} of the symmetry group G are given. The unit element of G_{SUB} is denoted e.

For $\varepsilon = -$, a Lagrangian field $u \in Y_{\mathcal{L}}$ is given and for an arbitrary smooth subset B of $\mathbb{R}^{s+1}$ the derivative in $g = e$ of the numerical function $g \longrightarrow I_u(g)$ on G_{SUB} will be computed

For $\varepsilon = +$ since $\mathcal{L} = \mathcal{L}_1 + \mathcal{L}_2$, $Y_{\mathcal{L}}$ is an affine subspace of the vector space Y. In particular if $\mathcal{L}_1 = 0$, $Y_{\mathcal{L}}$ is a <u>linear</u> subspace of Y and the derivative $I'_u(e)$ in the point $g = e$ of the following functional will be computed :

$$(3.2) \qquad G_{SUB} \ni g \longmapsto I_u(g) \in F^+_{STA}(Y_L)$$

If $\mathcal{L}_1 \neq 0$, the "derivative" $I'_u(e)$ in the point $g = e$ of the following "function"

$$(3.3) \qquad G_{SUB} \ni g \longrightarrow \tilde{I}_u(g) = \text{restriction of } I_u(g) \text{ to } Y_{\mathcal{L}}$$

will also be computed but the definition of $I'_u(e)$ needs now some precisions. In fact $\forall g \in G_{SUB}$, the form $I_u(g)$ on Y has two homogeneous components $I^1(g) : u \longrightarrow I^1_u(g)$ and $I^2(g) : (u,v) \longrightarrow I^2_{u,v}(g)$. The derivatives in the point $g = e$ of the following maps

$$g \longrightarrow \tilde{I}^1(g) = \text{restriction fo } I^1(g)$$
$$g \longrightarrow \tilde{I}^2(g) = \text{restriction of } I^2(g)$$

are denoted $I'^1(e)$ and $I'^2(e)$, and are written as restrictions to Y_L of multilinear and antisymmetric forms on Y. Finally we set $I'(e) = I'^1(e) + I'^2(e)$ and we write by extension $I'(e) \in F^+(Y_{\mathcal{L}})$.

(3.4) <u>Notations</u>

The derivatives in the point $g = e$ of the maps $g \longrightarrow g' \, x$ and $g \longrightarrow g'^{-1} x : G_{SUB} \longrightarrow \mathbb{R}^{s+1}$ are real linear maps $\text{Lie}(G_{SUB}) \longrightarrow \mathbb{R}^{s+1}$ denoted resp.

$$(3.5) \qquad X(x) : \delta g \longmapsto X(x) \, \delta g = (X^1(x) \, \delta g)_{1=0,\ldots s}$$

$$(3.6) \text{ and } \qquad X'(x) : \delta g \longmapsto X'(x) \, \delta g = (X'^1(x) \, \delta g)_{1=0,\ldots s}$$

For $\varepsilon = -$, let $u \in Y_{\mathcal{L}}$ be a given Lagrangian field. The derivative in the point $g = e$ of the map $g \longrightarrow g''(u(\cdot)) : G_{SUB} \longrightarrow Y$ is denoted

$$(3.7) \quad \Psi(u,x) : \delta g \longmapsto \Psi(u,x) \; \delta g = (\Psi_k(u,x) \; \delta g)_k \quad k = 1 \ldots d$$

Hence, the derivative in the point $g = e$ of the map $\alpha: g \longrightarrow gu = (g''(g^{l-1}x)) : G_{SUB} \longrightarrow Y$ is

$$(3.8) \quad U(u,x) : \delta g \longmapsto U(u,x) \; \delta g = (U_k(u,x) \; \delta g)_k$$

$$(3.9) \quad \text{with} \quad U_k(u,x) \; \delta g = \Psi_k(u,x) \; \delta g + \sum_1 (\partial_1 u_k(x)) X'^1(x) \; \delta g$$

Also for x arbitrary and given $\in \mathbb{R}^{s+1}$, the derivative in the point $g = e$ of

$$G_{SUB} \ni g \longrightarrow D_x((gu(x)) \in \mathbb{R}^{s+1}$$

is the product of α with the linear map $v \longrightarrow (D_x v)(x)$, i.e.

$$(3.10) \qquad \delta g \longrightarrow D_x(U(u,x) \; \delta g)$$

For $\varepsilon = +$, we assume moreover that $U(u,x) \; \delta g$ is linear with respect to $u \in Y$.

(3.11) <u>Current of the action of G_{SUB}</u>

For $\varepsilon = -$, the current of the action of G_{SUB} on the given Lagrangian field u is defined as the following differential form of degree s on R^{s+1}, depending linearely from $\delta g \in \text{Lie}(G_{SUB})$

$$(3.12) \qquad \theta(u,x) \; \delta g = \sum_\ell (\theta^\ell(u,x) \; \delta g) \; d_\ell x$$

with

$$(3.13) \qquad d_\ell x = dx^\circ \wedge \ldots \wedge dx^{\ell-1} \wedge dx^{\ell+1} \ldots \wedge dx^s$$

$$(3.14) \quad \text{and} \quad \theta^\ell(u,x) \; \delta g = \sum_k U_k(u,x) \; \delta g \; \frac{\partial \mathcal{L}}{\partial(\partial_\ell u_k)} + X^\ell(x) \; \delta g \mathcal{L}$$

Hence $\theta(u,x)$ is a differential form on $\mathbb{R}^{s+1}$ with values in the dual $(\text{Lie } G_{SUB})^*$ of the Lie algebra of G_{SUB}.

For $\theta = +$, the same formulas are used but with a different meaning. More precisely if $\mathcal{L}_1 = 0$, the formula (3.12) defines $\theta(u,x)$ as a différential form on $\mathbb{R}^{s+1}$ with values in $L(\text{Lie } G_{SUB}, F^+(Y_L))$. But now in the R.H.S. of (3.14), the term $(U_k(u,x) \, \delta g) \times (\partial \mathcal{L} / \partial (\partial_1 u_k))$ is not a product of numbers but an antisymmetrized product of forms on Y_L. If $\mathcal{L}_1 \neq 0$, Y_L is an affine vector subspace of Y. In this case, the R.H.S. of (3.14) is defined by restriction to Y_L of antisymmetric forms on Y. Hence $\theta(u,x)$ takes values in $L(\text{Lie } G_{SUB}, F^+(Y_L))$.

(3.15) <u>The action theorem</u>

> For $\varepsilon = -$, for a given Lagrangian field u and for an arbitrary smooth subset B of $\mathbb{R}^{s+1}$, the derivative in the point $g = e$ of the numerical function $I_u(g) = S_{g',B}(gu)$ on G_{SUB}, is the following linear form on $\text{Lie } (G_{SUB})$:

$$(3.16) \qquad \delta g \longrightarrow \delta I \; = \; \int_{\partial B} \theta(u,x) \; \delta g$$

> For $\varepsilon = +$, and for an arbitrary smooth subset B of $\mathbb{R}^{s+1}$, the derivative in the point $g = e$ of $I_u(g) : G_{SUB} \longrightarrow F^+(Y_L)$, is the linear map (3.16) with values in $F^+(Y_L)$

(3.17) <u>Lemma</u>

> The algebraic dual E^* of the vector space E is endoved with the weak topology. Then for an arbitrary C^1 function $Q : \mathbb{R}^{s+1} \longrightarrow E^*$ and for an arbitrary smooth subset B of $\mathbb{R}^{s+1}$, the derivative is the point $g = e$ of

$$(3.18) \qquad G_{SUB} \ni g \longrightarrow S(g) = \int_{g',B} Q(x)dx \in E^*$$

> is

$$(3.19) \qquad \delta g \longrightarrow \delta S = \int_B dx \sum_{k=0..s} \partial_k (Q(x) X^k(x) \, \delta g)$$

$$(3.19') \qquad\qquad = \int_{\partial B} \sum_k (Q(x) X^k(x) \, \delta g) \, d_k x$$

This last expression admits the following geometrical interpretation :
δS is the flux through the hypersurface ∂B of the vector field

on $\mathbb{R}^{s+1}$ constructed by tensor product of $Q(x) \in E^*$ with the infinitesimal displacement $X(x)\, \delta g$ of the point x.

Proof of the lemma

Putting $x = g'y$, $S(g)$ can be written

$$* \qquad S(g) = \int_B Q(g'(y)) \left| \det\, M_g(y) \right| dy$$

where $M_g(y)$ is a Jacobian matrix. Hence the jth row of $M_{g'}(y)$ is

$$\mathrm{grad}_y (g'y)^j = (\partial_0 (g'y)^j,\ \partial_1 (g'y)^j,\ \ldots,\ \partial_\delta (g'y)^{\dot{j}})$$

For example $g' \in \mathcal{P}_0$, $\det\, M_{g'}(y) = \det g' = 1$. In any case $M_e(y)$ is the identity matrix. Hence the derivative in $g = e$ of $g \longrightarrow \left| \det M_g(y) \right|$ is :

$$\delta_w \longrightarrow \Sigma_k\ \partial_k\ (X^k(x)\, \delta g)$$

Therefore in view of $\quad * \quad$:

$$\delta S = \int_B dx\, \Sigma_k \left[(\partial_k Q)(x)\, X^k(x)\, \delta g\ +\ Q(x)\, \partial_k (X^k(x)\, \partial g) \right]$$

$$= \int_B dx\, \Sigma_k\ \partial_k\ (Q(x) X^k(x)\, \delta g)$$

The expression $(3.19')$ of δS follows by application of Stockes's theorem.

Proof of the action theorem for $\epsilon = -$

A field map $u \in Y_{\mathcal{L}}$ is given. The derivative in the point $g = e$ of the numerical function $I_u(g)$ on G_{SUB} is the sum of three terms corresponding resp. to the g-dependancy of $(gu)(x)$, of $D(gu)(x)$ and of the integration domain $g'B$ of the integral (2.12). Hence using the lemma (3.17)

$$** \qquad \delta I = \int_B dx \left[\Sigma_k\, U_k(u,x)\, \delta g\, \frac{\partial \mathcal{L}}{\partial u_k}\ +\ \Sigma_{k1} \partial_l (U_k(u,x)\, \partial g)\, \frac{\partial \mathcal{L}}{\partial (\partial_1 u_k)} \right.$$
$$\left. +\, \Sigma_k\ \partial_k (X^k(x)\, \delta g \mathcal{L}) \right]$$

Hence replacing $\partial \mathcal{L} / \partial u_k$ by $\sum_\ell \partial_\ell \mathcal{L} / \partial(\partial_\ell u_k)$:

$$\delta I = \int_B dx \left[\sum_{k\ell} (\partial_\ell(U_k(u,x)\,\delta g)\, \partial\mathcal{L}/\partial(\partial_\ell u_k)) \right.$$
$$\left. + \sum_\ell \partial_\ell(X^\ell(x)\,\delta_g \mathcal{L}) \right]$$

Hence using Stockes's theorem

$$= \int_{\partial B} \sum_\ell \left(\sum_k U_k(u,x)\,\delta g\; \partial\mathcal{L}/\partial(\partial_\ell u_k) + X^\ell(x)\,\delta g \right) d_\ell x$$
$$= \int_{\partial B} \sum_1 (\theta^\ell(u,x)\,\delta g)\, d_\ell x$$

This proves the action theorem for $\varepsilon = -$. For $\varepsilon = +$, the function $I_u(g)$ on G_{SUB} takes values in $F^+(YL)$. The following lemma is needed.

(3.20) <u>Lemma</u>

> The derivative in the point $g = e$ of the function
>
> $$G_{SUB} \ni g \to J_g(u) = \int_B \mathcal{L}((gu)(x), D(gu)(x), x)dx$$
>
> with values in $F^+(Y_L)$ is the following linear map :
> Lie $(G_{SUB}) \to F^+(Y_L)$
>
> (3.21) $$\delta g \to \delta J = \int_B dx \left[\sum_k U_k(u,x)\,\delta g\, \frac{\partial \mathcal{L}}{\partial u_k} \right.$$
> $$\left. + \sum_{kl} \partial_1(U_k(u,x)\,\delta g)\; \partial\mathcal{L}/\partial(\partial_\ell u_k) \right]$$

In the last formula, the integral denotes a weak integral $\in F^+(Y_L)$ and the products $(U_k(u,x)\,\delta g)\; \partial\mathcal{L}/\partial u_k$ and $\partial_1(U_k(u,x)\,\delta g)\; \partial\mathcal{L}/\partial(\partial_1 u_k)$ are antisymmetrized products of forms. The proof is written only for $\mathcal{L} = \mathcal{L}_2$; the proof for $\mathcal{L} = \mathcal{L}_1$ is similar and the case $\mathcal{L} = \mathcal{L}_1 + \mathcal{L}_2$ follows by linearity.

Proof

The canonical basis of $\mathbb{C}^d$ and $(\mathbb{C}^{s+1} \times \mathbb{C}^d)$ are denoted resp ε_k and $\varepsilon_{1k'}$ with $k, k' = 1\ldots d$ and $1 = 0,1\ldots s$. The generie element of

$\mathbb{C}^M = \mathbb{C}^d \oplus (\mathbb{C}^{s+1} \times \mathbb{C}^d)$ can be written $u + t = \sum u_k \varepsilon_k + \sum_{k\ell} \varepsilon_{\ell k'} t_{\ell k'}$

For a fixed x, the bilinear form on $\mathbb{C}^M$ defined by $\mathcal{L}_2(\bullet,\bullet;x)$ is denoted (u,t) ; $(u',t') \longmapsto \mathcal{L}(u,t;u',t',x)$. Hence the value of the antisymmetric form J on the pair $(u;v) \in Y_L \times Y_L$ is using the notation $V\delta g = U(v,x)\delta g$:

$$\delta J(u,v) = \int_B \left[\mathcal{L}(U\delta g, D(U\delta g); v(x), Dv(x);x) \right.$$

$$+ \mathcal{L}(u(x),Du(x) ; V\delta g, D(V\delta g) ; x \Big] dx$$

$$= \int dx \left[\sum U_k \delta g \ (\varepsilon_k,0; v(x), Dv(x) ; x) \right.$$

$$+ \sum_{1k} \partial_1(U_k \delta_g) \mathcal{L}(0, \varepsilon_{1k}; v(x), D v(x) ; x)$$

$$+ \text{similar terms in} \ \ V_k \ \delta g \ \text{and} \ \partial_1(V_k \delta g) \Big]$$

$$= \int \frac{dx}{2} \sum_k \Big[U_k \delta g (\partial\mathcal{L} / \partial u_k)(v(x),Dv(x),x) - V_k\delta g (\partial\mathcal{L} / \partial u_k)(u(x),Du(x),x)$$

$$+ \ \frac{dx}{2} \sum_{1k} \Big[\partial_1(U_k\delta g)(\partial\mathcal{L}/\partial(\partial_1 u_k))(v(x),Dv(x),x)$$

$$- \partial_1(V_k \delta g)(\partial\mathcal{L}/\partial (\partial_1 u_k))(u(x),Du(x),x) \Big]$$

By definition of the antisymmetrized product of (antisymmetric) forms, this means that the antisymmetric form δJ is

$$\delta J(u) = \int \sum_k U_k \delta g (\partial\mathcal{L} / \partial u_k)(u(x),Du(x),x) \ dx$$

$$+ \sum_{1k} \int_B \partial_\ell(U_k(u,x) \delta g)(\partial\mathcal{L} / \partial (\partial_\ell u_k))(u(x),u(x),x) \ dx$$

Proof of the action theorem for $\varepsilon = +$.

A combined application of lemmas (3.14) and (3.20) shows that formula (3.19) is also valid for $\varepsilon = +$ with a different meaning. In fact the products in the R.H.S. of (3.19) are wedge products of antisymmetric forms on Y the product in the bracket are wedge products restricted to Y_L ; and the integral of form and is a weak integral of $F^+(Y_L)$ valued functions. Hence the arguments of the commuting case can be adapted to the anticommuting case.

(3.22) Corollary : canditates for conserved integrals

If the action of G_{SUB} preserves $\mathcal{L}$, then

$$(3.23) \qquad \forall B \qquad \int_{\partial B} \theta(u,x)\, \delta g = \int_{\partial B} \sum_{\ell} \theta^{\ell}(u,x)\, \delta g d_{\ell} x = 0$$

Applying this to cylindrical subsets :

$$B_n(t_o,t_1) = \left\{ x = (\vec{x},t) \in \mathbb{R}^{s+1} ; \left|\vec{x}\right| \leq n \text{ and } t_o < t < t_1 \right\}$$

and assuming the existence of limits for $n \to \infty$, the following inte-gral is time-independant :

$$(3.24) \qquad \int_{R^s} \theta^{\circ}(u,t,x)\, \delta g\, \vec{dx}$$

The corresponding conserved integral $\int_{R^s} \theta^{\circ}(u,t,x)\, \vec{dx}$ is an elelent of $(\text{Lie } G_{SUB})^*$ for $\mathcal{E} = -$ and an element of $L(\text{Lie } G_{SUB}, F^+(Y_L))$ for $\mathcal{E} = +$

(3.25) Examples of currents and of conserved integrals

a) The current $T(u,x)$ of energy impulsion is the current of the subgroup of translations $x \mapsto g'x = x - h$ in R^{s+1}, $h \in R^{s+1} = G_{SUB}$. In this particular case

$$X^{\ell}(x)\, \delta h = - \delta h^{\ell} \; ; \; X'^{\ell}(x)\, \delta h = \delta h^{\ell}$$

Since $\Psi = 0$ we have

$$U_k(u,x)\, \delta h = 0 + \sum_{\ell} (\partial_{\ell} u_k(x))\, \delta h^{\ell}$$

$$(3.26) \qquad T^{\ell}(u,x)\, \delta h = \sum_k \left(\sum_{\ell} \partial_{\ell} u_k(x)\, \delta h^{\ell} \right) \frac{\partial \mathcal{L}}{\partial(\partial_{\ell} u_k)} - \mathcal{L}\, \delta h^{\ell}$$

Hence $T^{\ell}(u,x)\, \delta h = \sum_j T_j^{\ell}(u,x)\, \delta h^j$ with

$$(3.27) \qquad T_j^{\ell}(u,x) = \sum_k \partial_j u_k(x)\, \frac{\partial \mathcal{L}}{\partial(\partial_j u_k)} - \mathcal{L}\, \delta_j^{\ell}$$

The contravariant components of $T(u,x)$ are $T^{\ell j} = g^{jj} T_j^{\ell}$. The corres-ponding conserved integral is $P(u) = (P^j(u))_j$ with

$$(3.28) \qquad P^j(u) = \int_{\mathbb{R}^s} T^{\circ j}(u;t,\vec{x})\, \vec{dx}$$

The Hamiltonian is $H(u) = P^\circ(u)$. For $\varepsilon = -$, u is given $\in Y_{\mathcal{L}}$, $P(u)$ is a vector and $P^\circ(u)$ is a scalar. But for $\varepsilon = +$, $P(u)$ $\in R^{s+1} \otimes F^+(Y_L)$ and the energy $P^\circ(u) \in F^+(Y_L)$ is an antisymmetric form on Y_L.

b) The complex vector space $\mathbb{C}^{d'}$ defines an underlying real vector space $(\mathbb{C}^{d'})_r$ of dimension $2d'$. The map

$$(3.29 \qquad (\mathbb{C}^{d'})_r \ni z \longrightarrow (z;\bar{z}) \in \mathbb{C}^{d'} \times \mathbb{C}^{d'}$$

defines a complexification of $(\mathbb{C}^{d'})_r$. Many times a field map $u : R^{s+1} \longrightarrow (\mathbb{C}^{d'})_r$ is replaced by the field map $(u,\bar{u}) : \mathbb{R}^{s+1} \longrightarrow \mathbb{C}^{d'} \times \mathbb{C}^{d'}$. Then the current J of charge is defined as the current of the group of first kind gauge transformations :

$$(3.30) \qquad (u,\bar{u}) \longrightarrow g''(u,\bar{u}) = (e^{-i\alpha} u,\ e^{+i\alpha} \bar{u})$$

In the present case dim $G_{SUB} = 1$, $X^\ell(x) = X'^\ell(x) = 0$ and

$$\Psi(u,x)\, \delta\alpha = (-iu\, \delta\alpha,\ \ldots;\ + i\,\bar{u}\,\delta\alpha,\ldots)$$

$$U_j(u,x)\, \delta\alpha = \begin{cases} -\, iu_j\, \delta\alpha & \text{for } j = 1 \ldots d' \\ +\, iu_j\, \delta\alpha & \text{for } j > d' \end{cases}$$

Hence the components of J are

$$(3.31) \qquad J^\ell(u,x) = \sum_j -i\, u_j\, \frac{\partial \mathcal{L}}{\partial(\partial_\ell u_j)} + i\, u_j\, \frac{\partial \mathcal{L}}{\partial(\partial_\ell \bar{u}_j)}$$

The corresponding conserved integral is the charge

$$(3.32) \qquad Q(u) = \int_{\mathbb{R}^s} J^\circ(u;t,\vec{x})\, d\vec{x}$$

The charge is a number for $\varepsilon = -$ and an antisymmetric form on Y_L for $\varepsilon = +$

(3.33) <u>In general</u> the usual expressions of currents and conserved integrals are valid for $\varepsilon = +$, at least if the products are written such that all terms containing $\mathcal{L}$ are written on the right hand side.

4 - EXAMPLE

Previous results can be used practically in the following way :

a) The usual formulas of E.L. and of Noether's theory are keeped without modification at least if products are written with the correct order of factors : see (3.33)

b) The meaning of these formulas is modified using anticommuting variables in the Lagrangian,for Fermi components of fields.

c) Conserved integrals are not scalars in the anticommuting case, but functionals with anticommuting arguments on infinite dimensional spaces.

We consider for example :

Ψ = Dirac spinor field = colomn vector

$\widetilde{\Psi} = \Psi^{*} \, \gamma^{\circ}$ = Dirac conjugate spinor field

= line vector

Usually,the Lagrangian of the free Dirac field is a quadratic function of the 8 components of $u = (\Psi, \Psi)$ and of the 16 components of $Du = (D\Psi, D\widetilde{\Psi})$:

$$(4.1) \quad \mathcal{L}(u,Du) = \frac{i}{2} \sum (\widetilde{\Psi} \gamma^{j}(\partial_{j}\Psi) - \partial_{j}\widetilde{\Psi} \gamma^{j} \Psi) - m \widetilde{\Psi} \Psi$$

The E.L. equations are

$$(4.2) \quad 0 = (\sum i \gamma^{j} \partial_{j} \Psi) - m \Psi = (\sum i \partial_{j}\widetilde{\Psi} \gamma_{j}) + m \widetilde{\Psi}$$

Using Noether's theorem the contravariant compoents of the energy impulsion tensor are

$$(4.3) \quad T^{k\ell}(u) = \frac{i}{2} \widetilde{\Psi}(x) \gamma^{k} \partial_{\ell}\Psi - \partial_{\ell}\widetilde{\Psi} \gamma^{k} \Psi$$

Using the Fourier decomposition of the field the energy momentum vector $P(u)$ has components $[2]$:

$$(4.4) \quad P^{1}(u) = \int \vec{dx} \, T^{o1}(u) = \int dk \, k^{1} \sum_{y=1,2} (a^{*+}_{y}(k) a^{-}_{y}(k) - a^{*-}_{y}(k) a^{+}_{y}(k))$$

Therefore the commutativity of arguments in the usual Lagrangian formalism produces an error of sign for the energy $P°(u)$. More precisely there is a discrepancy between the previous formalism and the corresponding informal quantum formalism. Now, keeping the same formulas, we assume that the Lagrangian $\mathcal{L}(u,Du)$ has anticommuting arguments. For example the term $\widetilde{\psi}\psi$ in (4.1) means now $\sum_j \widetilde{\psi}_j \wedge \psi_j$. In view of §2, the equations (4.2) can also be deduced by a principle of stationary anticommuting action. In view of §3, formulas (4.3) and (4.4) are also valid but with a different meaning : $P°(u)$ is a functional with anticommuting arguments hence

$$(4.5) \quad -a^{*-}(k)a^{+}(k) = -a^{*-}(k) \wedge a^{+}(k) = a^{+}(k) \wedge a^{*-}(k)$$

Hence the positivity of energy is restablished and there is non discrepancy between the new formalism and the corresponding quantum formalism of $[2]$ p125. In the free case or in the case of very weak interaction we pass from this new formalism to the corresponding quantum formalism by the symbolic calculus of $[3]$. This is possible since the symbol of a quantum operator is an usual functional for Boson components, and a functional with anticommuting arguments for Fermi-components of the field. This mathematical convention follows from ε-commutation relations of quantum fields and does not produce contradictions at the classical level, since unquantized Fermi fields are not physically observable. Note also that the present approach avoids the contradictions connected with the "Zitter-Bewegung" of electrons.

REFERENCES

[1]   F.A. BEREZIN

The method of second quantization Academic Press(1966)

[2]   N.N. BOGOLIUBOV and D.V.SHIRKOV

Introduction to the Theory of Quantized Fields
3th edition. John Wiley 1981

[3]   P. KRÉE

Lecture Notes in Mathematics n° 843 pp 373-404(1981)
Springer Verlag

[4]   P. KRÉE

Comptes Rendus Série I - t 292(6 Avriel 1981) pp 697-
698

[5]   P. KRÉE

Comptes Rendus t 292 (27 avril 1981) Série 1
pp 735-737

[6]   P. KRÉE

Colloque International du C.N.R.S. n° 307 (edition
du C.N.R.S. 1981) pp 435-460

[7]   J. SCHWINGER

Quantum Kinematics and Dynamics -- Benjamin . 1970

Institut de Mathématiques
Université de Paris 6
Place Jussieu - Paris 5e

HIGH-TEMPERATURE PATH METHOD
FOR A TWO-DIMENSIONAL RANDOM ISING MODEL

D. Merlini[+]

Mathematisches Institut

Ruhr-Universität Bochum

and

Fakultät für Physik

Universität Bielefeld

Germany

+ Present Adress: Fakultät für Mathematik, Ruhr-Universität,
4630 Bochum, Germany

Recently, there has been much interest in the investigation of phase transitions in spin lattice systems containing random parameters[1], randomly distributed ferromagnets[2] and spin glass systems[3,4,5,6]. The Hamiltonian of interest is usually written as

$$H = \sum_{\vec{x},\vec{y} \in \mathbb{Z}^{\nu}} J(\vec{x}-\vec{y}) s_{\vec{x}} s_{\vec{y}} - \sum_{\vec{x} \in \mathbb{Z}^{\nu}} h s_{\vec{x}}$$

where $s_{\vec{x}}$ is a spin variable (with values ± 1) and $J(\vec{x}-\vec{y})$ are, for different $\vec{x}-\vec{y}$, independent random variables distributed according to some probability measure μ. One difficulty in solving such models is given by the fact that, because the considered system is disordered, it is additionally necessary to average over μ the free energy. For models with neighbours interaction no exact solution has been found as of get. In a special case however [7], it has been shown recently that application of renormalization group techniques, yields absence of phase transitions. In this note we develope a rigorous compact formulation of the high-temperature cluster expansion for a two-dimensional random Ising model. Moreover we investigate the critical temperature of the model by the simple ratio method frequently employed in non-random spin systems. For a special class of "spin glass", it is found that the critical temperature is a monotonic decreasing function of the parameter describing the rate of disorder which varies from T_c (the Ising critical temperature) to $T = 0$ (symmetric case).

The random Ising model we consider here ("spin glass") is defined as follows: Let $\Lambda \subset \mathbb{Z}^2$ be a finite square box of $|\Lambda|$ sites. Let $\sigma_i = \pm 1$, $i \in \Lambda$, be an Ising spin variable and define $\sigma_A = \prod_{i \in A} \sigma_i$, $A \subset \Lambda$. Let B be any bond on the lattice i.e. $B = (i,j)$, (i,j) nearest-neighbours points on Λ. The Hamiltonian is defined by $H(J,\{t_B\}) = -\sum_B J \sigma_B t_B$, where the sum is over all bonds in Λ and t_B are independent random variables distributed according to some probability measure $P(t_B)$, so that the joint distribution is

$$d\mu = \prod_B P(t_B) dt_B \tag{1}$$

J is here a real constant (independent of B, σ_B, and t_B). We will restrict ourselves later essentially to the class of measures $\{\mu_p\}$ such that t_B takes the value $+1$ with probability p and the value -1 with probability $1-p$; in that case $\mu_p[\{t_B\}] = p^{N^+} \cdot (1-p)^{N^-}$, where $N^{\pm}$ is the number of t_B such that $t_B = \pm 1$. $(N^+ + N^- = 2|\Lambda|)$. We define

$$\overline{f(\{t_B\})}_p = \int d\mu_p(\{t_B\}) \, f(\{t_B\}) \tag{2}$$

for the average of some quantity with respect to μ_p. With $\beta J = \frac{1}{kT} J = K$ (k being Boltzmann's constant), the free energy per site is given by:

$$-\beta_{\Lambda}(\{t_B\},K) = \frac{1}{|\Lambda|} \ln Z_{\Lambda}(\{t_B\}) \tag{3}$$

where $Z_\Lambda(\{t_B\}) = \text{Trace } e^{-H(K,\{t_B\})}$, $H(K,\{t_B\}) = \beta^H(J,\{t_B\})$, the trace being with respect to the σ-variables.

With $x_B = \text{th}(Kt_B)$, the high temperature expansion reads[8]:

$$Z_\Lambda(\{t_B\}) = 2^{|\Lambda|} \prod_B \cosh(Kt_B) . \text{ Trace } \prod_B (1+x_B\sigma_B) . \tag{4}$$

Upper and lower bounds for the free energy are easily obtained, in fact:

$$Z_\Lambda(\{t_B\}) \leq \prod_B \cosh(K) . \text{ Trace } \prod_B (1+x_B\sigma_B) , \tag{5}$$

since only even graphs occours.

On the other hand by the definition:

$$Z\Lambda(\{t_B\}) = \text{Trace } e^{K\sum_B t_B \sigma_B} = \text{Trace } e^{\sum_B K(t_B-\alpha)\sigma_B + \sum_B K\alpha\sigma_B}$$

$$Z_\Lambda(\{t_B\}) \leq Z_{\Lambda\text{Ising}}(K\alpha) . \; e^{\sum_B K(t_B-\alpha)<\sigma_B>_{\text{Ising}}(K\alpha)} \tag{6}$$

from Jensen inequality, for any real α. Combining the two inequalities (5), (6), and choosing $\alpha = (2p-1)$ we get, for any $1/2 \leq p \leq 1$

$$-\beta f_{\text{Ising}}(K(2p-1)) \leq -\beta \overline{f_p(\{t_B\},K)} \leq -\beta f_{\text{Ising}}(K') \tag{7}$$

where $-\beta f_{\text{Ising}}(K')$ is the Ising free energy density with interaction parameter $J' = \frac{K'}{\beta}$. A cluster expansion for any p may be prooven easily by application of the Sherman theorem on paths (this is related to our previous use of the expansion to obtain a proof of inexistence of non-translationally invariant equilibrium states in the two-dimensional Ising model at low-temperature)[9,10]. The theorem say that the re partition function Z_r, is given by:

$$Z_r = Z_\Lambda \left(\prod_B \cosh(Kt_B)\right)^{-1} = e^{|\Lambda| \sum_c (-1)^{N_c} W(c)/\mu_c} \tag{8}$$

In eq. (8) c denotes any one-cycle on the lattice i.e. c is a connected closed trajectory starting at some point i, returning to i and having the property that at eve step no turn of $\pm\Pi$ degrees is allowed. N_c is the number of selfcrossing of c and μ_c its multiplicity, while $W(c) = \prod_{B \in C} x_B^{n_B}$, where n_B is the number of times the bond B accurs in c.

Assuming a measure μ_p giving the value 1 with probability p to t_B and the value -1 with probability $1-p$ as mentioned before, we obtain:

$$\overline{(x_B^{n_B})}_p = x_B^{n_B} \overline{(t_B^{n_B})}_p = \begin{cases} x_B^{n_B} (2p-1) & n_B \text{ odd} \\ x_B^{n_B} 1 & n_B \text{ even} \end{cases}$$

where $x = \text{th } K$.

Thus:

$$\lim_{|\Lambda|\to\infty} \frac{1}{|\Lambda|} \, (\overline{\ln \, \mathbb{Z}_r})_p = f_p^{\,odd} + f_p^{\,even} \tag{9}$$

where $f_p^{\,odd}$ is the sum of all c starting at some lattice point, where at least one bond $B \in C$ has n_B odd, while $f_p^{\,even}$ is the sum of all c starting at the same lattice point and such that n_B is even $\forall B \in C$. By the definition $f_p^{\,even} = f_{1/2}^{\,even}$ is independent of p and $f_{1/2}^{\,odd} \equiv 0$. The free energy for p appears as a perturbation around $p = {}^1\!/2$; this was to be expected since for $p = {}^1\!/2$ none of the ferromagnetic or anti ferromagnetic state is favored; in fact for $p = {}^1\!/2$, use of a symmetry property yields the result [6] that $\overline{<\sigma_x>} = 0 \; \forall x \subset \Lambda$ and all boundary conditions, while for $p > p_o > \frac{1}{2}$ use of refined Peierls arguments indicate the existence of a spontaneous magnetization at low temperature. Nevertheless , to our knowledge it is not known whether p_o may be arbitarily close to 1/2. We then conjecture, and give a numerical evidence below, that the critical temperature varies continuously between T_c and 0 with p, as p varies from 1 (pure Ising model) to $p = 1/2$.

We first consider the case $p = 1/2$ [12]. The question is then, whether the free energy is analytic; in particular if the specific heat is finite at all temeprature $T \neq 0$, since the absence of a symmetry breakdown of the state at $p = 1/2$, i.e. $\overline{<\sigma_x>} = 0 \; \forall x$, is by no means a sufficient condition for the absence of a singularity in the thermodynamic functions at $T \neq 0$. [13]

We notice that the set of all trajectories c which contribute to $f_{1/2} = f_{1/2}^{\,even} = f_{Ising} - f_1^{\,odd}$ is a subset (indeed very small) of all trajectories c which contribute to f_{Ising}, since n_B should be even $\forall B \in C$ and $\forall C$. Up to some order, we then obtain

$$f_{1/2} = -\frac{x^8}{2} - x^{12} + 4x^{14} - \frac{27}{4}x^{16} + 20x^{18} - 66x^{20} + \ldots \tag{10}$$

With eq. (10) and with the help of the result on the Ising model [12] (p = 1) we also obtain:

$$f_1^{\,odd} = x^4 + 2x^6 + 5x^8 + 12x^{10} + \frac{115}{3}x^{12} + \frac{883}{7}x^{14} + \frac{3981}{8}x^{16} + \frac{17448}{9}x^{18} + \ldots \tag{11}$$

The coefficients in $f_{1/2}$ increase very slowly, while those of $f_1^{\,odd}$ have small oscillations around those of the Ising model.

The series for the reduced specific heat, defined as $C_{1/2} = \partial^2 f_{1/2} / \partial K^2$ is given by:

$$C_{1/2} = -28y^3 + 34y^4 - 132y^5 + 884y^6 - 2460y^7 + 7956y^8 - 31920y^9 + \ldots \tag{12}$$

where $y = x^2$.

By the ratio method[12] we find that the convergence radius is given by $\dfrac{1}{x_c^2} = 0{,}989 \sim 1$,

so that $C_{1/2}$ is bounded for any temeprature $T > 0$; on the other hand writing $f_{1/2} = \Sigma a_n y^n$, $y = x^2$, it may easily be verified that the sequence $|a_n|^{1/n}$ converges rapidly to 1.

Within the accuracy of the numerical method, we have given a strong evidence that the model at $p = 1/2$ does not undergo a phase transition (except possibly at $T = 0$), a result which supports the one given previously using a renormalization technique [7].

Let us now consider the general case $1/2 < p < 1$. Here much more work is needed to compute 7 coefficients in the expansion (contrary to the case $p = 1/2$) but refined technique of the algebraic method could be employed[13].

Without too many efforts and using the Sherman theorem of closed trajectories (eq. 8), we have computed the series up to order 12 in x.

We obtain:

$$f_p^{\,odd} = x^4 \in^4 + 2x^6 \in^6 + x^8(7\in^8 - 2\in^6) + x^{10}(28\in^{10} - 4\in^4 - 12\in^8)$$

$$+ \; x^{12}(124\in^{12} - 62\in^{10} - 16\in^8 - 12\in^6 + \frac{13}{3}\in^4) + \cdots \tag{13}$$

where $x = \mathrm{th}\beta J$ and $\in = (2p-1)$.

We investigate the convergence radius of the series above which we identify with the inverse critical point; the coefficients are continuous function of p (polynomials) and the convergence radius for $f_p^{\,odd}$ should be a continuous function of p. Since the signum of the coefficients depends on $\in$ it is more convenient to introduce the new variable $S = \sinh 2\beta J$ instead of x. We then obtain:

$$f_p^{\,odd}(S) = \sum_{n=2}^{6} S^{2n}\, a_{2n}\, \frac{1}{2^4} \quad \text{where}$$

$$a_4 = \in^4$$

$$a_6 = \frac{\in^6}{2} - \in^4$$

$$a_8 = \frac{7}{8}\in^4 - \frac{3}{4}\in^6 + \frac{1}{16}(7\in^8 - 2\in^6)$$

$$a_{10} = -3/4\in^4 + \frac{27}{3^2}\in^6 - \frac{1}{8}(7\in^8 - 2\in^6) + \frac{1}{2^6}(28\in^{10} - 4\in^4 - 12\in^8)$$

$$a_{12} = \frac{165}{2^8}\in^4 - \in^6\frac{110}{128} + \frac{11}{64}(7\in^8 - 2\in^6) - (28\in^{10} - 4\in^4 - 12\in^8)\frac{5}{128}$$

$$+ \frac{1}{2^8}(124\in^{12} - 62\in^{10} - 16\in^8 - 12\in^6 + \frac{13}{3}\in^4).$$

If may be checked that the above series is alternating for $\forall 0 \leq \epsilon < 1$; for $\epsilon = 1 (p = 1)$ we know that the convergence radius is the Ising critical point such that $\rho = 1$. For different ϵ in the range $\frac{1}{100} < \epsilon < 1$, we have computed the sequence $\mu_n(\epsilon) = |a_n|^{1/n}$ and plotted $\mu_n(\epsilon)$ as function of $\frac{1}{n}$. We have then extrapolated the last 2 values to $n \to \infty$. Within the accuracy of the above numerical method, it may be said that the reciprocal convergent radius of the series $r(\epsilon)$ is a monotonic increasing function of ϵ. The plot is given in fig. 1.

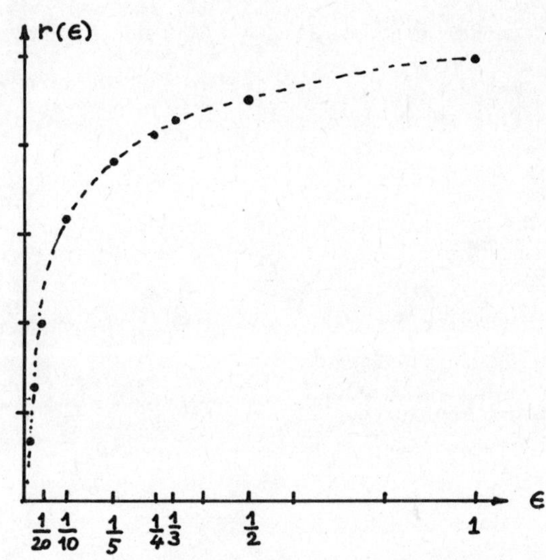

Fig. 1: Reciprocal convergence radius $r(\epsilon)$

It is thus found that the critical temperature $T_c(\epsilon)$ given by $\sin h2\beta_c \; J(\epsilon) = \frac{1}{r(\epsilon)}$ is decreasing with ϵ from the critical temperature $T_c(p = 1)$ to $T_c(p = {}^1\!/_2) = 0$.

Since for $p > 1/2$ the series has been computed only up to the order 12, a reasonable estimation of the critical index α for the specific heat anomaly is not expected. Moreover the analytical properties of f_p^{odd} as function of p are of interest. Finally, it may be remarked that although the spin glass at $p = 1/2$ does not have a phase transition at $T \; 0$, an exact solution for $f_{1/2}^{even}$ seems hard to be found.

In fact the local constraint n_B even $\forall B \in C$ and $\forall C$ appearing ni $f_{1/2}^{even}$ cannot be easily incorporated into the combinatorial method which leads to the solution of the two-dimensional Ising model.

References

1. L. A. Pastur and A. L. Figotin, Teoreticheskaya i Matematicheskaya Fizika Vol 35, $\underline{2}$, 193 (1978).

2. H. O. Georgii, Spontaneous magnetization of randomly diluted ferromagnets, preprint 1981, Universität Bielefeld.

3. I. Morgenstern and K. Binder, Phys. Rev $\underline{B}$, $\underline{22}$, 288 (1980).

4. Shang-Keng MA: Chin. J. Phys., $\underline{16}$, 2 (1978).

5. S. Goulard Rosa jr.: Phys. Lett. A, $\underline{76}$, 42 (1980).

6. J. E. Avron, Cr. Roepstorff and L. S. Shouman: Groundstate degeneracy and ferromagnetism in a spin glass, preprint (Princeton N. J. 1980).

7. R. B. Griffiths: Statistical mechanics Meeting, Rutgers N. J. (May 1980).

8. C. Gruber, A Hinterman and D. Merlini, Lectures Notes in Physics $\underline{60}$, Chapter 5, Springer Verlag (Berlin 1977).

9. D. Merlini, J. Stat. Phys. $\underline{21}$, 739 (1979).

10. D. Merlini, Lett. Nuovo Cimento $\underline{30}$, 15 (1981).

11. C. Gruber, J. Stat. Phys. $\underline{14}$, 811 (1976).

12. R. B. Griffiths, in Phase transition and critical Phenomena $\underline{Vol}$ 1, ed. C. Domb and M. S. Green, New York (1972).

13. D. Ruelle, Statistical mechanics, Benjamin (1967).

<u>STOCHASTIC PROCESSES IN MUSIC AND ART</u>

J.C. RISSET

Faculté des Sciences de Luminy et
Laboratoire de Mécanique et d'Acoustique du C.N.R.S.
13009 MARSEILLE - F R A N C E -

During the Luminy Symposium on Stochastic Processes, Daniel Arfib and I
presented a few instances of music and art based on stochastic processes. I can only
leave here a written trace of that presentation ; however the reader is referred to
a bibliography - and discography - that will direct him to detailed texts and to musical and visual examples.

RANDOMNESS, ORDER, AND STOCHASTIC MUSIC -

There is a certain arbitrariness inherent in many works of art, even those
which follow rather strict stylistic rules. Is it the infrangible right of the artist
to whims and creative freedom ? In any case, art has been long ago related to random processes. After Ma Sou-tchien, dices were used for musical composition in ancient
China. Musical dice games, permitting to "compose" music by assembling elements in a
way determined by the throwing of dices, were quite popular in the eighteen century
(one was even published under the name of Mozart). The technique of dripping, used in
the paintings of Masson and Pollock, is obviously somewhat random. Cage has used various chance operations in the composition of his music, some of which derived from
the I-Ching, an old chinese book of divination that yields random numbers by throwing coins or sticks, others derived from the observation of the imperfections of the
paper upon which he was writing.

As Pierce (1968) remarked, "since the coming of the computer, chaos has en-
tered music more scientifically". (Pierce has himself, with M.E. Shannon, performed
experiments on stochastic composition of music in 1949 ; so did D. Slepian in 1951).
Around 1956, several experiments were reported on the use of stochastic processes to
compose music. Pinkerton (1966) demonstrated the generation of simple melodies using
probability tables. This was pushed further by Olson and Belar (1961), who obtained
first-, second- and third order probability distributions for pitches in 11 tunes by
Stephen Foster ; Olson and Belar then obtained new tunes by generating successions of
random pitches with the same distributions : the new tunes were distinctly Stephen -
Foster-like. Barbaud performed similar experiments : he could generate music that
sounded like disorganized Mozart or Chopin from a statistical analysis of pieces by
these composers. Thus a computer, supplied with statistics describing the style of a
composer, can produce stochastic music with some similarity to the composer's style.
Brooks and al. (1957) studied the relevance of stochastic models in the analysis

and synthesis of simple music : they performed a statistical analysis of 37 tunes up
to an eighth order approximation ; then they used the results as tables of transition
probabilities for the synthesis of new tunes. They found that the best level of synthesis is around five or six : if the order of the synthesis is too low, it sounds
very random and does not resemble the original samples ; if it is too high, there is
too much duplication of the original samples - for instance one gets a "new" tune
consisting of parts of the original tunes assembled together. Even the best order
synthesis gives occasional "degenerate" tunes, for instance long ascending sequences.
Clearly Markov sequential probability methods are not sufficient to produce satisfactory tonal tunes ; sequential dependencies do not yield an acceptable long-range
structure.

The most systematic experiments were performed by Hiller and Isaacson (1959),
who used the Monte Carlo method to simulate musical composition with a computer. Numbers are associated to musical elements, e.g., to note pitches. Successive random
numbers are "filtered" through a set of rules : inadequate numbers are rejected until
one gets a string of numbers following the rules. These rules can either be statistical rules describing the distribution of pitches, or rules corresponding to those of
counterpoint. In the latter case, the computer produces counterpoints that are syn-
tactically correct, but often poorly organized and foreign to the spirit of counterpoint. Hiller and Isaacson's work has resulted in a musical composition, the "ILLIAC
Suite for string quartet", which reflects the experiment, and which is available on
record.

To get a better control of long-range structure, Tenney (1969) developed a
special process of "controlled random selection". Each successive note is selected
randomly and independently of other notes, but the range in which the random selection is made can be varied over the duration of the composition : the composer pres-
cribes this variation in a "score" which only controls the broad lines of the composition, leaving it to the random selection to fill in the details.

Xenakis (1955, 1963, 1971) is probably the most eloquent advocate of stochastic music. He claims (1955) that the effect of 12-tone organisational processes
is lost perceptually in a complex polyphony, hence that one should rather control
the aurally significant aspects of the sound, that is, statistical aspects like
overall densities. Thus several of his scores specify only statistical distributions,
and a random choice fills in the values of the parameters.

INFORMATION THEORY AND ANALYSIS OF MUSIC -

The above experiments imply the following model of musical composition :
composing consists of assembling the elements of a repertory of symbols (representing,
e.g., pitches) according to rules of selection and choice ; it thus extracts an order
from the chaos of multiple possibilities. This model is very close to the ideas of

Stravinsky (1947) who speaks of "the need that we feel to bring order out of chaos... All art presupposes a work of selection... To know how to discard, that is the great technique of selection".

Such a conception was reinforced by the development of information theory - and by bold extrapolations of its conclusions (Cf. Brillouin, 1956 ; Moles, 1958 ; Pierce, 1961 ; Risset, 1964). Schematically, the information rate (the negentropy) of a musical message is an important parameter : if it is very low, the music is dull, too predictable ; if it is too high, the music is not at all predictable and comple- tely unintelligible. White noise is the extreme case of unpredictability. The struc- ture of the music lowers the information rate, provided this structure is appreciated by the listener - who must have some implicit knowledge of the rules, i.e. of the musical language ; if this is not the case, the message is too original and unpre- dictable - like white noise. Indeed novel music has often been termed noise, from the time of Ars Nova to Mozart and the contemporaries.

Information theory has encouraged statistical studies, like those by Fucks (1962), who has computed statistical parameters related to the first-order skip - distributions (frequency versus pitch interval between successive notes). Fucks has tried to characterize the stylistic evolution of western music by statistical para- meters ; he finds that the curtosis $\chi = \dfrac{\mu_4}{(\sigma)^4}$ of the distribution of pitches in-

creases regularly with time, from a value around 2 in Palestrina and Willaert to 12 for Ravel and Prokoviev - but this evolution does not apply to 12-tone music, for which the curtosis is 3.5. This result may be significant, but it might be an abuse to conclude that it argues against the "naturalness" or "validity" of 12-tone music. Autocorrelation graphs for some melodies indicate that the Bach melodies are less correlated than the Beethoven melodies, and that those of Webern are highly uncorre- lated. These studies were extended by Gabura (1970), with more concern for the musi- cal context.

Autocorrelation fonctions are related to spectral densities by the Wiener- Khintchine relations. White noise (with a flat spectrum) is quite uncorrelated, while a signal with a $1/f^2$ spectral density is highly correlated. According to Voss and Clarke (1978), parameters associated with music tend to have a distribution with a $1/f$ spectral density ; Voss and Clark suggest that one should use a $1/f$ noise source for stochastic composition rather than imposing constraints upon a white noise source, in order to get an adequate amount of correlation (Cf. also Mandelbrot, 1977).

LIMITATIONS OF STOCHASTIC MODELS FOR MUSIC COMPOSITION AND APPRECIATION -
----
It has been argued that statistical analyses of music are not very signifi- cant musically, perhaps because they only reveal the statistical imprint of processes that are inherently non-statistical. One should certainly be cautious not to reduce

esthetic originality to statistical imprevisibility. And music appreciation is strongly influenced by context and previous musical experiences : it is not easy to take that in account in some kind of informational calculus. The musical signal seems too strongly organized to be grasped by an information measurement primarily designed to deal with events that are independent or only statistically related : it calls for an hypothetical formalism with some hierarchical structure, such that the "value" or "information" attached to an event should be contingent upon the presence of other elements.

Similarly, synthesis of music through the imposition of statistical constraints upon randomly chosen elements yields results that may be locally satisfactory, but that seen to wander aimlessly without conveying a sense of large form. An ergodic process goes nowhere. And filtering a random source by rules like those of counterpoint is not satisfactory either : as the composer Milton Babbitt remarked, the rules of counterpoint tell one what not to do ; they do not tell one what to do. Chomsky has shown that Markov sources were inadequate to generate certain classes of linguistic sequences : this applies to music, as the musicologist Herbert Schenker had already indicated fifty years earlier. Schenker had proposed, for the generation of tonal sequences, models which are very similar to transformational grammars : these models start from archetypal "Kernels" later "prolongated" by nested "prolongation" transformations, rather than resorting to randomly chosen elements later filtered by rules (Cf. Kassler, 1962). Applying the methodology of generative grammars to the study of music looks promising indeed (Lerdahl and Jackendoff, 1977).

STOCHASTIC PROCESSES IN SOUND -

At the level of the sound wave, it can be useful to apply random modulation to certain parameters of the sound. If one modulates the amplitude of a sine wave of frequency f by a random wave, one gets a band of noise whose spectrum is obtained by transporting the original spectrum S of the random wave around the f line. If S is broad, the ear perceives a colored noise, with a more or less clear pitch (clearer if S is narrow). If S is quite narrow, the effect can be perceived as amplitude fluctuations, or as the so-called chorus effect (it simulates the irregular beating due to the interactions of many closely-spaced frequency components, like the voices of singers in a chorus). This can help to make sound more natural and lively. In synthetized sound, imposing random variations to certain parameters from tone to tone helps avoiding the feeling of a mechanical performance (modulation by 1/f noise seems quite useful in this respect). However one should not believe than mere randomization is sufficient to make synthetic sounds natural and lively : many aspects of the sound must be carefully controlled, and the parameters must also be changed from tone to tone in a deterministic way (depending for instance of the frequency or of the amplitude of the tone).

There has been speculation on the way the ear could distinguish two simultaneous tones in unison. As demonstrated by Chowning (1980), Mc Adams et Mc Nabb, imposing two different random micromodulations to two tones enable the ear to discriminate between these "incoherent" tones, while the ear would fuse them if they were coherent (Chowning, 1980).

Sounds can be synthesized by computer through the computation of samples of the sound waveforms (Mathews, 1969). Xenakis (1971) has proposed methods to manufacture sounds from a direct stochastic control of the samples : however these methods do not appear to give essentially novel results (Smith, 1973).

RANDOM PROCESSES IN THE VISUAL ARTS -

Much of what has been said about music applies to the visual arts, although it is not as easy here to isolate elements and formulate rules. We shall only indicate here a few exemplary experiments.

Julesz (1975) has generated by computer many visual patterns made up of 100 x 100 regularly arranged dots of various shades of gray. If the shade of each dot is chosen randomly, one thus gets visual "noise". Julesz has produced many patterns with various ordering constraints, to study whether they were visually perceived. He found that the boundary between two areas with different n^{th} order probability distributions was apparent to the eye for n = 1 (case of a difference in average brightness), n = 2, but not for n = 3. Julesz has done other work on texture as well as on binocular cues for depth : in particular, he found that some differences are perceivable instantly, spontaneously, whereas others can be perceived only with scrutiny. In the course of his research, he has produced beautiful stochastic visual patterns (1971).

Noll was one of the pionners of the application of the computer to visual art. Although he did not use only stochastic processes, he did produce strong pictures from simple constraints imposed upon random choices (cf. Computers and Automation, 1965). He obtained his "computer composition with lines" through the following process : "the positions of the vertical and horizontal bars have been chosen at random with the constraint that the position must fall inside a circle. The length and width of the bars was chosen at random within a specified range. If the position of one bar fell within a parabolic region in the upper half of the circle, the length of the bar was shortened by a factor proportional to the distance of the position from the edge of the parabolic region". The motivation for this type of pattern came from Piet Mondrian's "Composition with lines" (1917). Actually Noll performed an interesting experiment : he presented copies of both Mondrian's picture and his picture to 100 subjects, asking them which one picture they preferred and which one they thought was produced by Mondrian. 59 percent of the subjects preferred the computer - generated picture ; only 28 percent were able to identify correctly the picture produced by Mondrian. Noll (1867) warns against hasty conclusions from this experiment - after all, Mondrian's painting was the inspiration for the algorithm.

Other artists have been stimulated by the use of stochastic processes : among them Lilian Schwartz, Stan Vanderbeek (working with K. Knowlton), Georg Nees. Vera Molnar has done many variations on the theme of the square by introducing various degrees of disorder, various amounts of visual noise : an instance is presented along this page (Cf. V. Molnar, 1980, 1981).

PARTING REMARKS -

Stochastic processes have helped to better understand what is specific in the creative activity of artists ; they also have directly inspired artists' work. Yet it does not look that one can describe completely works of art in terms of stochastic processes - or, for that matter, in term of any formalism : no description can exhaust the richness of art and music. Perhaps Gödel gives good reasons for that ; Goethe had already labeled theories as grey, compared with the evergreen tree of life.

REFERENCES -

- ARFIB (D.), (1977), Conception assistée par ordinateur en acoustique musicale. Thèse Ing. Dr., Université d'Aix-Marseille II.

- L'Art et l'Ordinateur (1975), IBM Informatique n° 13 (Reproduction of works by Noll, Nake, Nees, Molnar, Franke, etc...).

- BARBAUD (P.), (1960), Musique algorithmique. Esprit 28, (Janvier).

- BARBAUD (P.), (1966), Initiation à la composition musicale automatique. Dunod, Paris.

- BARBAUD (P.), French Gagaku, Mu Joken. On recording Inédits O.R.T.F., 995 025.

- BRILLOUIN (L.), (1956), Science and Information Theory. Academic Press, New-York.

- BROOKS, Jr.(F.P.), HOOKINS, Jr.(A.L.), NEUMANN (P.G.) & WRIGHT (W.V.), (1957), An experiment in musical composition, IRE Trans. on Electronic Computers EC-6, 175.

- CHOWNING (J.), (1980), Computer synthesis of the singing voice. In Sound Generation in winds,strings, computers (J.Sundberg, ed.), Stockholm : Royal Institute of Technology, 4-13.

- COHEN (J.E.), (1962), Information theory and music. Behavioral Science 7, 137-163.

- Computers and Automation (August 1965) : the annual computer art contest (works by NOLL & DILEONARDO).

- COQUART (R.) ed., (1981), Aspecten De computer in de Vsuelle Kunst-KSAC ; Bruxelles 1981 (Cf. MOLNAR (F.) : Création par ordinateur ou création assistée par ordinateur).

- Cybernetic Serendipity (1968) (Catalogue of a large exhibit on computer art, London).

- DEGEORGE (D.) & COOK (G.E.), (1968), Statistical music analysis : an objection to Fucks'- curtosis curve. I.E.E.E. Trans. on Inf. Th. IT 14, 152.

- DENES (P.B.) & MATHEWS (M.V.), (1968), Computer models for speech and music appreciation. Fall Joint Computer Conference, 319-327.

- FERENTZY (E.N.), (1965), Computer simulation of human behavior in music composition. Computational Linguistic Yearbook IV, Hungarian Academy of Sciences, Budapest.

- FUCKS (W.), (1962), Mathematical analysis of formal structure of music. IRE Trans. on Information Theory, IT-8, 225.

- GABURA (J.), (1970), Music style analysis by computer. In LINCOLN H.B., The Computer in Music, Cornell University Press, Ithaca & London.

- HILLER (L.A.), (1970), Music composed with computers - a historical survey. In LINCOLN H.B. editor, The Computer and Music, Cornell University Press, Ithaca & London, 1970, pp. 42-96.

- HILLER (L.A.) & BAKER (R.A.), (1967), Computer Music from the University of Illinois. Record Heliodor H/MS - 25053 - (Includes the ILLIAC Suite for string quartet).

- HILLER (L.A.) & ISAACSON (L.M.), (1959), Experimental music. McGraw Hill Book Company, Inc., New-York, Toronto, London.

- JULESZ (B.), (1971), Foundations of cyclopean perception. University of Chicago Press.

- JULESZ (B.), (1975), Experiments in the visual perception of texture. Sci. American, 232, 4.

- KASSLER (M.), (1962), A report of work, directed towards explication of Schenker's theory of tonality. Princeton University thesis.

- LERDAHL (F.) & JACKENDOFF (R.), (1977), Toward a formal theory of tonal music. J. of Music Theory, 21, 111-171.

- MANDELBROT (B.), (1977), Fractals : forme, chance and dimension. Freeman, San Francisco.

- MATHEWS (M.V.), (1969), The technology of computer music. M.I.T. Press, Cambridge, Mass.

- MOLES (A.), (1958), Théorie de l'information et Perception esthétique. Flammarion, Paris.

- MOLNAR (F.) ed., (1977), Dossier Arts Plastiques n° 1. Centre G. Pompidou, Paris.

- MOLNAR (V.), (1980), One per cent disorder (20 images). Wedgepress & Cheese, Bjerred, Sweden.

- MOLNAR (V.), (1981), The role randomness can play in visual art. Page (Computer Arts Society Quarterly, London), 47 (Jan.1981), 3.

- MOORER (J.A.), (1972), Music and computer composition. Commun. of the ACM, 15, 104-113.

- MOZART (W.A.), Mukatisches Würfelspiel. (Köchel 294). New edition : Guild Publications of Art & Music, 141 Broadway, New-York.

- NAKE (F.),(1957), Computer-grafik. Verlag Nadolski, Stuggart.

- NOLL (A.M.), (1966), Human or machine : a subjective comparison of Piet Mondrian's "Composition with lines" (1917) and a computer-generated picture. The Psychological Record, $\underline{16}$, 1-10.

- NOLL (A.M.), (1967), The digital computer as a creative medium. I.E.E.E. Spectrum (october 1967), pp. 89-95.

- OLSON (H.F.) & BELAR (H.), (1961), Aid to music composition employing a random probability system. J. Acoust. Soc. Am., $\underline{33}$, 1163.

- PIERCE (J.R.), (1961), An introduction to information theory (2^d edition, 1980 : Dover Publications, New-York).(Original title : Symbols, signals and noise).

- PIERCE (J.R.), (1968), Science, Art, and Communication. Clarkson N.Potter, New - York.

- RISSET (J.C.), (1964), Musique électronique et théorie de l'information. L'onde électrique, $\underline{44}$, 1055.

- SCHENKER (H.), (1921), Der Tonwille. Universal Edition, Vienna.

- STRAVINSKY (I.), (1947), Poetics of music. Harvard University Press, Cambridge, Mass.

- TENNEY (J.C.), (1969), Computer Music Experiments. Electronic Music Report, (Utrecht) 1, 23-60.

- VOSS (R.F.) & CLARKE (J.), (1978), "1/f noise" in music : Music from 1/f noise. J. Acoust. Soc. Am., $\underline{63}$, 258.

- XENAKIS (I.), (1954), La crise de la musique sérielle. Gravesaner Blätter 1, 15.

- XENAKIS (I.),(1962), Stochastic Music. Gravesaner Blätter 23/24, 156.

- XENAKIS (I.), (1963), Musiques formelles. La Presse Musicale, vol. 253-254, Editions Richard Masse, Paris.

- XENAKIS (I.), (1971), Formalized music. University of Indiana, Bloomington.

- XENAKIS (I.), (1965), Metastastis, Pithoprakta, Eonta. Record "Le Chant du Monde" LDX-A48368.

- YOUNGBLOOD (J.E.), (1958), Style as information. J. of Music Theory (Yale) $\underline{2}$, 24.

- ZARIPOV (R.K.), (1960), An algorithmic description of the music composing process. Doklady Akademia Nauk SSSR $\underline{132}$, 1283. English translation : Automation Express $\underline{3}$, (nov. 1960), 17.

HOPPING TRANSPORT IN DISORDERED ONE-DIMENSIONAL

LATTICE SYSTEMS: RANDOM WALK IN A RANDOM MEDIUM

W.R. Schneider
Brown Boveri Research Center
CH-5405 Baden, Switzerland

1. Introduction

Consider a particle moving stochastically on the one-dimensional lattice $\mathbb{Z}$. The probability to find the particle at time $t \geq 0$ on the site $n \in \mathbb{Z}$ is denoted by $P_n(t)$; we assume that

$$P_n(0) = \delta_{no} \quad , \tag{1.1}$$

i.e. the particle starts with certainty at the origin. The change in time of these probabilities is described by the master equation

$$\dot{P}_n(t) = \sum_{m \neq n} T_{nm} P_m(t) - \left(\sum_{m \neq n} T_{mn} \right) P_n(t) \tag{1.2}$$

where $T_{nm} \geq 0$ is the transfer rate from site m to site n, $m \neq n$. The two terms in (1.2) correspond to gains by hops ending in n and losses by hops starting in n. Globally, the gains and losses balance:

$$\sum_n \dot{P}_n(t) = 0 \quad . \tag{1.3}$$

In view of (1.1) this implies

$$\sum_n P_n(t) = 1 \tag{1.4}$$

for all times; the particle does not get lost, destroyed or trapped.

We restrict ourselves to the case of nearest neighbour hopping, i.e. $T_{nm} = 0$ if $|n-m| > 1$. Furthermore, we require the rate between n and m to be equal to the rate between m and n, i.e. $T_{nm} = T_{mn}$. To simplify notation, we set

$$W_n = T_{n,n+1} \qquad (= T_{n+1,n}) \quad . \tag{1.5}$$

This reduces (1.2) to

$$\dot{P}_n = W_{n-1}(P_{n-1} - P_n) + W_n(P_{n+1} - P_n) \quad . \tag{1.6}$$

In ordered systems the transfer rates W_n have fixed non-negative values. If the system is invariant against all translations all W_n have the same fixed value $w > 0$. A Laplace transform

$$\tilde{P}_n(s) = \int_0^\infty dt\, e^{-st}\, P_n(t) \tag{1.7}$$

in time and a Fourier transform in space lead from (1.6) to

$$s \sum_n \tilde{P}_n\, e^{-ink} - 1 = 2w(\cos k - 1) \sum_n P_n\, e^{-ink} \tag{1.8}$$

or

$$\sum_n \tilde{P}_n\, e^{-ink} = \{s + 2w(1 - \cos k)\}^{-1} \quad . \tag{1.9}$$

This yields immediately

$$\tilde{P}_n(s) = (2w)^{-1} (\lambda^2 + 4\lambda)^{-1/2} \{1 + \lambda - (\lambda^2 + 4\lambda)^{1/2}\}^{|n|} \tag{1.10}$$

with $2w\lambda = s$ as well as

$$\sum_n P_n\, e^{-ink} = e^{-2wt}\, e^{2wt \cos k} \tag{1.11}$$

from which

$$P_n(t) = e^{-2wt}\, I_{|n|}(2wt) \tag{1.12}$$

is read off, I_κ denoting the modified Bessel function of index κ. Differentiating (1.11) twice with respect to k and setting k = 0 yields the mean square displacement

$$R^2(t) = \sum_n n^2\, P_n(t) = 2wt \quad . \tag{1.13}$$

Similarly, we obtain from (1.9)

$$D(s) = \frac{1}{2} s^2 \sum_n n^2 \tilde{P}_n(s) = w \quad , \tag{1.14}$$

the diffusion function, which in this simple case is a constant.

In the above example only the motion of the particle is stochastic whereas the medium in which the motion takes place is deterministic (i.e. its properties do not contain any randomness). This is an idealization; real media are always to a certain degree disordered (lattice defects, impurities etc.). A simple way to introduce disorder is to assume that the transfer rates W_n are random variables. For the sake of simplicity we further assume that they are all independent and equally distributed according to a probability measure ν (as $W_n \geq 0$ the support of ν has to be in $\mathbb{R}_+$).

In view of (1.16) all the site probabilities $P_n(t)$ as well as quantities derived thereof like $R^2(t)$ and $D(s)$ become random variables. As such they do not represent observable quantities. The latter are obtained by taking the expectation defined by

$$E(F) = \int \prod_{n \in \mathbb{Z}} d\nu(w_n) \, F(\{w_n\}) \tag{1.15}$$

where F is a measurable function on $\mathbb{R}_+^{\mathbb{Z}}$.

In Section 2 we develop the necessary formalism to deal with (1.6). In particular, we show that

$$E(\tilde{P}_0(s)) = \iint_{\mathbb{R}_+^2} d\mu_s(x) d\mu_s(y) \, (x + y + s)^{-1} \quad , \qquad s \geq 0 \quad , \tag{1.16}$$

where μ_s is a probability measure on $\mathbb{R}_+$ satisfying a homogeneous linear integral equation [1]. We sketch in Section 3 proofs for existence and uniqueness of this measure and exhibit continuity properties [2]. Section 4 is devoted to the behavior of μ_s as s tends to zero [3], [4]. Partial translation invariance is reintroduced in Section 5 by requiring $W_{n+L} = W_n$ as a means to study finite size effects. The large L behaviour of the static diffusion function is discussed [4], [5]. Finally, we would like to point out that a variety of physical problems admit a description in terms of our random walk model (possibly after a suitable transformation); a comprehensive review may be found in [6].

2. Formalism

We define X_n for $n = 0,1,2,\ldots$ by the following infinite continued fraction:

$$X_n = \cfrac{1}{\cfrac{1}{W_n} + \cfrac{1}{s + \cfrac{1}{\cfrac{1}{W_{n+1}} + \cdots}}} \qquad \qquad (2.1)$$

Obviously, X_n does not depend on W_k, $k < n$. Furthermore, X_n and X_{n+1} are related by

$$X_n = \{W_n^{-1} + (s + X_{n+1})^{-1}\}^{-1} \qquad . \qquad (2.2)$$

Setting $W_n' = W_{-n-1}$ for $n = 0,1,2,\ldots$ and replacing $\{W_k\}$ by $\{W_k'\}$ in (2.1) yields X_n' having analogous properties.

With the help of these auxiliary variables we define

$$\tilde{P}_0 = (s + X_0 + X_0')^{-1} \qquad \qquad (2.3)$$

and

$$\tilde{P}_n = \tilde{P}_0 \prod_{k=1}^{n} X_{k-1}(X_k+s)^{-1} \qquad , \qquad n = 1,2,\ldots \qquad . \qquad (2.4)$$

By putting primes on X_{k-1} and X_k we obtain $\tilde{P}_{-n}$. One verifies easily that

$$s\,\tilde{P}_n - \delta_{no} = W_{n-1}(\tilde{P}_{n-1} - \tilde{P}_n) + W_n(\tilde{P}_{n+1} - \tilde{P}_n), \qquad n \in \mathbb{Z} \qquad , \qquad (2.5)$$

is satisfied which is the Laplace transform of (1.6).

If W_0, $W_1,\ldots$ are equally distributed random variables the same is true for X_0, $X_1,\ldots$ as is seen from (2.1). The independence, however, is not transferred as (2.2) shows. Nevertheless, as already noted, W_n and X_{n+1} are independent. Hence, (2.2) implies

$$\mu_s(B_x) = \iint_{A_{s,x}} d\nu(y)d\mu_s(z) \qquad \qquad (2.6)$$

for the common probability measure μ_s of X_0, $X_1,\ldots$. Here, the notation

$$\sigma(B_x) = \int_{B_x} d\sigma(x') \qquad , \qquad B_x = [0,x) \qquad , \qquad (2.7)$$

σ a probability measure on $\mathbb{R}_+$, has been used and the set $A_{s,x}$ is given by

$$A_{s,x} = \{(y,z) \in \mathbb{R}_+^2 \mid \{y^{-1} + (s+z)^{-1}\}^{-1} < x\} \quad . \tag{2.8}$$

Anticipating that the integral equation (2.6) has a (unique) solution, we obtain from (2.3)

$$\mathrm{Prob}(\tilde{P}_0(s) < c) = \iint\limits_{(s+x+y)^{-1}<c} d\mu_s(x)d\mu_s(y) \tag{2.9}$$

and, as a consequence, the representation (1.16) of $E(\tilde{P}_0(s))$ by means of μ_s.

3. Uniqueness, Existence and Continuity

Consider the map $\psi_s : \mathbb{R}_+^2 \to \mathbb{R}_+$ with

$$\psi_s(y,z) = \{y^{-1} + (s+z)^{-1}\}^{-1} \quad . \tag{3.1}$$

From its graph the following decomposition of the set $A_{s,x}$, introduced in (2.8) into disjoint subsets is obtained:

$$A_{s,x} = A_{\infty,x} \cup C_{s,x} \quad . \tag{3.2}$$

The two components are given by

$$A_{\infty,x} = B_x \times \mathbb{R}_+ \quad , \qquad B_x = [0,x) \tag{3.3}$$

$$C_{s,x} = \{(y,z) \mid 0 \le z < \frac{yx}{y-x} - s \; , \qquad x \le y < \phi_s(x)\} \tag{3.4}$$

with

$$\phi_s(x) = \begin{cases} sx/(s-x) & , \quad x < s \\ \infty & , \quad x \ge s \end{cases} \quad . \tag{3.5}$$

The decomposition (3.2) and Fubini's theorem imply that (2.6) may be rewritten as

$$f_s = g + K_s f_s \tag{3.6}$$

where

$$f_s(x) = \mu_s(B_x) \quad , \qquad g(x) = \nu(B_x) \tag{3.7}$$

and

$$(K_s f)(x) = \int_x^{\phi_s(x)} d\nu(y) f\left(\frac{yx}{y-x} - s\right) \quad . \tag{3.8}$$

The linear operator K_s may be defined on the Banach space

$$B_\alpha = L^1(\mathbb{R}_+, \rho_\alpha) \quad , \qquad d\rho_\alpha(x)/dx = \alpha(1+x)^{-1-\alpha} \quad , \qquad 0 < \alpha < 1 \quad . \tag{3.9}$$

Note that the original domain of K_s, consisting of the functions f with $f(x) = \sigma(B_x)$ for some probability measure σ on $\mathbb{R}_+$, is total in B_α (i.e. the linear span is dense).

A simple calculation yields

$$\| K_s f \|_\alpha \le \alpha \int_0^\infty d\nu(y) \int_0^\infty dz \, k_s(y,z) \, |f(z)| \tag{3.10}$$

with equality holding for non-negative f, where

$$k_s(y,z) = y^2(s+y+z)^{-2} \{1 + y(s+z)/(s+y+z)\}^{-1-\alpha} \quad . \tag{3.11}$$

For $y > 0$, $z > 0$ we have the estimate

$$k_s(y,z) < (1+z)^{-1-\alpha} \tag{3.12}$$

leading to

$$\| K_s f \|_\alpha < \| f \|_\alpha \quad , \qquad f \ne 0 \quad . \tag{3.13}$$

Obviously, (3.13) implies that (3.6) has at most one solution $f_s \in B_\alpha$ for arbitrary $g \in B_\alpha$, $g \ne 0$.

Consider now the sequence

$$f_s^{(n)} = \sum_{m=0}^n K_s^m g \quad , \qquad n = 0,1,2,\ldots \quad . \tag{3.14}$$

Using

$$f_s^{(n)} = g + K_s f_s^{(n-1)} = f_s^{(n-1)} + K_s^n g \qquad (3.15)$$

one shows inductively that

(i) $f_s^{(n)}(x)$ is isotonic in x

(ii) $f_s^{(n)}(x)$ is left-continuous

(iii) $f_s^{(n)}(x) = 0$ for $x \le 0$

(iv) $f_s^{(n)}(x) \to 1$ as $x \to \infty$

and

$$0 \le f_s^{(0)}(x) \le f_s^{(1)}(x) \le \cdots \qquad \le 1 \quad . \qquad (3.16)$$

Consequently, $\lim_{n \to \infty} f_s^{(n)}(x) = f_s(x)$ exists pointwise and the properties (i)-(iv) are transferred to f_s. Hence, there is a (unique) probability measure μ_s which reproduces f_s when integrated over B_x. By Lebesgue's dominated convergence theorem $f_s^{(n)}$ converges also in B_α to f_s. This implies in view of (3.15) and the boundedness of K_s that (3.6) is satisfied.

Explicit solutions of (2.6) may be obtained for

(1) $\nu = \delta_w$ (Dirac measure at $w \ge 0$): The solution of (2.6) is

$$\mu_s = \delta_{a(s)} \quad , \quad a(s) = \frac{1}{2} \{(4ws + s^2)^{1/2} - s\} \quad . \qquad (3.17)$$

This corresponds to the case where all W_n in (1.6) have the same fixed value w.

(2) $s = 0$: The solution of (2.6) is

$$\mu_s = \delta_0 \quad . \qquad (3.18)$$

(3) $s = \infty$: The solution of (2.6) is

$$\mu_\infty = \nu \quad . \qquad (3.19)$$

Actually, f_s tends in B_α to g as s tends to infinity.

We turn now our attention to the continuity properties of f_s and μ_s. In very much the same way as (3.10) we also obtain

$$\|K_s f - K_t f\|_\alpha = \alpha \int_0^\infty d\nu(y) \int_0^\infty dz\{k_s(y,z) - k_t(y,z)\} f(z) \tag{3.20}$$

for non-negative f and $0 \leq s < t$. By Lebesgue's dominated convergence theorem this tends to zero as t tends to s. The extension to arbitrary f is trivial as K_s is positivity preserving. Combined with $\|K_s f\|_\alpha \to 0$ as $s \to \infty$ we obtain the result that the map $s \to K_s$ from $\mathbb{R}_+ \cup \{\infty\}$ to the bounded operators of $\mathcal{B}_\alpha$ is strongly continuous. As $\|K_s\| \leq 1$ this implies that each term in (3.14) depends continuously on s. Now, $f_s^{(n)} \leq f_0^{(n)}$ as is easily shown by induction. Hence, $f_s^{(n)}$ converges uniformly to f_s; therefore, f_s also depends continuously on s. By partial integration we obtain

$$\mu_s(g) := \int g(x) \, d\mu_s(x) = - \int g'(x) f_s(x) dx \tag{3.21}$$

for arbitrary $g \in C_o^1(\mathbb{R})$, the set of continous function having compact support and a continuous derivative, g'. Hence, there is a constant $C(g) < \infty$ such that

$$|\mu_s(g) - \mu_t(g)| \leq C(g) \|f_s - f_t\|_\alpha \quad , \tag{3.22}$$

i.e. $\mu_s(g)$ is continous in s. This is still true for $g \in C_o(\mathbb{R})$, the set of continuous functions with compact support, as $C_o^1(\mathbb{R})$ is dense in $C_o(\mathbb{R})$ with respect to the sup-norm. In the terminology of probabilists this means that the map $s \to \mu_s$ from $\mathbb{R}_+ \cup \{\infty\}$ to the probability measures on $\mathbb{R}_+$ is vaguely continuous.

4. Asymptotic Behaviour

We have seen that $\mu_s(f)$ with $f \in C_0(\mathbb{R})$ depends continuously on s. As is easily seen this is still true for the larger class of continuous functions f satisfying $f(x) \to 0$ as $x \to \infty$. Especially, we have

$$\lim_{s \downarrow 0} \mu_s(f) = f(0) \quad , \tag{4.1}$$

as $\mu_0 = \delta_0$. Choosing $f(x) = \exp(-px)$, $p > 0$, yields

$$\lim_{s \downarrow 0} \tilde{\mu}_s(p) = 1 \tag{4.2}$$

where the tilde denotes Laplace transform. We may obtain more information on how this limit is reached if we specify the behaviour of the probability measure ν near zero. It is intuitively clear that many small transfer rates will cause a slow-down of the speed with which the particle is diffusing to infinity. To make this statement precise ν is supposed to satisfy the two conditions A1 and A2 as well as one of the conditions C1, C2, C3:

(A1) $\nu(\{0\}) = 0$

This condition excludes percolation (i.e. a non-zero probability of vanishing transfer rates) and implies that ν_I with

$$\nu_I(B_x) = 1 - \nu(B_{1/x}) \tag{4.3}$$

is a probability measure; it describes the distribution of the random variables $M_n = W_n^{-1}$.

(A2) $\tilde{\nu}_I \in L^1(\mathbb{R}_+)$

This condition is of technical nature; it guarantees the existence of integrals involving $\tilde{\nu}_I$ and permits the interchange of limits and integrations. Note that in view of (4.3)

$$\tilde{\nu}_I(p) = \int_0^\infty d\nu(y)\, e^{-p/y} \tag{4.4}$$

holds.

(C1) ν_I has a finite first moment:

$$a = \int_0^\infty x \, d\nu_I(x) = \int_0^\infty x^{-1} \, d\nu(x) < \infty \quad . \tag{4.5}$$

(C2) There exist constants $a > 0$ and $c > 0$ such that

$$\int_0^c x^{-2} \, |g(x) - ax| \, dx \quad < \infty \tag{4.6}$$

$$(\text{recall } g(x) = \nu(B_x)) \quad .$$

(C3) There exist constants $a > 0$ and α, $0 < \alpha < 1$

$$\lim_{x \downarrow 0} \; g(x) x^{\alpha - 1} = a^{1-\alpha} \tag{4.7}$$

We shall denote by C1, C2, C3 the classes of probability measures satisfying (A1), (A2) and (C1), (C2), (C3), respectively. These classes are disjoint and correspond to an increasing singular behaviour near zero. Simple but typical examples are

$$\frac{d\nu}{dx} = \begin{cases} 1 & , & 1 \le x \le 2 & : \; C1 \\ 1 & , & 0 \le x \le 1 & : \; C2 \\ (1-\alpha)x^{-\alpha} , & 0 < x \le 1 & : \; C3 \end{cases} \tag{4.8}$$

with $d\nu/dx = 0$ outside the indicated interval.

Defining the function h by

$$h(x) = x^{1/2} \quad , \quad (-2x/\ell n x)^{1/2} \quad , \quad x^{1/(2-\alpha)} \tag{4.9}$$

for C1, C2, C3, respectively, and setting $\varepsilon(s) = h(as)/a$ yields

$$\lim_{s \downarrow 0} \; s^{-1} \varepsilon(s) \, \{\tilde{\nu}_I(0) - \tilde{\nu}_I(\varepsilon(s)p)\} = \phi(p) \tag{4.10}$$

with

$$\phi(p) = p \quad , \quad p \quad , \quad \Gamma(\alpha)p^{1-\alpha} \tag{4.11}$$

for C1, C2, C3, respectively.

The Laplace transform $\tilde{\mu}_s$ of μ_s satisfies

$$\tilde{\mu}_s(p) = \iint d\nu(y)\, d\mu_s(z) e^{-p\psi_s(y,z)} \tag{4.12}$$

as follows from (2.6), (2.7) and (3.1). Using the identity

$$1-e^{-p/b} = p^{1/2} \int_0^\infty dq\, q^{-1/2}\, J_1(2\sqrt{pq})e^{-qb} \tag{4.13}$$

$(J_\kappa = $ Bessel function of index κ) leads to

$$\tilde{\mu}_s(p) = 1-p^{1/2} \int_0^\infty d\mu_s(z) \int_0^\infty dq\, q^{-1/2}\, J_1(2\sqrt{pq})e^{-q/(z+s)}\, \tilde{\nu}_I(q) \tag{4.14}$$

Rescaling the variables p, q, z as follows

$$p \to p/\varepsilon(s) \quad , \quad q \to q\,\varepsilon(s) \quad , \quad z \to z\,\varepsilon(s) \tag{4.15}$$

and setting

$$\tilde{\pi}_s(p) = \tilde{\mu}_s\,(p/\varepsilon(s)) \tag{4.16}$$

yields for the probability measure π_s

$$\tilde{\pi}_s(p) = 1-p^{1/2} \int_0^\infty d\pi_s(z) \int_0^\infty dq\, q^{-1/2}\, J_1(2\sqrt{pq})\, \exp\{-q/(z+s/\varepsilon(s))\}\, \tilde{\nu}_I(q\varepsilon(s)). \tag{4.17}$$

In the limit $s\downarrow 0$ we obtain for $\pi = \pi_0$ using (4.10)

$$p^{1/2} \int_0^\infty d\pi(z) \int_0^\infty dq\, q^{-1/2}\, J_1(2\sqrt{pq})e^{-q/z}\, (\phi(q) - qz^{-2}) = 0 \quad . \tag{4.18}$$

In view of (4.11) this reduces to

$$p \int_0^\infty d\pi(z)\, (z^2-1)e^{-pz} = 0 \tag{4.19}$$

for C1 and C2 with the solution

$$\pi = \delta_1 \quad . \tag{4.20}$$

For C3 (4.18) may be rewritten as

$$p\tilde{\pi}(p) = \int_0^\infty d\pi(z)p^{1/2} \int_0^\infty dq\, q^{-1/2}\, J_1(2\sqrt{pq})e^{-q/z}\, \Gamma(\alpha)q^{1-\alpha} \tag{4.21}$$

or, in terms of the Mellin transform $\hat{\pi}$

$$\hat{\pi}(s) = \int_0^\infty d\pi(x)x^{s-1} = \Gamma(1-s)^{-1} \int_0^\infty dp\ p^{-s}\ \tilde{\pi}(p) \qquad , \tag{4.22}$$

as

$$\hat{\pi}(s) = \Gamma(\alpha)\Gamma(s+1-\alpha)\Gamma(s+1)^{-1}\ \hat{\pi}(s+2-\alpha) \qquad . \tag{4.23}$$

The solution of this difference equation is [3]

$$\hat{\pi}(s) = \frac{\beta}{\Gamma(\beta)}\ \gamma^{1-s}\ \frac{\Gamma(\beta(s-1))\Gamma(\beta s)}{\Gamma(s-1)} \tag{4.24}$$

with

$$\beta = (2-\alpha)^{-1} \qquad , \qquad \gamma = \left[\beta^2\Gamma(\alpha)\right]^\beta \qquad . \tag{4.25}$$

By inverting the Mellin transform we obtain for the density of π

$$\frac{d\pi(x)}{dx} = \frac{\beta}{\Gamma(\beta)}\ \gamma\ H_{12}^{20}\ \left(\gamma x\ \Big|\ \begin{matrix}(-1,1)\\(-\beta,\beta)\quad(0,\beta)\end{matrix}\right) \tag{4.26}$$

where H_{pq}^{mn} denotes the general Fox function [7], one of the generalizations of Gauss's hypergeometric functions. For a discussion of $d\pi/dx$ and its properties see [3].

From (1.16) and (4.16) we obtain the behaviour of $E(\tilde{P}_0(s))$ as $s\downarrow 0$: Rewriting (1.16) in terms of π_s yields

$$E(\tilde{P}_0(s)) = \varepsilon(s)^{-1} \iint d\pi_s(x)d\pi_s(y)(x+y+s/\varepsilon(s))^{-1} \tag{4.27}$$

leading to

$$\lim_{s\downarrow 0}\ E(\tilde{P}_0(s))/\{a\ f(as)\} = 1 \tag{4.28}$$

where

$$f(x) = \tfrac{1}{2}x^{-1/2} \quad , \quad 2^{-3/2}(-\ell n\ x/x)^{1/2} \quad , \quad c_\alpha\ x^{-\beta} \tag{4.29}$$

for C1, C2, C3, respectively. The constant C_α is given by

$$C_\alpha = \iint d\pi(x)d\pi(y) \ (x+y)^{-1} \qquad (4.30)$$

whereas the exponent β is defined in (4.25).

From general Abelian and Tauberian theorems for the inverse Laplace transform we obtain from (4.28)

$$\lim_{t\to\infty} E(P_0(t))/g(t/a) = 1 \qquad (4.31)$$

where

$$g(x) = (4\pi x)^{-1/2}, \quad (\ln x/8\pi x)^{1/2}, \quad \frac{C_\alpha}{\Gamma(\beta)} x^{-\beta(1-\alpha)} \qquad (4.32)$$

for C1, C2, C3, respectively.

So far we did not succeed to obtain similar results for $E(\tilde{P}_n(s))$ or $E(P_n(t))$, n arbitrary. Nevertheless, the behaviour of these quantities may be discussed with the help of a scaling hypothesis [3], borne out by Monte Carlo simulations.

5. Finite Size Effects

To simulate systems of finite length L (L integer) we impose the following periodicity condition on the transfer rates W_n of the infinite system (1.6):

$$W_{kL+j} = W_j \quad , \quad j = 1,2,\ldots,L, \qquad k \in \mathbb{Z} \quad . \qquad (5.1)$$

By introducing the quantities

$$R_j(s,q) = \sum_k e^{-iq(kL+j)} \ \tilde{P}_{kL+j}(s) \qquad (5.2)$$

the infinite system of equations (2.5) is reduced to the following finite system

$$sR_j - \delta_{jL} = W_{j-1}(e^{-iq}R_{j-1} - R_j) + W_j(e^{iq}R_{j+1} - R_j) \qquad (5.3)$$

where $j = 1,2,\ldots,L$.

A tedious calculation yields for the diffusion constant

$$D = \lim_{s \downarrow 0} D(s) \tag{5.4}$$

with $D(s)$ defined in (1.14) the simple result $D = D_L$ with

$$D_L = L \left(\sum_{j=1}^{L} W_j^{-1} \right)^{-1} . \tag{5.5}$$

This may be interpreted as the conductivity of a system of conductances $W_1, W_2, \ldots, W_L$ in series.

For a disordered system we assume $W_1, W_2, \ldots, W_L$ to be independent equally distributed random variables; their distribution is described by a probability measure ν of class C1, C2, C3, respectively, introduced in Section 4. The question arises how the expectation $E(D_L)$ behaves for large L [5].

Let σ_L denote the probability measure associated with the random variable

$$y_L = B_L \left(\sum_{j=1}^{L} W_j^{-1} \right)^{-1} \tag{5.6}$$

where

$$B_L = aL , \qquad aL \, \ell nL , \qquad aL^{1(1-\alpha)} \tag{5.7}$$

for C1, C2, C3, respectively. Then, adapting the procedure of Section 4, one shows that σ_L tends to σ_∞ as $L \to \infty$ where

$$\sigma_\infty = \delta_1 \tag{5.8}$$

for C_1 and C_2, and

$$\frac{d\sigma_\infty(x)}{dx} = \rho \Gamma(\alpha)^\rho \; H_{11}^{10} \left(\Gamma(\alpha)^\rho x \; \Big|\; \begin{matrix} (-1,1) \\ (-\rho,\rho) \end{matrix} \right), \quad \rho = (1-\alpha)^{-1} \tag{5.9}$$

for C_3. Hence, $E(D_L)$ behaves asymptotically for large L as follows for C1, C2, C3, respectively,

$$E(D_L) \sim a^{-1} , \quad (a \, \ell nL)^{-1} , \quad a^{-1} \Gamma(1+\rho) \Gamma(\alpha)^{-\rho} L^{-\alpha\rho} . \tag{5.10}$$

REFERENCES

[1] Bernasconi, J., Alexander, S., Orbach, R.: Phys.Rev.Lett.41, 185-187 (1978)

[2] Schneider, W.R.: Submitted to Comm. Math.Phys.

[3] Bernasconi, J., Schneider, W.R., Wyss, W.: Z.Phys. B37, 175-184 (1980)

[4] Schneider, W.R., Bernasconi, J.: Lecture Notes in Physics 153 , pp.389
     Berlin, Heidelberg, New York: Springer 1982

[5] Bernasconi, J., Schneider, W.R.: Phys.Rev.Lett. 47, 1643-1647 (1981)

[6] Alexander, S., Bernasconi, J., Schneider, W.R., Orbach, R.: Rev.Mod.Phys.
     53, 175-198 (1981)

[7] Braaksma, B.L.J.: Compos.Math. 15, 239-341 (1963)

LARGE SCALE BEHAVIOR OF EQUILIBRIUM TIME CORRELATION FUNCTIONS
FOR SOME STOCHASTIC ISING MODELS

Herbert Spohn
Department of Mathematics
Rutgers University
New Brunswick, NJ 08903

1. Introduction

Systems in thermal equilibrium are analysed experimentally through scattering
experiments. Such a scattering experiment measures the intensity of light (or of
other probes such as X-rays, neutrons or electrons depending on the experimental
situation) scattered by the system with a transfer of momentum

$$k = k_s - k_i$$

and a transfer of energy

$$\omega = \omega_s - \omega_i \quad ,$$

cf. Figure.

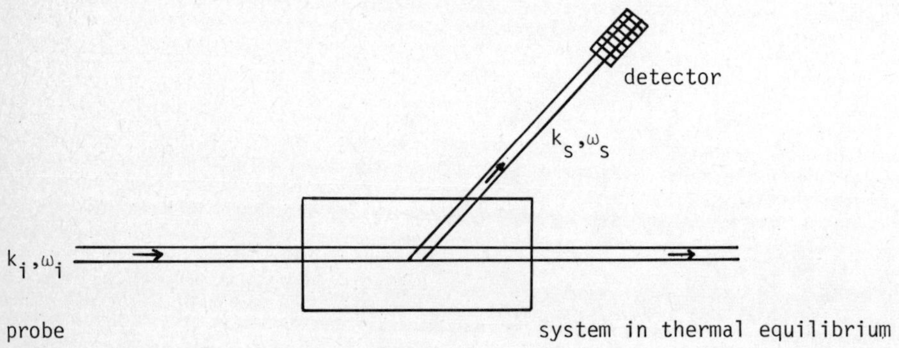

Figure

A narrow light beam with incident wave number (momentum) k_i and frequency (energy)
ω_i is scattered into wave number k_s and frequency ω_s. The detector records the distribution of the frequencies ω_s.

The interaction between the system and the probe is weak. Therefore, simple second order perturbation theory shows that the scattering intensity is proportional to the so-called dynamical structure factor $S(k,\omega)$.[1,2] [To my knowledge there is no proof of this widely used relationship, even in the case the fluid is assumed to be non-interacting.] If our system is a fluid, then $S(k,\omega)$ is the space-time Fourier transform of the density-density time correlation function. For a magnetic system, probed with neutrons, $S(k,\omega)$ would be the space-time Fourier transform of the spin-spin time correlation function.

I hope that these remarks explain why Statistical Mechanics has made a large effort to understand and to predict the structure of equilibrium time correlation functions. Of course, except for trivial systems, there is no way to compute these functions explicitly. So one has to rely on a more qualitative theory. Of particular interest is then the behavior of $S(k,\omega)$ for small k and ω, equivalently, the behavior of the time correlation functions for large separation in space-time.

In this note, I will stay within the mathematically rather safe ground of stochastic Ising models. I will try to explain the mathematically more accessible part of the expected qualitative picture. Since the exposition is short by necessity, I recommend to the reader the book by Forster.[1] I review a few recent rigorous results and include an argument which proves the exponential decay of correlation functions in space-time for a class of stochastic Ising models at high temperature. I will provide brief outlooks on the hydrodynamics of stochastic lattice gases and on linearized hydrodynamics for fluids.

2. Single Spin Flip Dynamics / Memory Function Formalism

Single spin flip dynamics goes back to R. Glauber.[3] There exists a well developed mathematical theory due to F. Spitzer, R.L. Dobrushin, R.A. Holley, T.M. Liggett, D.W. Stroock and many others. I refer the reader to the survey article by Liggett.[4] The memory function formalism has been introduced by R. Zwanzig[5] and H. Mori.[6]

I want to consider spins on the d-dimensional lattice Z^d. The spin σ_x at site $x \in Z^d$ takes the values ± 1. σ denotes a spin configuration. The space of all possible spin configurations is then $\{-1,1\}^{Z^d} \equiv \Omega$. Let

$$c(x,\sigma) \geq 0 \tag{2.1}$$

be the rate that, given the configuration σ, the spin at site x flips from σ_x to $-\sigma_x$. We assume that the rates are translation invariant, i.e. $c(x + a, \tau_a\sigma) = c(x,\sigma)$, where τ_a denotes the shift by $a \in Z^d$. The generator of the dynamics is given by

$$(Lf)(\sigma) \;=\; \sum_x c(x,\sigma)\,(\partial_x f)(\sigma) \tag{2.2}$$

acting on strictly local functions $f: \Omega \to R$. Here $(\partial_x f)(\sigma) = f(_x\sigma) - f(\sigma)$ where $_x\sigma$ the configuration σ with spin at site x reversed, i.e.

$$(_x\sigma)_y \;=\; \begin{cases} \sigma_y & \text{for } x \neq y \\ -\sigma_y & \text{for } x = y \end{cases} \;\; .$$

If the flip rates $c(x,\sigma)$ are sufficiently local, i.e. if $c(x,\sigma)$ depends only little on spins far away from x, then L determines a unique Markov semigroup

$$T_t \;=\; e^{Lt} \;, \quad t \geq 0 \;, \tag{2.3}$$

on $C(\Omega)$, the space of bounded continuous functions on Ω. In turn, it follows then from the general theory of Markov processes that T_t determines a unique Markov jump process σ_t with state space Ω. The canonical path space of this process is the space of functions $t \mapsto \sigma_t$ $(\in \Omega)$ which are right continuous and have left limits.

Let μ be a probability measure on Ω which is time-invariant. If $<\cdot>$ denotes its expectation, then this means

$$<f> \;=\; <T_t f> \tag{2.4}$$

for all $t \geq 0$ and continuous functions f on Ω. Equivalently, for all $t \geq 0$ and every bounded set $\Lambda \subset Z^d$

$$\langle T_t \prod_{x \in \Lambda} \sigma_x \rangle \equiv \langle \prod_{x \in \Lambda} \sigma_{x,t} \rangle = \langle \prod_{x \in \Lambda} \sigma_x \rangle \qquad (2.5)$$

We assume that μ is invariant under translations.

As explained in the introduction in many cases the object of interest is the magnetization fluctuation field

$$\{ \sigma_{x,t} - \langle \sigma_{x,t} \rangle \mid x \in Z^d, \, t \in R\} \qquad (2.6)$$

in the stationary state $\langle \cdot \rangle$. We think of $\sigma_{x,t} - \langle \sigma_{x,t} \rangle$ as a <u>random field</u> in space-time. $\sigma_{x,t} - \langle \sigma_{x,t} \rangle \equiv \xi_{x,t}$ is stationary in space-time. (By time invariance $\xi_{x,t}$ is defined for all t.) We inserted here "fluctuation" because $\xi_{x,t}$ gives the deviation, in space-time, of the magnetization from its average value $\langle \sigma_{x,t} \rangle$. In a scattering experiment where the probe couples to the magnetization one measures directly the dynamical structure factor

$$S(k,\omega) = \sum_x e^{ikx} \int dt \, e^{i\omega t} (\langle \sigma_{0,0} \sigma_{x,t} \rangle - \langle \sigma_{0,0} \rangle \langle \sigma_{x,t} \rangle) \ . \qquad (2.7)$$

Therefore one would like to know the covariance of the fluctuation field,

$$\langle \xi_{x,t} \, \xi_{y,s} \rangle = \langle \sigma_{x,t} \, \sigma_{y,s} \rangle - \langle \sigma_{x,t} \rangle \langle \sigma_{y,x} \rangle \ , \qquad (2.8)$$

in particular its behavior for large separation $|x-y|$ and $|t-s|$.

Which fluctuation field one wants to investigate is model-dependent. For example, if one associates an energy $H(\sigma)$ with the configuration σ, say

$$H(\sigma) = - \sum_{|x-y|=1} \sigma_x \, \sigma_y \ , \qquad (2.9)$$

then one may want to consider the energy fluctuation field

$$\{ \sum_{|e|=1} (-\sigma_{x,t} \, \sigma_{x+e,t} + \langle \sigma_{x,t} \, \sigma_{x+e,t} \rangle) \mid x \in Z^d, t \in R\} \ . \qquad (2.10)$$

In particular, the fields corresponding to the locally conserved quantities and to the order parameters play a distinguished role.

The memory function formalism now rewrites the covariance (2.8) in a form which is physically more suggestive. For this purpose let us consider the Hilbert space $\mathcal{H} = L^2(\Omega,\mu)$ with the scalar product $\langle f|g \rangle = \langle f^* g \rangle - \langle f^* \rangle \langle g \rangle$. Let P be the orthogonal projection onto the subspace linearly spanned by all $\{\sigma_x \mid x \in Z^d\}$. (As before σ_x also denotes the function $\sigma_x : \sigma \mapsto \sigma_x$.) Explicitly P is given by

$$(Pf)(\sigma) = \sum_{x,y} \sigma_x \chi^{-1} (x-y) < \sigma_y|f> . \qquad (2.11)$$

Here $\chi(x-y) = <\sigma_x| \sigma_y>$ and χ^{-1} is the matrix inverse. Let $\hat{\chi}(k,z)$ be the Fourier-Laplace transform of the covariance matrix,

$$\hat{\chi}(k,z) = \sum_x e^{ikx} \int_0^\infty dt e^{-zt} <\sigma_x| T_t \sigma_0> , \qquad (2.11)$$

$k \in [-\pi,\pi]^d$ and $\mathrm{Re}\ z > 0$. After some rearrangements one obtains the following identity

$$\hat{\chi}(k,z) = \hat{\chi}(k) \left[z - \frac{\omega(k)}{\hat{\chi}(k)} - \frac{1}{\hat{\chi}(k)} M(k,z) \right]^{-1} \qquad (2.13)$$

with

$$\hat{\chi}(k) = \sum_x e^{ikx} (<\sigma_x\sigma_0> - <\sigma_x> <\sigma_0>) ,$$

$$\omega(k) = \sum_x e^{ikx} <\sigma_x| L \sigma_0> , \qquad (2.14)$$

$$M(k,z) = \sum_x e^{ikx} <\sigma_x| LQ (z - QLQ)^{-1} QL \sigma_0> ,$$

$$Q = 1 - P .$$

(2.13) is one of the basic results of the memory function formalism.

$\omega(k)$ is called the frequency coefficient and $M(k,z)$ the generalized transport coefficient. Note that $\omega(k)$ is given by static correlation functions on which one supposes to have some information. All the dynamics is burried in the generalized transport coefficient $M(k,z)$.

The name memory function formalism comes from writing (2.13) in time as

$$\frac{\partial}{\partial t} \hat{\chi}(k,t) = \frac{\omega(k)}{\hat{\chi}(k)} \hat{\chi}(k,t) + \int_0^t ds \frac{1}{\hat{\chi}(k)} M(k,t-s) \hat{\chi}(k,s) , \qquad (2.15)$$

$$\hat{\chi}(k,t=0) = \hat{\chi}(k) ,$$

with

$$\hat{\chi}(k,t) = \sum_x e^{ikx} <\sigma_x| T_t \sigma_0 > ,$$

$$M(k,t) = \sum_x e^{ikx} <\sigma_x| LQ e^{QLQt} LQ \sigma_0> . \qquad (2.16)$$

The second term on the right of (2.15) is called the memory term, because it expresses how $\hat{\chi}(k,t)$ depends on its own past.

Why should the simple rewriting (2.13) of the covariance matrix be of any value? The main point is that with some physical understanding of the dynamics the form (2.13) allows one to <u>guess</u> the behavior of $\hat{\chi}(k,z)$ for small k and z, i.e., the behavior of the covariance matrix (2.8) for large $|x-y|$ and $|t-s|$. This guess is particularly useful in situations where the large distance behavior is not determined by a single field, as pretended here, but by several fields, as e.g. for a fluid where mass, velocity and energy density field have to be considered simultaneously. In this case, the generalization of (2.13) contains considerable information on the general structure of $\hat{\chi}(k,z)$ for small k and z, in particular on its symmetry properties and on the allowed couplings between the various fields. In addition (2.13) is a convenient starting point for phenomenological theories which aim at predicting $\hat{\chi}(k,z)$ over a large range of k,z-values and not only its asymptotic behavior $k,z \to 0$. To actually compute either $\hat{\chi}(k,z)$ or the transport coefficients which enter the asymptotic k,z-behavior is another story. Such computations are in general based on perturbation expansions whose asymptotic validity relies on the availability of a small parameter.

When "reading off" the asymptotic behavior of $\hat{\chi}(k,z)$ from (2.13) we have to distinguish several cases.

(A) Let us assume that the static ($<\cdot>$) correlation functions decay fast such that

$$\lim_{k \to 0} \hat{\chi}(k) \;=\; \chi \tag{2.17}$$

exists.   Then

$$\hat{\chi}(k,z) \;\overset{\sim}{=}\; \chi \left[z - \frac{1}{\chi}\, \omega(k) \;-\; \frac{1}{\chi}\, M(k,z) \right]^{-1}. \tag{2.18}$$

(A1) Let us assume (A). If also $\omega(0)$ is finite and if the time decay is sufficiently rapid such that $M(k,z)$ tends to a finite limit as $k,z \to 0$, then

$$\hat{\chi}(k,z) \;\overset{\sim}{=}\; \frac{\chi}{z + \lambda} \tag{2.19}$$

with $\lambda > 0$ for small k and z. The magnetization decays exponentially to its average value with decay rate λ. Typically, in this case μ will be the only stationary state of T_t.

In the following section we will prove the exponential decay in space-time for a certain class of spin flip models.

(A2) Let us assume (A). In addition let us assume that the dynamics conserves the magnetization. [This is impossible for the spin flip models defined above. But if we modify the dynamics in such a way that also pairs of spins can exchange their values with a certain rate, then we can consider a dynamics such that at each transition the (total) magnetization remains unchanged. Systems with a conservation law are more naturally discussed in the language of stochastic lattice gases. We will discuss in detail these models in Section 5.] Typically, in this case T_t has a one-parameter family of stationary states, at least one for each value in the allowed range of the mean magnetization. A simple computation, cf. Section 5, shows that because of conservation

$$\omega(k) \stackrel{\sim}{=} -(k,D_o k),$$

$$M(k,z) \stackrel{\sim}{=} -(k,\hat{M}(k,z)k) \quad . \tag{2.20}$$

Because our system is not necessarily isotropic, D_o and $\hat{M}(k,z)$ are $d \times d$ matrices. We use the shorthand $(k,D_o k) = \sum_{i,j=1}^{d} k_i (D_o)_{ij} k_j$. If $\hat{M}(k,z)$ tends to a finite limit as $k,z \to 0$, then

$$\tilde{\chi}(k,z) \stackrel{\sim}{=} \frac{\chi}{z + (k,Dk)} \quad , \tag{2.21}$$

$D>0$, for small k,z. D is the so-called bulk diffusion coefficient (matrix). Therefore $<\xi_{x,t}\, \xi_{y,s}> \stackrel{\sim}{=} \chi \; e^{-t(\nabla,D\nabla)}(x,y)$. The covariance matrix is approximately given by the fundamental solution of the diffusion equation. Physically, this result is quite convincing. After all, because of the conservation law, any excess magnetization cannot simply disappear. It has to diffuse out slowly.

  We will discuss the case of a locally conserved field more thoroughly in Section 5

(B) Let us assume that the static correlation functions decay slowly. In some cases, one has a small k-behavior as

$$\chi(k) = \frac{c}{k^2} \quad . \tag{2.22}$$

Then

$$\tilde{\chi}(k,z) \stackrel{\sim}{=} \frac{c}{k^2} [z - \frac{k^2}{c} \omega(k) - \frac{k^2}{c} M(k,z)]^{-1} \quad . \tag{2.23}$$

(B1) Let us assume (B). If $\omega(k)$ and $M(k,z)$ have finite limits as $k,z \to 0$, then

$$\hat{\chi}(k,z) \stackrel{\sim}{=} \frac{1}{k^2} \frac{c}{z + D_1 k^2} \quad , \qquad (2.24)$$

$D_1 > 0$, for small k,z. So again one obtains a diffusive behavior. However, in this case, it is the result of slowly decaying static correlation and not of a conservation law.

In Section 4 we will explain that the voter model in dimension $d \geq 3$ is an example for (B1).

(B2) Let us assume (B). If the magnetization is conserved, then by the same reasoning as before one would expect that

$$\hat{\chi}(k,z) \stackrel{\sim}{=} \frac{1}{k^2} \frac{c}{z + D_2 k^4} \quad , \qquad (2.25)$$

$D_2 > 0$, for small k,z. This kind of behavior is predicted for the Heisenberg ferromagnet in three dimensions at low temperature, cf. [1, Chapter 8].

To make these crude arguments more convincing one would have to go into more detail and by necessity a more complicated picture would emerge. I only want to draw as a main conclusion that for the large scale behavior of the covariance matrix (2.8) one has to take into account
   (i)   decay properties of the static correlation functions
   (ii) conservation laws.

3. Exponential Decay of Correlation Functions in Space-Time for a Class of Spin Flip Models with No Local Conservation Law and at High Temperature

For a class of spin flip models Holley and Stroock[7,8] show that a uniform exponential decay in time implies exponential decay of the spatial correlation functions of the (unique) stationary state. It is therefore not surprising that as a corollary exponential clustering in space-time follows.

We describe here a set of sufficient conditions. They are by no means optimal.

We assume that the flip rates are translation invariant and are of finite range, i.e.

$$\partial_x \, c(y, \sigma) = 0 \quad \text{for } |x-y| \geq M. \tag{3.1}$$

Let $\chi_F(\sigma) = \prod_{x \in F} \sigma_x$ for any bounded set $F \subset Z^d$ with the convention $\chi_\phi(\sigma) = 1$. We assume that

$$c(x,\sigma) = \sum_F \hat{c}\,(x,F)\,\chi_F(\sigma) \tag{3.2}$$

such that for some $0 < \alpha < 1$

$$\sum_{F \neq \phi} |c(x,F)| \leq \alpha\,\hat{c}(x,\phi) \equiv \alpha\,c \ . \tag{3.3}$$

Then T_t has a unique invariant measure μ. Since the flip rates are translation invariant, so is μ. Let f depend only on finitely many spins. Then there exists a constant $C(f)$ such that

$$\sup_\sigma |(T_t f)(\sigma) - \,<f>\,| \leq C(f)\,e^{-2c(1-\alpha)t} \tag{3.4}$$

with $<f> = \int d\mu f$, [4, Theorem 1.3.10].

<u>Theorem 1.</u> Let $\Lambda \subset Z^d$ be bounded and let f,g depend only on spins in Λ , i.e. $\partial_x f = 0 = \partial_x g$ for $x \in \Lambda$. Then there exists a constant $\kappa > 0$, not dependent on f, g, and constants A(f), A(g) such that for all $t \in R_+$, $\quad x \in Z^d$

$$|\,<g T_t \tau_x f>\, - \,<g><f>\,| \leq A(f)A(g)\,e^{-\kappa(t + |x|)}. \tag{3.5}$$

<u>Proof:</u> Let $t \geq \theta_0 \, |x|$ for some $\theta_0 > 0$. Then by (3.4)

$$|\,<g\,T_t \tau_x\,f>\, - \,<g><f>\,|$$

$$= | \langle \tau_{-x} g \, T_t f \rangle - \langle \tau_{-x} g \rangle \langle f \rangle |$$

$$\leq \|g\| \; C(f) \; e^{-2c(1-\alpha)t} \leq \|g\| \; C(f) \; e^{-[c(1-\alpha)t + c(1-\alpha)\theta_0 |x|]} . \qquad (3.6)$$

Therefore it remains to show that we can choose θ_0 such that (3.5) holds for $t \leq \theta_0 |x|$. Let L be such that $\Lambda \subset [-L,L]^d$. Let $\Lambda \subset \tilde{\Lambda}$ and let ρ be the distance of Λ from the complement of $\tilde{\Lambda}$. We set $|x| = \rho + 2L$. For a configuration σ we denote by ξ its part in $\tilde{\Lambda}^c$. Let $\mu^{\tilde{\Lambda},\xi}$ on $\{-1,1\}^{\tilde{\Lambda}}$ be the measure μ conditioned on the configuration ξ in $\tilde{\Lambda}^c$ and let $T_t^{\tilde{\Lambda},\xi}$ be the Markov semigroup of the spin flip dynamics in $\tilde{\Lambda}$ with the configuration ξ in $\tilde{\Lambda}^c$ frozen in. Then by [7, (4.7) Theorem]

$$| \int d\mu \; g \; T_t \, \tau_x \, f - \int d\mu \; g \quad d\mu \; f |$$

$$= | \int d\mu \; \tau_{-x} g \; [\int d \, \mu^{\tilde{\Lambda},\xi} \, T_t f - \int d\mu \; f \,] |$$

$$\leq \|g\| \; \sup_{\xi} \; | \int d\mu^{\tilde{\Lambda},\xi} \, T_t f - \int d\mu \; f \, |$$

$$\leq \|g\| \; \sup_{\xi} \{ \, | \int d \, \mu^{\tilde{\Lambda},\xi} \, (T_t f - T_t^{\tilde{\Lambda},\xi} f)| + | \int d\mu^{\tilde{\Lambda},\xi} \; f - \int d\mu \; f \, |\} . \qquad (3.7)$$

The second term is exponentially small in ρ. For the first we use that

$$|T_t f \, (\sigma) - T_t^{\tilde{\Lambda},\xi} \, f(\sigma) | \leq A'(f) \; e^{Ct} \; \frac{(Ct)^{N+2}}{(N+2)!} \qquad (3.8)$$

with $C > 0$ a constant and $N = [\rho/M] - 1$, where $[\rho/M]$ denotes the integer part of ρ/M. Let us set $t < \theta|x|$. Then using Stirling's formula

$$|T_t \, f(\sigma) - T_t^{\tilde{\Lambda},\xi} f(\sigma)|$$

$$\leq A''(f) \; \exp[\; |x| \; (C\theta + \frac{1}{M} \log M \, C\theta + \frac{1}{M} \,] \; . \qquad (3.9)$$

Let us take now θ_0 sufficiently small such that the [ ]-term in the exponent is less than $-2\alpha'$, $\alpha' > 0$, for all $0 \leq \theta \leq \theta_0$. Then for all $t \leq \theta_0 \, |x|$

$$| \langle g \; T_t \, \tau_x f \rangle - \langle g \rangle \langle f \rangle | \leq$$

$$A(f)A(g) \; e^{-2\alpha''|x|} \leq A(f)A(g) \; \exp[-(\frac{\alpha''}{\theta_0} t + \alpha'' \, |x| \,)] \; \blacksquare \qquad (3.10)$$

The flip rates can be arranged in such a way that the invariant measure μ is a Gibbs state for some potential. Holley and Stroock translated the conditions (3.1) and (3.3) into a condition on the potential and found a high temperature condition.

I find the following example rather instructive. Let the dimension $d = 1$ and let us assume the flip rates

$$c(x,\sigma) = 1 + \delta\sigma_{x-1}\,\sigma_{x+1} - \frac{1}{2}(1 + \delta)\gamma\,(\sigma_{x-1}\sigma_x + \sigma_x\sigma_{x+1})\;. \tag{3.11}$$

The positivity of the flip rates requires $|\gamma| \le 1$, $|\delta| \le 1$. The nearest neighbor Gibbs state

$$Z^{-1}\;\exp[\beta(\gamma)\;\sum_x \sigma_x\sigma_{x+1}] \tag{3.12}$$

with the inverse temperature determined by γ is time-invariant. The condition (3.3) requires then

$$|\delta| + |(1 + \delta)\gamma| < 1\;. \tag{3.13}$$

One expects exponential clustering in space-time for $|\delta| \ne 1$. For $\delta = 1$ flips which conserve energy are forbidden. Therefore in addition to (3.12) the measures concentrated on the configuration $(\ldots, 1, 1, -1, -1, 1, 1, \ldots)$ and its shift are time-invariant. In principle, this should not destroy the exponential clustering of equilibrium time correlation functions. But less "crude" methods than perturbation seems to be required to prove it. For $\delta = -1$ all energy non-conserving flips are forbidden. Now the energy is locally conserved. The Gibbs states (3.12) for every temperature are invariant. In this case space-time correlation functions are diffusive and exponential clustering definitely fails.

4. Slowly Decaying Static Correlations / the Voter Model

We discuss the, to my knowledge, only example for (B1). For more details I refer to [9 - 13].

The voter model is a spin flip model with flip rates

$$c(x,\sigma) \;=\; 2d \;-\; \sum_{|x-y|=1} \sigma_x \sigma_y \;. \tag{4.1}$$

We assume dimension $d \geq 3$. Then T_t has a one-parameter family of extremal time-invariant states, $\langle \cdot \rangle_m$. They are also invariant under translations in space and are parametrized by the magnetization m, $|m| \leq 1$,

$$\langle \sigma_x \rangle_m \;=\; m \;. \tag{4.2}$$

The static covariance is given by

$$\langle \sigma_x \sigma_y \rangle_m \;-\; m^2 \;=\; (1 - m^2)\; p(x-y) \;. \tag{4.3}$$

Here $p(x)$ is the probability that a symmetric random walker on Z^d who starts at $x \in Z^d$ will ever hit the origin. For Brownian motion in R^d, given it starts at x, $|x| > 1$, the probability to ever hit the unit ball is exactly $|x|^{-d+2}$. $p(x)$ has the same large distance behavior, i.e.

$$p(x) \;\simeq\; (1 - p(e))\; |x|^{-d+2} \tag{4.4}$$

for large $|x|$, $|e| = 1$. Equivalently, in Fourier space

$$\hat{p}(k) \;\simeq\; (1 - m^2)(1 - p(e))\; k^{-2} \tag{4.5}$$

for small k. The static correlation functions decay slowly. Bramson and Griffeath[11] show that $\langle \cdot \rangle_m$ scaled appropriately looks like a Gaussian measure on R^d with covariance matrix $(1 - m^2)(1 - p(e))\; \Delta^{-1}(x,y)$ (mass zero free field).

The time displaced covariance can be computed explicitly. One obtains

$$\langle \sigma_{x,t} \sigma_{y,s} \rangle \;-\; m^2 \;=\; (1 - m^2) \sum_{x'} e^{2|t-s|\Delta}(x,x') p(x'-y) \tag{4.6}$$

with Δ the discrete Laplacian on Z^d, $(\Delta f)(x) = \sum_{|e|=1} (f(x+e) - f(x))$. The magnetization time correlation function decays diffusively because of (4.4).

The duality of the voter model to annihilating random walks allows one to obtain results which go beyond the covariance of the fluctuation field. Let $f \in \mathcal{S}(R^d)$, the Schwartz space of rapidly decreasing functions. We define the scaled magnetization fluctuation field by

$$\xi^{\varepsilon}(f,t) \;=\; \varepsilon^{((d/2)+1)} \sum_{x} f(\varepsilon x) \, (\sigma_{x,\varepsilon^{-2}t} - m) \tag{4.7}$$

in the state $\langle \cdot \rangle_m$. The power of the prefactor is dictated by the scale invariance of (4.5). Note that because of slow static decay the fluctuations are huge by usual standards. For given spatial scale the time scale is a consequence of the scale invariance of the diffusion equation.

<u>Theorem 2</u> [12,13) Let $\xi(f,t)$ be the Gaussian random field over R^{d+1} with mean zero and covariance

$$\langle \xi(f,t)\, \xi(g,s) \rangle \;=\; (1 - m^2)(1 - p(e)) \int_{R^d} dk \, \hat{f}(k)\hat{g}(k) \, \frac{1}{k^2} \, e^{-2|t-s|k^2} . \tag{4.8}$$

Then

$$\lim_{\varepsilon \to 0} \xi^{\varepsilon}(f,t) \;=\; \xi(f,t) \tag{4.9}$$

in the sense of weak convergence of path measures on $D([0,\infty), \mathcal{S}'(R^d))$.

5. Dynamics of a Locally Conserved Field

In a ferromagnet which is strongly coupled to its termal surroundings the magnetization will hardly be conserved. However if we think of a lattice gas where particles jump stochasticly, clearly, the number of particles is conserved. In this section we will switch then to the more convenient lattice gas language and will define the dynamics in such a way that the number of particles is conserved (Kawasaki dynamics[14,15]).

In most physical systems conservation laws are of great importance. A stochastic lattice gas is one of the simplest models where the effect of such a conservation law can be investigated.

There seem to be at least two classes of physical systems which are a reasonable realization of a stochastic lattice gas. One of them is the AB-alloy where the particles correspond to one of the species.[16] The other one is the so-called super-ionic conductors.[17] An example is silverbromide. Bromium ions form a rigid lattice and, when suitable manufactured, the silver ions partially occupy the interlattice spacings. The silver ions are rather mobile and jump due to thermal activations from lattice spacing to lattice spacing provided it is empty.

I want to consider particles on the d-dimensional lattice Z^d. There is at most one particle per lattice site. $n_x = 1$ corresponds to site x occupied and $n_x = 0$ to site x empty. n denotes a particle configuration. The space of all particle configurations is then $\{0,1\}^{Z^d} = \Omega$. Let

$$c(x,y,n) \geq 0 \qquad\qquad (5.2)$$

be the rate that, given the configuration n, the occupations at site x and y are interchanged. Therefore $c(x,y,n) = c(y,x,n)$. If $n_x = 1$ and $n_y = 0$, then $c(x,y,n)$ is the rate for the particle at x to jump to y given the configuration n. If either both $n_x = 0 = n_y$ or both $n_x = 1 = n_y$, then we may set $c(x,y,n) = 0$. We assume that the rates are invariant under translations, i.e. $c(x+a, y+a, \tau_a n) = c(x,y,n)$ for all $a \in Z^d$. The generator of the dynamics is given by

$$(Lf)(n) = \frac{1}{2} \sum_{x,y} c(x,y,n) \, (\partial_{xy} f)(n) \qquad\qquad (5.3)$$

acting on strictly local functions $f: \Omega \to R$. Here $(\partial_{xy} f)(n) = f(_{xy}n) - f(n)$ where $_{xy}n$ is the configuration n with occupations at sites x and y interchanged, i.e.

$$(_{xy}n)_z = \begin{cases} n_x & \text{for } z = y \\ n_y & \text{for } z = x \\ n_z & \text{for } z \neq x,y \end{cases} .$$

If the jump rates $c(x,y,\eta)$ are sufficiently local, then L determines a unique Markov semigroup

$$T_t = e^{Lt} \ , \quad t \geq 0 \tag{5.4}$$

on $C(\Omega)$. In turn T_t determines then a unique Markov process η_t with state space Ω. The canonical path space of this process is the space of functions $t \mapsto \eta_t$ (ϵ Ω) which are right continuous and have left limits.

In the following we will assume that L satisfies the <u>condition of detailed balance</u> with respect to some measure μ on Ω, i.e.

$$<g \, T_t f> \ = \ <f \, T_t g > \tag{5.5}$$

for all $f,g \ \epsilon \ C(\Omega)$. T_t defines then a semigroup of self-adjoint contractions on $L^2(\Omega,\mu)$. If the jump rates $c(x,y,\eta)$ are of finite range, then (5.5) is equivalent to

$$<g \, L \, f> \quad = \quad <f \, L \, g> \tag{5.6}$$

for all strictly local f,g. Note that in particular μ is time-invariant. A measure μ which satisfies (5.5) is called reversible for T_t.

5.1 Current, Bulk Diffusion and Green-Kubo Formula

The aim of this section is to establish the connection between the bulk diffusion coefficient and the current-current time correlation function. Under the assumptions (i) to (iii) stated below the computations can be rigorized up to the stage where the existence of the limit $k,z \to 0$ is assumed.

In trying to emphasize the main features let me restrict myself to the simplest case. I will assume the following:

(i) $c(x,y,\eta) = 0$ for $|x - y| > 1$.

This means that only nearest neighbor jumps are allowed.

(ii) $c(x,y,\eta)$ is of finite range and translation invariant.

(iii) L satisfies detailed balance with respect to some measure μ . μ is translation invariant and has exponentially decaying correlations.

Let $\rho = \mu(\eta_x)$ be the average density. To emphasize the density dependence we write $< \cdot >_\rho$ for μ .

In most applications the following situation is met: The energy of a configuration η is

$$H(\eta) = \beta \sum_{|x-y|=1} \eta_x \eta_y \ . \tag{5.7}$$

One chooses the jump rates such that they satisfy detailed balance with respect to the Gibbs state $Z^{-1}e^{-H}$,

$$c(x,y,\eta) \;=\; c(x,y,_{xy}\eta)\,\exp[-H(_{xy}\eta) + H(\eta)] \qquad . \tag{5.8}$$

Then $<\cdot>_{\rho}$ is constructed as the infinite volume limit of $Z^{-1}e^{-H}$ with prescribed density ρ, $0 \le \rho \le 1$. Because of the possible occurrence of a phase transition, for some values of ρ $<\cdot>_{\rho}$ may not be extremal and would then violate (iii).

The density field is given by $\{n_{x,t} \mid x \in Z^d, t \in R\}$. Since the density is locally conserved, there exists a corresponding current: To each bond (x,y), $|x-y| =1$, we associate the (integrated) current

$j(x,y,[t_1,t_2])$ = number of particles which cross the bond from x to y minus the number of particles which cross the bond from y to x during the time span $[t_1,t_2]$.

$\{j(x,y,[t_1,t_2]) \mid |x-y|=1, 0 \le t_2 - t_1 < \infty\}$ is the <u>current random field</u>. It is defined for any history in path space. $j(x,y,[t_1,t_2])$ is a discrete vector field and satisfies $j(x,y,[t_1,t_2]) = -j(y,x,[t_1,t_2])$. We have the obvious local conservation law

$$\sum_{x \in \Lambda} (n_{x,t} - n_{x,s}) \;+\; \sum_{\substack{x \in \Lambda \\ y \notin \Lambda}} j(x,y,[s,t]) \;=\; 0 \tag{5.9}$$

for any bounded region Λ and finite time interval $[s,t]$. (5.9) is the usual definition of a locally conserved field. Its value inside the bounded region Λ changes only by a flow through the surface of Λ.

Let $\Lambda = \{x\}$ and let us average (5.9) given the system starts in the arbitrary initial state $\bar{\mu}$. Then one obtains

$$E_{\bar{\mu}} (j(x,y,[s,t])) \;=\; \int_s^t dt' \int \bar{\mu}(d\eta)\, T_{t'} j(x,y,\eta) \;, \tag{5.10}$$

where

$$j(x,y,\eta) \;=\; c(x,y,\eta)(n_x - n_y) \;, \quad |x-y| =1 \;. \tag{5.11}$$

We used here that

$$L\, n_x \;=\; -\sum_{|x-y|=1} j(x,y) \;. \tag{5.12}$$

$j(x,y,\eta)$ is called the <u>current function</u>. It is a function on the space of configurations Ω and when integrated over the state of the system at time t it gives the

<u>average</u> current through the bond (x,y). We note that $j(x,y) = -j(y,x)$ and that, because of the translation invariance of the jump rates, $j(x+a,y+a) = \tau_a j(x,y)$ for all $a \in Z^d$.

In Section 2 we argued that the density time correlations behave like the fundamental solution of the diffusion equation for large separation $|x-y|$ and $|t-s|$,

$$<\eta_{x,t}\eta_{y,s}>_\rho - \rho^2 \; \tilde{=} \; \chi_\rho \; e^{-|t-s| \; (\nabla,D(\rho)\nabla)}(x,y) \; . \tag{5.13}$$

Here χ_ρ is the static compressibility,

$$\chi_\rho = \sum_x [<\eta_x\eta_0>_\rho - \rho^2]. \tag{5.14}$$

Let us define the scaled <u>density fluctuation field</u>

$$\xi^\varepsilon (f,t) = \varepsilon^{d/2} \sum_x f(\varepsilon x) (\eta_{x,\varepsilon^{-2}t} - \rho) \; , \tag{5.15}$$

$f \in \mathcal{S}(R^d)$, in the state $<\cdot>_\rho$. The time scale is a consequence of the scale invariance of the diffusion equation.

<u>Conjecture 1</u>. For some positive $d \times d$ matrix $D(\rho)$ the covariance of the density fluctuation field has the limit

$$\lim_{\varepsilon \to 0} <\xi^\varepsilon(f,t) \; \xi^\varepsilon(g,s) >_\rho = \chi_\rho \int dk \; \hat{f}(k)\hat{g}(k) \; e^{-|t-s|(k,D(\rho)k)} . \tag{5.16}$$

We will obtain now an expression for the bulk diffusion coefficient $D(\rho)$. The Fourier-Laplace transform of (5.16) reads

$$\lim_{\varepsilon \to 0} \varepsilon^2 \hat{\chi}_\rho (\varepsilon k,\varepsilon^2 z) = \frac{\chi_\rho}{z + (k,D(\rho)k)} \tag{5.17}$$

provided this limit exists. We use now (2.13). For the first summand one has

$$\lim_{\varepsilon \to 0} \varepsilon^{-2} \; \hat{\omega}(\varepsilon k) = \lim_{\varepsilon \to 0} \varepsilon^{-2} \sum_x e^{i\varepsilon kx} <\eta_x|L\eta_0>$$

$$= -\frac{1}{2} \sum_{i=1}^d <c(0,e_i)(\eta_{e_i} - \eta_0)^2>_\rho \; k_i^2 \; . \tag{5.18}$$

e_i is the unit vector in the positive direction of the i-th coordinate axis. We used that by detailed balance for $|e| = 1$

$$\langle \eta_x \, c(0,e)(\eta_e - \eta_0) \rangle_\rho = \begin{cases} 0 & \text{for } x \neq 0, e \\[2mm] - \langle \eta_0 \, c(0,e)(\eta_e - \eta_0) \rangle_\rho & \text{for } x = e \\[2mm] - \langle \eta_e \, c(0,e)(\eta_e - \eta_0) \rangle_\rho & \text{for } x = 0 \, . \end{cases} \qquad (5.19)$$

For the generalized transport coefficient we obtain, using translation invariance and detailed balance,

$$M(k,z) = \int_0^\infty dt \, e^{-zt} \sum_x e^{ikx} \langle \, L\eta_x | \, Q e^{QLQt} \, Q \, L\eta_0 \, \rangle$$

$$\qquad (5.20)$$

$$= \int_0^\infty dt \, e^{-zt} \sum_x e^{ikx} \sum_{i,j=1}^d \langle ((\tau_{e_i}-1) \, j \, (x, x+e_i)) \, Q \, e^{QLQt} Q \, (\tau_{e_j}-1) j(0,e_j) \rangle_\rho .$$

Therefore

$$\lim_{\varepsilon \to 0} \varepsilon^{-2} M(\varepsilon k, \varepsilon^2 z) = \int_0^\infty dt \sum_x \sum_{i,j=1}^d \langle j(x,x+e_i) \, Q \, e^{QLQt} \, Qj(0,e_j) \rangle_\rho \, k_i k_j \, .$$

$$\qquad (5.21)$$

Again by detailed balance

$$\sum_x \langle j(x,x+e_i) \, Q \, e^{QLQt} \, Qj(0,e_j) \rangle_\rho = \sum_x \langle j(x,x+e_i) \, e^{Lt} \, j(0,e_j) \rangle_\rho \, .$$

Therefore the diffusion matrix is given by

$$D_{ij}(\rho) = \chi_\rho^{-1} \Big\{ \tfrac{1}{2} \langle c(0,e_i)(\eta_{e_i} - \eta_0)^2 \rangle_\rho \delta_{ij} - \int_0^\infty dt \sum_x \langle j(x,x+e_i) \, e^{Lt} j(0,e_j) \rangle_\rho \Big\} \quad (5.22)$$

$$\equiv D_{ij}^{(0)}(\rho) - D_{ij}^{(f)}(\rho) \, ,$$

$i,j = 1, \ldots , d$. $D^{(0)}$ is given by static correlation functions whereas $D^{(f)}$ involves the dynamics of the lattice gas. In analogy to fluids (5.22) is called the <u>Green - Kubo formula</u> for the bulk diffusion matrix.

There is a rewriting of (5.22) in a form which apparently is valid for any many particle system. For this purpose we define the <u>current - current correlation function</u> in equilibrium,

$$\langle j(x, x+e_i, dt) \, j \, (0, e_j, dt') \rangle_\rho \, ,$$

$$\qquad (5.23)$$

where dt, dt' are small time intervals. From the definition of the current field we conclude that (5.23) is a signed measure in $dt \times dt'$ on R^2. By stationarity it depends only on $t - t'$. For this signed measure on the real line we write then

$$\langle j(x,x+e_i,dt)\; j(0,e_j,0)\rangle_\rho \; . \tag{5.24}$$

By the conservation law (5.9)

$$\sum_x \int dt\; \langle \eta_{x,t}\eta_{0,0}\rangle_\rho \; \frac{\partial^2}{\partial t^2}\; \psi(t,x)$$

$$= \sum_x \int \; \sum_{i,j=1}^{d} \; \langle j(x,x+e_i,dt)\; j(0,e_j,0)\rangle_\rho \, (\psi(t,x)-\psi(t,x+e_i)-\psi(t,x-e_j)+\psi(t,x+e_i-e_j)) \tag{5.25}$$

for any smooth function ψ . Twice partial integration on the left and Fourier transforming leads to

$$\iint dk\; \hat\psi(t,k)\; \sum_x e^{ikx}\; \sum_{i,j=1}^{d}\; \langle j(x,x+e_i,dt)\; j(0,e_j,0)\rangle_\rho \, (1-e^{ik_i})(1-e^{-ik_j})$$

$$= \int dk\; \hat\psi(0,k)\; \sum_{i,j=1}^{d}\; \langle c(0,e_i)(\eta_{e_i}-\eta_0)^2\rangle_\rho \, \delta_{ij}\, (1-e^{ik_i})(1-e^{-ik_j})$$

$$- \int dt \int dk\; \hat\psi(t,k)\; \sum_x e^{ikx}\; \sum_{i,j=1}^{d}\; \langle j(x,x+e_i)e^{L|t|}\; j(0,e_j)\rangle_\rho \, (1-e^{ik_i})(1-e^{-ik_j}) \; . \tag{5.26}$$

Therefore

$$\langle j(x,x+e_i,dt)\; j(0,e_j,0)\rangle_\rho = \tag{5.27}$$

$$\delta(t)\delta_{x,0}\, \langle c(0,e_i)(\eta_{e_i}-\eta_0)^2\rangle_\rho \, \delta_{ij} - dt\; \langle j(x,x+e_i)e^{L|t|}\; j(0,e_j)\rangle_\rho$$

and

$$D_{ij}(\rho) = (2\,\chi_\rho)^{-1} \int dt \sum_x \langle j(x,x+e_i,dt)\; j(0,e_j,0)\rangle_\rho \; . \tag{5.28}$$

The transport coefficient associated with a locally conserved field equals the space-time integral over the correlation function of its associated current. This is the reason why besides the correlation functions of the locally conserved fields their current-current correlation functions are of major interest.

Since in our model of a lattice gas particles have no inertia, the current correlation function has a δ-contribution at $t=0$. This is reflected in the cusp of $\langle \eta_{x,t}\eta_{0,0}\rangle_\rho$ at $t=0$. For a mechanical system with a smooth interaction potential such a term is not present. A δ-contribution at $t=0$ occurs however for systems with hard core. It may happen that the spatial sum over the smooth part of the current correlation function vanishes. In this case, $D(\rho) = D^{(0)}(\rho)$ is given purely in terms of static correlation functions. If $c(x,y,\eta)$ depends only on the nearest neighbors of the bond (x,y), then in one dimension all jump rates for which $D^{(f)}(\rho)=0$ have been computed.[18] Finally note that

$$\sum_x \langle j(x,x+e_i)\, e^{Lt}\, j(0,e_j)\rangle \;\geq\; 0 \tag{5.29}$$

as a matrix. Therefore one has the bounds

$$0 \;\leq\; D(\rho) \;\leq\; D^{(o)}(\rho) \;. \tag{5.30}$$

5.2 Conductivity, Linear Response to a Small External Electric Field

We still make the same assumptions on the jump rates as in Section 5.1. We imagine that our lattice gas is placed in a uniform electric field $E = (E_1,\ldots,E_d)$. The field favors jumps in its direction. Therefore the dynamics is given by the new, field dependent jump rates

$$c(x,x+e_i,\eta,E) \;=\; c(x,x+e_i,\eta)\big((1+E_i)\eta_x + (1-E_i)\eta_{x+e_i}\big), \tag{5.31}$$

$i=1,\ldots,d$, $|E_i|\leq 1$. Whether the definition (5.31) models well the effect of an external electric field has to come from physics. At least for small E (5.31) is quite reasonable and seems to be universally adopted. The generator and current depend now on E. This will be indicated by a subscript E.

Let us assume that for each E there exists a translation invariant stationary state $\langle\cdot\rangle_{\rho,E}$ for the jump rates (5.31) such that $\langle\eta_x\rangle_{\rho,E} = \rho$ and such that $\langle\cdot\rangle_{\rho,E}$ depends continuously on E. In particular, $\lim_{E\to 0}\langle\cdot\rangle_{\rho,E} = \langle\cdot\rangle_{\rho}$. (It may happen that $\langle\cdot\rangle_{\rho,E}$ does not depend on E.) For the field pointing in the i-direction the steady state current along the j-direction is

$$\langle j_E(0,e_j)\rangle_{\rho,E_i} \;=\; \langle c(0,e_j)(\eta_0 - \eta_{e_j})\rangle_{\rho,E_i} + E_i\,\langle c(0,e_j)(\eta_0 - \eta_{e_j})^2\rangle_{\rho,E_i}\;. \tag{5.32}$$

The static conductivity is defined as

$$\sigma_{ij} \;=\; \frac{\partial}{\partial E_i}\,\langle j_E(0,e_j)\rangle_{\rho,E_i}\Big|_{E=0}\;. \tag{5.33}$$

The derivative of the second term in (5.32) is $\delta_{ij}\langle c(0,e_j)(\eta_0-\eta_{ej})^2\rangle_{\rho}$. To obtain the derivative of the first term we compute formally for any local function f

$$\langle f \rangle_{\rho,E} \;=\; \langle e^{L_E t} \, f \rangle_{\rho,E} \tag{5.34}$$

$$\langle e^{Lt} \, f \rangle_{\rho,E} \;+\; \int_0^t ds \; \langle (L_E - L) \, e^{Ls} \, f \rangle_{\rho,E} \quad .$$

As $t \to \infty$

$$\langle f \rangle_{\rho,E} \;-\; \langle f \rangle_\rho \;=\; \int_0^\infty dt \; \langle (L_E - L) \, e^{Lt} \, f \rangle_{\rho,E} \quad . \tag{5.35}$$

Now

$$\langle (L_{E_i} - L) \, e^{Lt} \, f \rangle_\rho$$

$$= E_i \sum_x \langle c(x,x+e_i)(n_x - n_{x+e_i}) \partial_{x,x+e_i} \, e^{Lt} \, f \rangle_\rho$$

$$= -2 E_i \sum_x \langle c(x,x+e_i)(n_x - n_{x+e_i}) \, e^{Lt} \, f \rangle_\rho \tag{5.36}$$

by detailed balance. Therefore the derivative of the second term in (5.32) is
given by

$$-2 \int_0^\infty dt \sum_x \langle c(x,x+e_i)(n_x - n_{x+e_i}) \, e^{Lt} \, c(0,e_j)(n_0 - n_{e_j}) \rangle_\rho \quad . \tag{5.37}$$

Adding both terms we obtain

$$\sigma_{ij} \;=\; 2 \chi_\rho \, D_{ij}(\rho) \quad . \tag{5.38}$$

The static conductivity is proportional to the diffusion constant. This result is
valid for a large class of physical systems.

More generally we may consider a space-time periodic electric field

$$E(x,t) \;=\; E \, e^{ikx} e^{i\omega t} \tag{5.39}$$

and study the response of the system _linearly_ in E. We insert (real and imaginary
part of) the electric field in (5.31). Let us assume that the system is initially
in the state $\langle \cdot \rangle_\rho$. For $E \neq 0$ this state will evolve in time to $\langle \cdot \rangle_\rho(t)$. We compute the average current through the bond $(x,x+e_j)$ at time t, $\langle j(x,x+e_j) \rangle_\rho(t)$,
linearly in E. Then as $t \to \infty$ the current follows the field with the same frequency
but possibly some delay (phase shift), i.e.

$$\langle j(x,x+e_j) \rangle_\rho(t) \;\cong\; \hat{\sigma}_{ij}(k,\omega) \, E_i \, e^{ikx} \, e^{i\omega t} \tag{5.40}$$

for large t and small E. (5.40) defines the frequency and wave vector dependent conductivity $\hat{\sigma}(k,\omega)$. Since in general $\hat{\sigma}(k,\omega)$ is complex, it contains the amplitude and the phase shift of the response current. By a computation identical to the one before one obtains

$$\hat{\sigma}_{ij}(k,\omega) \;=\; <c(0,e_j)(n_0 - n_{e_j})^2>_\rho \; \delta_{ij}$$

$$-2 \int_0^\infty dt\, e^{i\omega t} \sum_x e^{ikx} \; <j(x,x+e_i)\, e^{Lt}\, j(0,e_j)>_\rho \quad . \tag{5.41}$$

(5.41) yields a further interpretation of the static part of the diffusion coefficient. In an oscillatory external electric field a part of the induced current follows the applied field <u>instantaneously</u>. The strength of this part is determined by the static contribution to the diffusion coefficient. The other part of the induced current follows the applied field with some delay which is determined by the dynamical contribution to the diffusion coefficient.

5.3 Density Field of a Lattice Gas at Infinite Temperature

We assume the jump rates

$$c(x,y,n) \;=\; \begin{cases} (n_x - n_y)^2 & \text{for } |x-y| = 1 \\[2em] 0 & \text{otherwise.} \end{cases} \tag{5.42}$$

Independently of the configuration particles jump to nearest neighbor sites whenever the final site is empty. This model is also known as the simple symmetric exclusion process. The equilibrium states are the translation invariant Bernoulli measures $<\cdot>_\rho$, i.e. each site independently is occupied with probability ρ and empty with probability $1 - \rho$, $0 \le \rho \le 1$. For this model the density time correlation function can be computed explicitly. One obtains

$$\sum_x e^{ikx} (< n_{x,t}\, n_{0,0} >_\rho - \rho^2)$$

$$= \rho(1-\rho)\, \exp[-2t\, (\sum_{j=1}^{d} (1 - \cos k_j))] \quad . \tag{5.43}$$

The density time correlations are diffusive because of the local conservation law.

Since the simple exclusion process is dual to random walks with exclusion, one can even analyze the full density fluctuation field.

Theorem 3 [19] . Let $\xi(f,t)$ be the Gaussian random field on R^{d+1} with mean zero and covariance

$$\langle\xi(f,t)\,\xi(g,s)\rangle = \rho(1-\rho)\int_{R^d} dk\ \hat{f}(k)\hat{g}(k)\ e^{-|t-s|k^2}\ . \tag{5.44}$$

Then

$$\lim_{\varepsilon\to 0}\ \xi^\varepsilon(f,t) = \xi(f,t) \tag{5.45}$$

in the sense of weak convergence of path measures on $D([0,\infty], \mathcal{S}'(R^d))$.

The total (i.e. summed over space) current correlation function is $\delta(t)\,\delta_{ij}$. There is no dynamical contribution to the diffusion coefficient. The diffusion coefficient is independent of the density .

6. Outlook: Hydrodynamics for Stochastic Lattice Gases

So far we have restricted ourselves to the time-dependent density fluctuations in equilibrium. In this section I will try to place this problem within a somewhat wider context.

To be specific let me assume (i), (ii) and (5.8) together with (5.7) of Section 5.1. Furthermore I will assume that (iv) $|\beta| \leq \beta_o$ where β_o is small enough such that the static high temperature expansion converges. Then for every ρ, $0 \leq \rho \leq 1$, $<\cdot>_\rho$ is the unique Gibbs state for the nearest neighbor potential defined by (5.7). (iii) of Section 5.1 is a consequence of our assumptions. If β is such that a phase transition may occur, the general picture has to be modified considerably.

I want to consider an <u>initial value problem</u> with an initial state where the density of particles varies slowly on the scale set by the lattice. To be specific I assume that the initial state μ^ε is a product measure. Particles at each site are independent and

$$\mu^\varepsilon(\eta_x) = \rho(\varepsilon x) \tag{6.1}$$

with ρ: $R^d \to [0,1]$ a continuous function. Note that in the state μ^ε the density field $n^\varepsilon(f,0) = \varepsilon^d \sum_x f(\varepsilon x)\eta_x$ has a limit as $\varepsilon \to 0$,

$$\lim_{\varepsilon \to 0} n^\varepsilon(f,0) = \int dx\, f(x)\rho(x) \tag{6.2}$$

in probability. The initial density field tends to the deterministic density $\rho(x)$.

If the scale of density variation remains large compared to the lattice spacing also at later times, then one may consider the density $\rho(x,t)$ approximately as a smooth function. By local conservation of mass

$$\frac{\partial}{\partial t}\, \rho(x,t) + \text{div}\, \vec{j}\,(x,t) = 0 \;, \tag{6.3}$$

where $\vec{j}(x,t)$ is the mass current per unit area per unit time. In equilibrium, the current $\vec{j}(x,t)$ is zero. Expanding locally in small deviations from equilibrium one expects then on phenomenological grounds that

$$\vec{j}_i(x,t) = - \sum_{j=1}^{d} D_{ij}\,(\rho(x,t))\, \frac{\partial}{\partial x_j}\, \rho(x,t) \;, \tag{6.4}$$

i.e. the current is proportional to the density gradient (Fick's law). The proportionality constant $D(\rho)$ will depend, in general, on the local density. Combining (6.3) and (6.4) we find that the "macroscopic" density is governed by the nonlinear diffusion equation

$$\frac{\partial}{\partial t}\, \rho(x,t) \;=\; \sum_{i,j=1}^{d} \frac{\partial}{\partial x_i}\, D_{ij}(\rho(x,t))\, \frac{\partial}{\partial x_j}\, \rho(x,t) \;. \tag{6.5}$$

This is the <u>hydrodynamic equation</u> for our lattice gas. At that stage we have no way to guess what the diffusion matrix $D(\rho)$ as a function of ρ would be. From a phenomenological point of view $D(\rho)$ has to be given.

To put our simple argument into a precise conjecture we consider the scaled density field

$$n^{\varepsilon}(f,t) \;=\; \varepsilon^{d} \sum_{x} f(x)\, \eta_{x,\varepsilon^{-2}t} \tag{6.6}$$

with initial state μ^{ε}.

<u>Conjecture 2</u> . Under the above assumptions, the following limit exists,

$$\lim_{\varepsilon \to 0} \; n^{\varepsilon}(f,t) \;=\; \int dx\, f(x)\, \rho(x,t) \tag{6.7}$$

in probability. $\rho(x,t)$ is the solution of the non-linear diffusion equation (6.5) with initial conditions $\rho(x,0) = \rho(x)$. The function $\rho \to D_{ij}(\rho)$ is given by (5.22).

To identify $D(\rho)$ in (6.5) with the $D(\rho)$ given by (5.22) one argues as follows: $\mu_0(f) \equiv \langle f\eta_0\rangle_\rho/\rho$ is the equilibrium distribution of particles given that there is a particle at the origin. Therefore $\langle \eta_{x,t}\eta_{0,0}\rangle_\rho/\rho$ is the average density of particles at time t at site x given the system starts initially in the state μ_0. In the state μ_0 the density deviates on a macroscopic scale only little from the uniform density. Therefore it should be safe to compute the average density at time t by the non-linear diffusion equation (6.5) <u>linearized</u> around the constant density ρ. If this is assumed, then $\langle \eta_{x,t}\eta_{0,0}\rangle_\rho$ is governed by the <u>linear</u> diffusion equation with diffusion matrix $D(\rho)$. The whole Section 5 aimed at to argue that $D(\rho)$ is given by (5.22).

Of course, there is no reason to consider time-dependent density fluctuations only in the equilibrium state. In particular, since according to Conjecture 2 the density in the initial value problem has a deterministic limit as $\varepsilon \to 0$, it is meaningful also to consider fluctuations around this time-dependent deterministic density. Let us define the scaled density fluctuation field by

$$\xi^{\varepsilon}(f,t) \;=\; \varepsilon^{d/2} \sum_{x} f(\varepsilon x)[\eta_{x,\varepsilon^{-2}t} - \mu_{\varepsilon}(T_{\varepsilon^{-2}t}\,\eta_x)] \tag{6.8}$$

with initial state μ^ε. Under our assumptions it is expected that $\xi^\varepsilon(f,t)$ tends to a Gaussian field as $\varepsilon \to 0$. Let us first amplify Conjecture 1 in this direction.

<u>Conjecture 3.</u> Let $\xi(f,t)$ be the Gaussian random field with mean zero and covariance (5.16), where $D(\rho)$ is given by (5.22). Let $\xi^\varepsilon(f,t)$ be the density fluctuation field in the equilibrium state $<\cdot>_\rho$. Then

$$\lim_{\varepsilon \to 0} \xi^\varepsilon(f,t) = \xi(f,t) \ . \tag{6.9}$$

Let $\xi(f,t) = \int dx\, f(x)\, \xi(x,t)$. Then the Gaussian field $\xi(x,t)$ is governed by the stochastic partial differential equation

$$d\,\xi(x,t) = \sum_{i,j=1}^{d} \frac{\partial}{\partial x_i} D_{ij}(\rho) \frac{\partial}{\partial x_j} \xi(x,t)dt + R_\rho^{\frac{1}{2}} dW_t(x). \tag{6.10}$$

Here R_ρ is the positive linear operator defined by the quadratic form

$$\int dx\, g(x)(R_\rho f)(x) = 2 \sum_{i,j=1}^{d} \int dx\, \chi_\rho\, D_{ij}(\rho)\, (\frac{\partial}{\partial x_i} g)(x)(\frac{\partial}{\partial x_j} f)(x) \tag{6.11}$$

and $dW_t(x)$ is white noise over $R^d \times R$. The fluctuations around a non-equilibrium state are assumed to be governed by the "minimal" extension of (6.10).

<u>Conjecture 4</u> . Let $\xi^\varepsilon(f,t)$ be the density fluctuation field with initial state μ^ε . Then

$$\lim_{\varepsilon \to 0} \xi^\varepsilon(f,t) = \xi(f,t) \tag{6.12}$$

exists. $\xi(f,t) = \int dx\, f(x)\, \xi(x,t)$ is an infinite-dimensional, non-homogeneous Ornstein-Uhlenbeck process governed by the stochastic partial differential equation

$$d\xi(x,t) = A_{\rho(t)}\, \xi(x,t)dt + (R_{\rho(t)})^{\frac{1}{2}} dW_t(x) \tag{6.13}$$

with Gaussian initial conditions with mean zero and covariance $<\xi(f,0)\xi(g,0)> = \int dx\rho(x)(1-\rho(x))\, f(x)g(x)$. Here $A_{\rho(t)}$ is the linear operator obtained by linearizing (6.5) at $\rho(x,t)$, i.e.

$$(A_{\rho(t)}\xi)(x) = \sum_{i,j=1}^{d} \frac{\partial}{\partial x_i} \frac{\partial}{\partial x_j} D_{ij}(\rho(x,t))\, \xi(x) \ , \tag{6.14}$$

and $R_{\rho(t)}$ is the positive linear opeartor defined by the quadratic form

$$\int dx \, g(x) \, (R_{\rho(t)}f)(x) = 2 \sum_{i,j=1}^{d} \int dx \, \chi_{\rho(x,t)} D_{ij}(\rho(x,t)) \left(\frac{\partial}{\partial x_i} g\right)\left(\frac{\partial}{\partial x_j} f\right)(x). \qquad (6.15)$$

For $\rho(x,t) \equiv \rho$ (6.13) reduces to (6.10). For the lattice gas at infinite temperature discussed in Section 5.3 Conjectures 2 to 4 are proved.[19,20]

7. Outlook: Linearized Hydrodynamics for a Fluid in Thermal Equilibrium

In principle, the analysis of equilibrium time correlation functions explained here is also applicable to a real fluid. (Historically, it was after all a fluid for which the method was developed.) There are however several complications which I list briefly.

We consider an infinite system of classical particles in equilibrium at inverse temperature β and density ρ. We assume ρ to be sufficiently small in order to be in the fluid phase. The particles interact via the central pair potential v. The dynamics is given by Newton's equation of motion

$$\frac{d}{dt} q_i(t) = v_i(t)$$

$$\frac{d}{dt} v_i(t) = \sum_{j \neq i} - \text{grad } V(q_j(t) - q_i(t)). \qquad (7.1)$$

For potentials V which are sufficiently repulsive near the origin and decay sufficiently rapidly at infinity (7.1) has a unique, global solution for a set of initial data of full equilibrium measure.[21-23] In particular, equilibrium time correlation functions exist.

(i) In contrast to the stochastic lattice gas, a fluid has <u>five</u> locally conserved fields, namely the mass density, the three components of the velocity density and the energy density. Therefore instead of a single correlation function we have now to consider the matrix of correlation functions $\langle \xi_\alpha(x,t)\xi_{\alpha'}(x',t')\rangle$, $\alpha,\alpha''=1,\ldots, 5$, with $\xi_1(x,t)$ the mass density fluctuations, etc. One expects that for large separation $|x-x'|$ and $|t-t'|$ $< \xi_\alpha(x,t) \, \xi_{\alpha'}(x',t')$ equal approximately the fundamental solution of the equations of <u>linearized hydrodynamics</u>,[1,2] i.e. the equations obtained by linearizing the usual non-linear hydrodynamic equations at mass density $\rho = $ const, velocity density $u=0$ and energy density $e = $ const.

(ii) The linearized hydrodynamic equations have in addition to the diffusive terms drift contributions, i.e. contributions of the form ik in Fourier space. Physically, they correspond to sound waves. One therefore has <u>two time scales</u>. On a short time scale (Euler time scale of the order ε^{-1}) one obtains a reversible flow, whereas on a longer time scale (Navier-Stokes time scale of the order ε^{-2}) one obtains in addition diffusive damping.

(iii) Last, not least, the underlying dynamics is deterministic. This makes the problem of deriving linearized hydrodynamics for a fluid, although without question well-defined, rather hopelessly difficult in view of present day mathematical machinery.

The only deterministic interacting particle model for which the validity of linearized hydrodynamics is proved in full is the fluid of hard rods in one dimension.[24] For this model also the validity of non-linear hydrodynamics governing the evolution of the velocity density fields is proved.[25]

<u>Acknowledgments</u>

It is my pleasure to thank S. Goldstein, C. Kipnis, J.L. Lebowitz, and E. Presutti for many constructive discussions on stochastic particle systems.

<u>References</u>

1) D. Forster, Hydrodynamic Fluctuations, Broken Symmetry, and Correlation Functions. Benjamin, Reading, Mass., 1975.
2) P. Resibois, M. DeLeener, Classical Kinetic Theory of Fluids. John Wiley, New York, 1977.
3) R. Glauber, J. Math. Phys. $\underline{4}$, 294 (1963).
4) T.M. Liggett, The Stochastic Evolution of Infinite Systems of Interacting Particles. Lecture Notes in Mathematics 598, Springer, Berlin, 1977.
5) R. Zwanzig, Statistical Mechanics of Irreversibility. In: Lectures in Theoretical Physics, vol. 3 (Boulder Lectures), Interscience, New York, 1961.
6) H. Mori, Progr. Theor. Phys. $\underline{34}$, 423 (1965).
7) R.A. Holley, D.W. Stroock, Comm. Math. Phys. $\underline{48}$, 249 (1976).
8) R.A. Holley, D.W. Stroock, Z. fur Wahrscheinlichkeits theorie verw. Gebiete $\underline{35}$, 87 (1976).
9) P. Clifford, A. Sudbury, Biometrika $\underline{60}$, 581 (1973).
10) R.A. Holley, T.M. Liggett, Annals of Prob. $\underline{3}$, 643 (1975).
11) M. Bramson, D. Griffeath, Annals of Prob. $\underline{7}$, 418 (1978).
12) R.A. Holley, D.W. Stroock, Annals of Math. $\underline{110}$, 333 (1979).
13) E. Presutti, H. Spohn, Hydrodynamics of the Voter Model, preprint.
14) K. Kawasaki, Phys. Rev. $\underline{142}$, 164 (1966).
15) K. Kawasaki, Kinetics of Ising Models. In: Phase Transitions and Critical Phenomena. Eds. C. Domb, M. Green, vol. 2. Academic Press, New York, 1972.
16) K. Binder, M.H. Kalos, J.L. Lebowitz and J. Marro, Advances in Coloid and Interface Science $\underline{10}$, 173 (1979).
17) W. Dietrich, P. Fulde, I. Peschel, Adv. Phys. $\underline{29}$, 527 (1980).
18) H. Singer, I. Peschel, Z. Physik $\underline{B39}$, 333 (1980).
19) A. Galves, C. Kipnis, H. Spohn, Fluctuation theory for the symmetric simple exclusion process, preprint in preparation.
20) A. Galves, C. Kipnis, C. Marchioro, E. Presutti, Comm. Math. Phys. $\underline{81}$, 127 (1981).
21) O.E. Lanford, Time Evolution of Large Classical Systems. In: Dynamical Systems, Theory and Applications, ed. J. Moser. Lecture Notes in Physics 38, Springer, Berlin, 1975.
22) C. Marchioro, A. Pelligrinotti, E. Presutti, Comm. Math. Phys. $\underline{40}$, 175 (1975).
23) E. Presutti, M. Pulvirenti, B. Tirozzi, Comm. Math. Phys. $\underline{47}$, 81 (1976).
24) H. Spohn, Annals of Physics, in press.
25) C. Boldrighini, R.L. Dobrushin, Yu.M. Sukhov, Hydrodynamics of one-dimensional hard rods, preprint.

QUANTUM DISSIPATION AND STOCHASTIC PROCESSES

R.Vilela Mendes
CFMC — Instituto Nacional de Investigação Científica
Av.Gama Pinto, 2 — 1699 Lisboa Codex — Portugal

ABSTRACT

A general scheme to quantize dissipative and volume preserving dynamics is reviewed.

The operator solutions of two examples are used to derive the stochastic differential equations obeyed by these processes.

1. INTRODUCTION

Consideration of dissipative effects in a quantum context is useful in the description of heavy ion scattering, fission, threshold phenomena in lasers and tunneling in macroscopic systems [1-3].

In Ref.4 it was shown that the problem of quantizing dissipative systems is reducible, through embedding, to the quantization of volume preserving dynamics. Generalizing, for (non-Hamiltonian) volume-
-preserving dynamics, the usual Heisenberg picture operator quantization a general scheme was obtained to quantize arbitrary nonconservative systems.

After a brief review of these results in section 2,   the operator solutions of two examples treated by this method are used to characterize the stochastic processes associated to the quantum volume preserving systems. Time dependence of the diffusion coefficients seems to be the main characteristic of these processes.

2. QUANTIZATION OF DISSIPATIVE AND VOLUME PRESERVING DYNAMICS

The quantization scheme developed for dissipative systems is based in the following results:

Theorem 1 [5] : In an N-dimensional (N even) C^∞-manifold M one can find for every $x \in M$ a neighbourhood Ω of x, a Riemannian metric g and N-1 sympletic forms ω_i on Ω such that every vector field X defined on   the nbd can be decomposed into one gradient and N-1 Hamiltonian fields.

Theorem 2 [4] : For any dissipative vector field X in R^N (characterized on a neighbourhood by one gradient and N-1 Hamiltonian functions) there is a volume preserving vector field in R^{2N} (characterized by N hamiltonian functions, N-1 of which are the same as for X) which on restriction to R^N coincides with X.

From these results the following rules emerge to quantize non-
-conservative systems

(1)   Use Theorem 1 to decompose the dynamical system into gradient and
      Hamiltonian components:
(2)   Embed the N-dimensional system into a 2N-dimensional volume
      preserving dynamics.
(3)   Quantize the volume preserving dynamical system.

To implement the last step in the program one notices that according to the Theorem 1 a volume preserving classical dynamical system in R^N

$$\dot{x}_i = X_i(x) \qquad\qquad (\partial_i X_i = 0) \tag{2.1}$$

can be written

$$\dot{x}_i = \sum_{k=1}^{r} \{x_i, H^{(k)}\}^k \tag{2.2}$$

where $\{\ \}^k$ is the Poisson bracket associated to the symplectic form ω_k and $k \leq N-1$. In general the dynamical evolution brings into play several distinct symplectic structures and not just one as in the Hamiltonian systems.

Several quantization schemes are now possible. For example in the scheme of quantization through deformations[6] one would have to consider the deformations of the r distinct Poisson Lie algebras[7].

If instead one decides to generalize the usual Heisenberg picture operator formalism an operator representation $T_k(x_i)$ must be chosen for each ω_k-structure and the observable x_i is associated to an r-tuple of operators $\{T_k(x_i) ; i = 1...r\}$. The quantum brackets $[\ ]^k$ for the operator r-tuples are

$$[A,B]^k = \{T_i T_k^{-1} [T_k(A), T_k(B)] ; \quad i = 1...r\} \tag{2.3}$$

With the definition (2.3) and symmetrizing all products the equation of motion for operators (in the Heisenberg picture) is

$$\dot{A} = \sum_{k=1}^{r} \frac{1}{i} [A, H^{(k)}]^k \tag{2.4}$$

The states are r-tuples of functions $\{\psi_i(x) ; i = 1...r\}$ and

$$A \{\psi_i(x)\} = \{T_i(A)\psi_i(x)\}$$

The expectation values are

$$<A> = \sum_i <A>_i = \sum_{i=1}^{r} (\psi_i, T_i(A)\psi_i) \tag{2.5}$$

The eqs. (2.3-5) together with the operator representation at $t = 0$ define a (Heisenberg picture) quantization for volume preserving dynamics.

3. PROBABILISTIC INTERPRETATION AND STOCHASTIC PROCESSES

To derive the stochastic differential equations associated to the volume preserving quantum systems one needs the configuration space probability density $\rho(x,t)$. Because the Schrödinger picture does not exist for $r > 1$ one cannot simply compute $\rho(x,t)$ as the modulus square $|\psi(x,t)|^2$ of the wave function. Instead let $\underset{\sim}{x}_t$ be the (time dependent) Heisenberg picture position operator. By the spectral theorem

$$<e^{i\alpha \underset{\sim}{x}_t}> = \int e^{i\alpha x} d(\psi, E_x(t)\psi) = \int e^{i\alpha x} \rho(x,t) dx$$

Hence

$$\rho(y,t) = \frac{1}{2\pi} \int dx \ e^{-i\alpha y} {}_{<e}{}^{i\alpha x_{\sim}t}{}_{>} \tag{3.1}$$

For the volume preserving system associated to the damped harmonic oscillator[4]

$$\dot{x} = p \ ; \quad \dot{p} = -x-\lambda p \ ; \quad \dot{z} = 0 \ ; \quad \dot{w} = \lambda w$$

one operator representation of x_t is

$$T_1(x_{\sim t}) = e^{-\lambda t/2}(x \cos \Omega t + (\frac{1}{i} \frac{\partial}{\partial x} + \frac{\lambda x}{2})\frac{\sin \Omega t}{\Omega}) \tag{3.2}$$

where $\Omega = (1 - \lambda^2/4)^{1/2}$. Restricting ourselves to states $\begin{vmatrix} \psi(x) \\ 0 \end{vmatrix}$ and defining

$$\phi(y,t) = \exp\left\{- \frac{i}{2} (x^2 - \frac{\partial^2}{\partial x^2} + \frac{\lambda}{2} \{x, \frac{1}{i} \frac{\partial}{\partial x}\}_+)t\right\} \psi(x) \Bigg|_{x = e^{\lambda t/2}y}$$

one obtains from (3.1) and (3.2)

$$\rho(y,t) = e^{\lambda t/2}|\phi(y,t)|^2 \tag{3.3}$$

For the eigenstates of the constant of motion $\phi_1 = (p^2+x^2+\lambda xp)w/2$

$$\psi_n^{(1)} = c_n \ \exp\{-i\lambda/2-\sqrt{1-\lambda^2/4}\}\frac{x^2}{2} \ H_n\left[(1-\lambda^2/4)^{1/4} x\right]e^{iz}$$

one obtains a time dependent density distribution that becomes sharper as time flows,

$$\rho_n(y,t) = |c_n|^2 \ e^{\lambda t/2}e^{-\Omega y^2 e^{\lambda t}} \ H_n^2\left[\sqrt{\Omega} \ y \ e^{\lambda t/2}\right]$$

From (3.3) it also follows that $\rho(y,t)$ obeys the equation

$$\frac{d}{dt} \rho(y,t) = - \frac{\partial}{\partial y}(b^+\rho) + \frac{e^{-\lambda t}}{2} \frac{d^2}{dy^2}\rho \tag{3.4}$$

with

$$b^+(y,t) = \frac{e^{-\lambda t}}{2} \{\frac{\nabla\phi}{\phi} + \frac{\nabla\phi^*}{\phi^*}\} + \frac{e^{-\lambda t}}{2i} \{\frac{\nabla\phi}{\phi} - \frac{\nabla\phi^*}{\phi^*}\} \tag{3.5}$$

Therefore in the stochastic differential equation corresponding to this process

$$dx(t) = b^+(x,t)dt + e^{-\lambda t/2}dW(t)$$

both the drift and the diffusion coefficient are time-dependent.

For the quantum analogue of a dynamical system with a classical limit cycle[4]

$$\dot{x} = p + \lambda x(K - x^2 - p^2) \ ; \quad \dot{p} = -x + \lambda p(K - x^2-p^2)$$

$$\dot{z} = 0 \qquad\qquad ; \quad \dot{w} = -2\lambda(K - 2x^2 - 2p^2)w$$

let $X(t)$ and $L(t)$ be the operators

$$X(t) = \exp\left\{\tfrac{i}{2}\left(x^2 - \tfrac{\partial^2}{\partial x^2}\right)t\right\} \times \exp\left\{-\tfrac{i}{2}\left(x^2 - \tfrac{\partial^2}{\partial x^2}\right)t\right\} \tag{3.6}$$

$$L(t) = \left\{\frac{K\,e^{2K\lambda t}}{K^2 + \left(x^2 - \tfrac{\partial^2}{\partial x^2}\right)(e^{2K\lambda t}-1)}\right\}^{1/2} \tag{3.7}$$

Then the T_1 operator representation of $\underset{\sim}{x}_t$ is

$$T_1(\underset{\sim}{x}_t) = \tfrac{1}{2}(X(t)L(t) + L(t)X(t)) \tag{3.8}$$

For functions of $\underset{\sim}{x}_t$ it is convenient to define a conventional ordering

$$\langle e^{i\alpha\cdot\underset{\sim}{x}_t}\rangle \doteq \sum_n \frac{(i\alpha)^n}{n!}\,\tfrac{1}{2}\,\langle X^n(t)L^n(t) + L^n(t)X^n(t)\rangle \tag{3.9}$$

For the eigenstates $\left|\psi_n(x)\right\rangle$ of the constant of motion
$$\underset{\sim}{\phi}_2 = (p^2 + x^2)(p^2 + x^2 - K)\underset{w}{\overset{0}{}}$$

$$\psi_n^{(2)} = c_n\,\exp(iz - x^2/2)\,H_n(x)$$

one obtains from (3.1) and (3.9)

$$\rho_n(y,t) = |c_n|^2\,\ell_n^{1/2}\,e^{-\ell_n y^2}\,H_n^2\left[\ell_n^{1/2}y\right] \tag{3.10}$$

where
$$\ell_n = \{K + (2n + 1)(e^{2K\lambda t} - 1)\}/(K\,e^{2K\lambda t})$$

In the limit $t \to \infty$ one obtains the quantum limit cycle probability distribution

$$\lim_{t\to\infty}\rho_n(y,t) = |c_n|^2\{(2n+1)/K\}^{1/2}\,e^{-(2n+1)y^2/K}\,H_n^2\left[\{(2n+1)/K\}^{1/2}y\right] \tag{3.11}$$

Although the limiting average "energy" is the same in all cases

$$\lim_{t\to\infty}\langle\tfrac{1}{2}(p^2(t) + x^2(t))\rangle_n = K/2$$

eq. (3.11) shows that there are infinitely many quantum limit cycle distributions and the one that is actually approachad as $t \to \infty$ depends on the initial condition.

Because of the operator nature of the $L(t)$ factor in $\underset{\sim}{x}_t$ one cannot in general derive a simple Fokker-Planck equation as in the case of the damped harmonic oscillator, eq. (3.4). However for states $\phi(x) = \sum_n d_n\,\psi_n^{(2)}$ with d_n strongly peaked around a particular n one finds the following approximate equations

$$\frac{d}{dt}\rho_n(y,t) = -\frac{\partial}{\partial y}(b_n^+\rho_n) + \nu_n(t)\rho_n \tag{3.12}$$

where the diffusion coefficient is $\nu_n(t) = \ell_n^{-1/2}(t)/2$. This means that in this processes the time-dependent diffusion coefficient depends on the initial condition. Taking the $t \to \infty$ limit one concludes that on the quantum limit cycles the process has an effective Planck constant $\hbar\sqrt{\dfrac{K}{2n+1}}$.

REFERENCES

|1| R.W. HASSE; Rep. Prog. Phys. <u>41</u>, 1027 (1978)

|2| K.W.H. STEVENS; Phys. Lett <u>75A</u>, 463 (1980)

|3| A.O. CALDEIRA and A.J. LEGGETT; Phys. Rev. Lett. <u>46</u>, 211 (1981)

|4| R.V. MENDES; Marseille preprint CPT-81/P.1306, to appear in Phys. Rev. D

|5| R.V. MENDES and J.T. DUARTE; J. Math. Phys. <u>22</u>, 1420 (1981)

|6| F. BAYEN, M.FLATO, C.FRONSDAL, A.LICHNEROWICZ, D.STERNHEIMER; Ann. of Physics <u>111</u>, 61 (1978)

|7| R.V. MENDES and J.F. RODRIGUES; to appear.

<u>L I S T   O F   C O N T R I B U T O R S   A N D   P A R T I C I P A N T S</u>

L. ACCARDI	Cagliari	R.L. HUDSON	Nottingham
M. AIZENMANN	Princeton	R. JANCEL	Paris VII
S. ALBEVERIO	Bochum	G. JONA-LASINIO	Rome and IHES
T. AREDE	Marseille	W. KARWOWSKI	Wroclaw
A. BADRIKIAN	Clermont-Ferrand	W. KIRSCH	Bochum
J. BELLISSARD	Université de Provence and CPT, CNRS Marseille	J. KLAUDER	Bell Laboratories
F. BENTOSELA	Aix-Marseille II and CPT, CNRS Marseille	R. KOTECKY	IHES
		P. KREE	Paris VI
A.M. BERTHIER	Paris VI	C. MACEDO	Coimbra
J. BERTRAND	Paris VII	R. MARRA	Salerne
Ph. BLANCHARD	Bielefeld	F. MARTINELLI	Bochum
F. CALHEIROS	Marseille	D. MERLINI	Bochum
P. COLLET	Ecole Polytechnique, Paris	S. MIRACLE-SOLE	Saragosse and CPT CNRS Marseille
Ph. COMBE	Aix-Marseille II and CPT, CNRS Marseille	L. MORATO	Padoue
		Ch. PFISTER	Lausanne
I. DAUBECHIES	Bruxelles	G. RIDEAU	Paris VII
I. DAVIES	Edimburgh	J.C. RISSET	Marseille
G.F. DE ANGELIS	Salerne and CPT, CNRS Marseille	R. RODRIGUEZ	Aix-Marseille II and CPT, CNRS Marseille
D. DE FALCO	Salerne and Princeton	P. RUGGIERO	Salerne
S. DE GREGORIO	Rome	W. SCHNEIDER	Baden-Dätwill
G.F. DELL'ANTONIO	Rome	E. SCOPPOLA	Rome
C. DEWITT-MORETTE	Austin	J.B. SHLOSMAN	Moscou
D. DOHRN	Salerne	M. SIRUGUE	CPT, CNRS Marseille
M.D. DONSKER	Courant Institute, New York	M. SIRUGUE-COLLIN	Univ. de Provence and CPT, CNRS Marseille
D. DÜRR	Bochum	T. SPENCER	New York
K.D. ELWORTHY	Coventry	H. SPOHN	IHES
G.G. EMCH	Rochester	R.F. STREATER	London
W. FARIS	IHES	L. STREIT	Bielefeld
A. FORMOSO	Université de Provence and CPT, CNRS Marseille	D. TESTARD	Avignon and CPT, CNRS Marseille
J. FRÖHLICH	IHES	A. TRUMAN	Edimburgh
M. FUKUSHIMA	Osaka	B. TIROZZI	Rome
W. GARCZYNSKI	Wroclaw	R. VILELA-MENDEZ	Lisbonne
F. GUERRA	Rome	T.T. WU	Harvard University
R. HØEGH-KROHN	Oslo and Marseille		

Lecture Notes in Physics